T0317740

Broadband Optical Access Networks and Fiber-to-the-Home

Broadband Optical Access Networks and Fiber-to-the-Home

Systems Technologies and Deployment Strategies

Edited by

Chinlon Lin

Center for Advanced Research in Photonics
Chinese University of Hong Kong

John Wiley & Sons, Ltd

Copyright © 2006   John Wiley & Sons Ltd, The Atrium, Southern Gate, Chichester,
West Sussex PO19 8SQ, England

Telephone (+44) 1243 779777

Email (for orders and customer service enquiries): cs-books@wiley.co.uk
Visit our Home Page on www.wiley.com

All Rights Reserved. No part of this publication may be reproduced, stored in a retrieval system or transmitted in
any form or by any means, electronic, mechanical, photocopying, recording, scanning or otherwise, except under
the terms of the Copyright, Designs and Patents Act 1988 or under the terms of a licence issued by the
Copyright Licensing Agency Ltd, 90 Tottenham Court Road, London W1T 4LP, UK, without the permission
in writing of the Publisher. Requests to the Publisher should be addressed to the Permissions Department,
John Wiley & Sons Ltd, The Atrium, Southern Gate, Chichester, West Sussex PO19 8SQ, England, or
emailed to permreq@wiley.co.uk, or faxed to (+44) 1243 770620.

Designations used by companies to distinguish their products are often claimed as trademarks. All brand names
and product names used in this book are trade names, service marks, trademarks or registered trademarks of their
respective owners. The Publisher is not associated with any product or vendor mentioned in this book.

This publication is designed to provide accurate and authoritative information in regard to the subject matter
covered. It is sold on the understanding that the Publisher is not engaged in rendering professional services.
If professional advice or other expert assistance is required, the services of a competent professional should
be sought.

Other Wiley Editorial Offices

John Wiley & Sons Inc., 111 River Street, Hoboken, NJ 07030, USA

Jossey-Bass, 989 Market Street, San Francisco, CA 94103-1741, USA

Wiley-VCH Verlag GmbH, Boschstr. 12, D-69469 Weinheim, Germany

John Wiley & Sons Australia Ltd, 42 McDougall Street, Milton, Queensland 4064, Australia

John Wiley & Sons (Asia) Pte Ltd, 2 Clementi Loop #02-01, Jin Xing Distripark, Singapore 129809

John Wiley & Sons Canada Ltd, 6045 Freemont Blvd, Mississauga, ONT, L5R 4J3

Wiley also publishes its books in a variety of electronic formats. Some content that appears in print may not be
available in electronic books.

Library of Congress Cataloguing-in-Publication Data

Broadband optical access networks and fiber-to-the-home: systems
technologies and deployment strategies/edited by Chinlon Lin.
    p. cm.
Includes bibliographical references and index.
ISBN-13: 978-0-470-09478-5 (cloth : alk. paper)
ISBN-10: 0-470-09478-8 (cloth : alk. paper)
1. Optical fiber subscriber loops.   2. Broadband communication
systems. I. Lin, Chinlon.
TK5103.592.O68B73 2006
004.6′4–dc22   2006006411

British Library Cataloguing in Publication Data

A catalogue record for this book is available from the British Library

ISBN-10: 0-470-09478-8
ISBN-13: 978-0-470-09478-5

Typeset in 10/12pt Times by Thomson Digital
Printed and bound in Great Britain by CPI Antony Rowe, Eastbourne
This book is printed on acid-free paper responsibly manufactured from sustainable forestry
in which at least two trees are planted for each one used for paper production.

*"Let there be **Light**, and there was **Light**."*
– Genesis 1:3

*"I have set my **Rainbow** in the clouds."*
– Genesis 9:13

This book is dedicated to

All the engineers and scientists worldwide who worked in laser photonics and lightwave technologies for optical fiber communications over the last 40 years, for they are the ones who have collectively and successfully made the modern global information infrastructure based on worldwide optical fiber telecommunications networks a reality, which in turn transformed the human communications and ushered the entire human civilization into the new era of ubiquitous broadband.

Contents

11 Integrated Broadband Optical Fibre/Wireless LAN Access Networks

Foreword

The huge successes during the last decade of the Web and digital video (first as DVDs, then as satellite and cable entertainment, now as video recording, HDTV, and flat-panel displays) have changed people's habits and their demands for service delivery. Consequently, consumer adoption of broadband access to facilitate use of the Internet for knowledge, commerce, and entertainment is following these same patterns, outpacing the ability of some technologies to keep up. A growing number of service providers – old and new – are turning to solutions capable of exploiting the full potential of optical fiber for service delivery: 'FTTH: fiber to the home.' In the laboratory for 30 years, and successfully meeting the technical objectives of worldwide field trials throughout the 1980s, FTTH was nevertheless stymied by high costs. In recent years, however, costs have fallen as service demand rose dramatically so that FTTH has become cost-effective in more and more situations. The ongoing pressures to reduce costs of integrated circuits and other technologies for mass-market products, especially PCs, Ethernet data networking, and digital video, combined with similar progress in electro-optics and fiber have all contributed to driving down the cost of FTTH. Ongoing operational cost savings from moving fiber to the very edges of the network are better understood and also improve the business case. In light of the rising consumer demand for so-called 'triple-play' services (digital video, broadband Internet access, and fully featured voice services including Internet delivery), the time is right to understand what the new FTTH systems and strategies offer.

This book you are about to read addresses the need. Chinlon Lin and two dozen eminent authors, representing the spectrum from developers and manufacturers to network and service providers, treat the systems aspects of FTTH comprehensively. This includes passive optical networks, standardization of PONs which has accelerated product availability and deployment, the ongoing evolution to deliver a Gigabit per second Ethernet, and the growing trend to migrate to dense wavelength-division multiplexing. Business and operational issues, trends, and evolution strategies are also discussed. In organizing this book, Chinlon Lin has brought with him his broad perspective. Having a career path at Bell Labs, Bellcore (now Telcordia Technologies), and Jedai Broadband, he worked intimately with the R and D of the underlying optical technologies and architectures, as well as some of the deployment strategies. His past interactions ranged from telecommunications companies to broadband cable service providers. In the chapters of this book, recognized experts provide international perspectives along these lines for the reader to develop their own knowledge base and business perspective of FTTH access networks, including even the popular use of wireless once the premises are reached.

I know you will find this book timely, offering accurate and authoritative information for those interested in the topics of broadband optical access and Fiber to the Home. Enjoy it!

Paul W. Shumate
Executive Director of the IEEE LEOS
(Lasers and Electro-Optics Society)
January 2006

Preface and Overview

Over the last decade the world has seen a great transformation in telecommunications, and its impacts on the human civilization in digital information transmission and storage, media, and entertainment. This is in many ways due to the advances in computer networking and the emergence of global Internet, which were made possible by the various advances in the high-speed, high-capacity broadband telecommunications technologies. While satellite and wireless microwave transmissions played important roles, without the invention of Lasers and optical fiber communication systems, current ubiquitous global telecom infrastructure interconnecting the global Internet over many continents, such transformations would have been impossible.

Prof. Charles H. Townes was honored by UC Berkeley in October 2005 in a special symposium with lectures by many fellow Nobel Laureates, when he reached 90 years' birthday anniversary. His invention of the idea of Lasers in 1958, and Dr Ted Maiman's first demonstration of the first laser, a Ruby Laser at 694.3 nm in 1960, pioneered the beginning of a brand new era in Modern Photonics. Forty-five years later in 2005, human society continues to benefit enormously from numerous novel applications of Lasers and Modern Photonics. One can easily find modern examples such as laser traps and tweezers for basic physical and biomedical sciences, ultra-high-capacity optical storage technologies like Blu-ray DVD and HD DVD for data and HDTV, large-screen display LCD and OLED TV, laser-based eye-sight correction techniques, Biophotonic Sensors, Nonlinear Multiphoton and SHG Bioimaging, etc. as well as the future potential of 3D HDTV.

Dr Charles K. Kao first demonstrated the potential of low loss in silica-glass optical fibers for long-distance telecommunications in 1966, and evangelized his idea in Europe, Japan, and the US afterwards. Dr Kao's idea was soon realized by Corning and AT&T Bell Labs researchers in 1970 with the demonstration of low-loss silica optical fibers. Some 35 years later in 2005, the impact of optical fiber for telecommunications worldwide is amazingly beyond simple description. The entire world is now connected with the speed of light through the global optical fiber telecommunication networks, via both the terrestrial and the undersea optical fiber communications links. This has in turn made possible many new possibilities, notably the global Internet, which has indeed transformed the human civilization in a new and fundamental way forever. Dr Kao has since been honored with many awards including the Marconi Fellowship Award, the Japan Prize, the Draper Prize (with Dr John MacChesney of Bell Labs and Dr Robert Maurer of Corning), to name a few, and was selected to be the *Man of the Century* in Asia in 2001 by the *Asia Week* magazine.

Whenever I travel on a Boeing or Airbus jumbo-jetplane flying over the Pacific or Atlantic Oceans, I am *always* amazed by the human ingenuity and collective success in the high-tech industry in the last century. Would Wright Brothers, two bicycle shop mechanics, when they achieved the first successful human flight in the air for 12 seconds over 40 m in 1903, have ever thought of the possibility that, 100 years later, 14-hour long transcontinental air travels over 10,000 km are 'routine' and taken for granted? Would Prof. Townes (and Prof.

N. G. Basov and Prof. A. M. Prokhorov) have ever thought about the possibility that, 45 years later, Lasers and associated photonics technologies have changed the world in such a fundamental way in every aspect of human lives? Would Prof. Kao have ever thought of the possibility that, 35 years later, long-haul optical fiber telecommunications would be so ubiquitous and make possible the global Internet and the global information infrastructure (GII) spanning the continents, through the deployment of undersea optical fiber systems using the low-loss single-mode fibers, dense wavelength-division-multiplexing (DWDM) and erbium-doped fiber amplifier (EDFA) technologies in the oceans?

This is human ingenuity and collective wisdom and contribution at work. The Masters and the Pioneers are special and unique people in that they pointed to the way and led the great adventure, and therefore deserve our special recognition and respect. What is also extremely important is that many others (hundreds or thousands) followed with enthusiasm and top-notch research and development effort, making the practical impacts so amazing, so impressive, and so long-lasting. Furthermore, whenever I thought about the simple and elegant concept of WDM, the rainbow principle, I was reminded that perhaps Sir Issac Newton should be credited with the first WDM concept: with his discovery of the multi-colored nature of the white light with his prism dispersion experiment, the first WDM demultiplexing experiment. Also, note that today's global telecommunications and information infrastructure is nothing if we were not communicating at the speed of light. So, a most fundamental and interesting question is: who invented the light, decided the speed of light, and the colors of the rainbow, and put it up in the sky?

Looking into the future, what would be the impact of Optical Fiber Communications in the next 10 years? The answer is: a Broadband Society with the ultimate Broadband Optical Access.

The future of the modern broadband information society will indeed rely on the optical fiber-based access infrastructure – an optical fiber-based access network, using novel photonics systems technologies to reach the ultimate human society's bandwidth needs. Even with xDSL and Cable Modem providing the current-generation high-speed Internet access, and wireless LAN of Wi-Fi and WiMAX providing the wireless last-link connection, optical fibers are essential for all these access platforms. It is generally agreed that eventually the ultimate broadband services with HDTV and high-definition multimedia streaming (and two-way HD video communications) would require Gb Ethernet to the home, so Fiber to the Home (FTTH), the apartments and the offices, etc. would sooner or later be needed. It is also essential that broadband fiber-based in-home networks be developed to accommodate the interconnection of emerging 'broadband intelligent information appliances.' All these broad-band FTTH and home networks would make it possible to have the ultimate information superhighway reaching inside the homes or apartments, and to usher the entire human society into the Broadband Future.

The root of the great telecom industry transformation we are witnessing now actually started way back in the mid-1980s when AT&T divestiture in 1984 was forced upon AT&T and seven RBOC were separated from AT&T. And I witnessed it first hand. Over the last 30+ years since graduating from UC Berkeley, I have worked at Bell Labs, AT&T Bell Labs, Bellcore, and Tyco Submarine Systems (now Tyco Telecom) R and D Labs, all the while living in Holmdel, New Jersey where Lucent Bell Labs (soon to be Alcatel-Lucent Labs?) still is in 2006. At Bellcore, which was designed to be the Bell Labs of RBOCs, we began to look at FTTH on behalf of RBOCs and working closely with technical expert teams from

RBOCs. In 1986, 20 years ago, on the first Bell South FTTH trial site, an up-scale new development ('Greenfield') outside Orlando, Florida, a team of Bellcore people were meeting with technical experts from seven RBOCs' R and D groups, to discuss the FTTH technology tradeoffs and various techno-economic issues. I remember seeing Dr Paul W. Shumate, Jr., the primary *FTTH Evangelist* whom we all know and respect, now the IEEE LEOS Executive Director, enthusiastically taking a photo of fiber being installed all way to the kitchen (FTTK, Fiber to the Kitchen). These were wonderful days full of FTTH pioneering spirit and enthusiasm, but unfortunately the FTTH system was too expensive then, and during the late 1980s and early 1990s, there was no Internet as it is today, and broadband 'killer applications' were nowhere to be found to justify the cost of significant FTTH deployments. Nevertheless, the research work at Bellcore in the late 1980s and early 1990s has indeed focused on broadband optical access networks, using DWDM (which I called HD WDM then, for high-density WDM), for high-capacity date and video distribution. Early researchers using DWDM for local access distribution included Dr Chuck Brackett and his team (Dr Stu Wagner was one of them) at Morristown, NJ, working on the LamdaNet experiments, and my research team at Red Bank, NJ, working on DWDM-based data and video distribution networks, which later led to EDFAs for RF analog video systems (work done mainly by Dr Winston Way) and hybrid AM/64QAM analog/digital video distribution experiments over lightwave systems. During this time Bell Labs researcher Dr Anders Olsson demonstrated the first really small-channel-spacing WDM point-to-point long-distance optical fiber transmission experiment which was later termed DWDM. All these have helped to pioneer the field of DWDM in the late 1980s and early 1990s, which then blossomed into practical commercial development of DWDM systems around the 1995 time frame by Ciena for SPRINT. The rest is history, well known to most of us.

With DWDM and EDFA technologies making great impacts in the global long-haul fiber networks over the last 10 years or so today there are excellent well-established fiber backbone network infrastructures worldwide. But the FTTH and optical fiber access networks had to wait until 2003–2004 when FTTH began to regain the new momentum. The various PON architectures were studied and technologies standardized by FSAN led by NTT, British Telecom, Bell South, France Telecom, and Deutsche Telekom initially and others who joined later. The FTTH systems technologies became more mature and cost-effective for Greenfield deployments, as compared with lower-bandwidth and distance-limiting copper-based technologies. While there are still no obvious 'killer applications,' the recent emergence of video/multimedia-rich Internet made it necessary to have higher-speed broadband Internet access, and the triple-play services are gradually becoming a reality. In Korea, Korea Telecom (KT) started nation-wide full deployment of DSL technologies, becoming an early leader of broadband deployment. In Japan, NTT, facing the competition from new players like Yahoo!BB in Japan, started significant FTTH deployment and gained much expertise and experience. In the US, in ADSL's competition against the Cable Modem services offered by the Cable MSO's, major Telcos such as SBC and Verizon began to seriously consider the FTTH deployment beyond the current copper-based DSL services, to provide broadband services with much higher bandwidths. Certainly, all these key players have their different strategies, timelines, roadmaps, and techno-economic considerations, but based on the available information and analyses, competition has indeed been the main driving force for most of the FTTH push in the last few years.

It is perhaps, therefore, the appropriate time to have this book on FTTH and Broadband Optical Access Technologies in 2006. FTTH and broadband access technologies are difficult subjects to have a comprehensive and cohesive treatment, unlike the long-haul backbone fiber networks where the photonics systems technologies are much more challenging, as very often in broadband access it is not just the technology that matters. Very often the technology, the architecture and the economics of FTTH are intermingled in a complex manner and the deployment strategies are also critically dependent on the services for now and for the future, as well as the practical issues of existing infrastructure investment, service take rate, governmental policy and regulatory issues, the geographical makeup, future service upgrade path, and very importantly, the degree and nature of the competitions. This book therefore tries to look at FTTH and broadband optical access from the views and experiences of (1) key service providers in different leading broadband countries, (2) FTTH equipment suppliers' technology experts who are very familiar with the technology and techno-economic strategy considerations, (3) key technical and strategic designers of broadband communities deploying FTTH, and (4) active researchers in the broadband optical access field looking into the next-generation optical access technology options and the competitive issues of the ultimate broadband access.

This book has 12 contributed chapters. We are fortunate that these expert contributors from Japan, USA, Sweden, Korea, UK, the Netherlands, and China agreed to share with us their experience and insights. This is certainly not a comprehensive list but is indeed a very good representative list in 2005–2006. Every chapter has its unique view and focus, but also with some overlaps with others, so that the readers can appreciate learning about the unique strategies and concerns of various leading broadband players, as well as seeing a common theme throughout the many chapters. Note that Japan, Korea, and Sweden are among the leading broadband services providing countries. The US would be the largest FTTH deployment countries soon because of SBC and Verizon's FTTH commitment. US is also where many broadband communities are being set up by the state or municipal authorities (e.g., in the State of Utah's Utopia Project), instead of the traditional telecom or Cable TV operators. China has also just started the FTTH trials, but its potential rapid ramp-up and sheer size for FTTH deployment is something to really watch in the next few years.

The titles and contributors of the 12 chapters in this book are as follows:

(1) Hiromichi Shinohara and Tetsuya Manabe of NTT, Tokyo, Japan wrote the first chapter on 'Broadband Optical Access Technologies and FTTH Deployment in NTT.' It summarized NTT's past experience with optical access technologies, current vision for the new optical broadband generation, and the strategic thinking in their FTTH deployments to meet the emerging broadband services needs. It also discussed the competitive forces in Japan driving the rapid deployment of FTTH. It is clear that NTT's world-leading FTTH effort is in response to the strong competition in Japan, without having the 'killer applications' requiring high-bandwidth which was thought to be essential to justify the FTTH deployment.

(2) Eugene Edmon, Kent Mccammon, Renee Estes, and Julie Lorentzen of SBC (now the new AT&T) California, USA, wrote the second chapter on 'Today's Broadband Fiber Access Technologies and Deployment Considerations at SBC.' In this chapter they declared that 'There is no doubt that the age of Fiber to the Home (FTTH) has arrived.' They then discussed their strategies and practical deployment considerations at SBC, in

which both FTTH for the Greenfield deployment and Fiber-to-the-Node or Neighbor-hood (FTTN) for the brownfield deployment were summarized. Both BPONs and GPONs are being considered for current and future deployments. Note for video services SBC will use a switched digital video (SDV) architecture with IPTV delivery, in sharp contrast to that of Verizon's FTTH strategy, in which the more traditional RF video overlay architecture is being deployed to accommodate the legacy systems and terminals. In this way, SBC will have a totally IP centric access network, but not the hybrid with both IPTV and RF video distribution, so a new digital set-top box will be needed for the new IPTV video services.

(3) Claus Popp Larsen, Örjan Mattsson & Gunnar Jacobsen from ACREO of Stockholm, Sweden contributed Chapter 3, 'FTTH – The Swedish Perspective', and presented a comprehensive view of the broadband services and market trends in the Swedish FTTH boom. Note that at the time of their writing, 'almost 20 % of all broadband connections in Sweden are of FTTH type, which is the highest share in world.' The chapter also briefly described key Swedish broadband players and their roles in the broadband Sweden. A short discussion of PON and P2P Ethernet-based broadband access architecture was also presented. Two interesting FTTH deployment examples, repre-senting both greenfield and brownfield FTTH communities in or near Stockholm, were also discussed.

(4) Kilho Song and Yong-Kyung Lee of Korea Telecom (KT), Seoul, contributed Chapter 4, 'Broadband Access Networks and Services in Korea.' The chapter started with a historical account of their Fiber-in-the-Loop (FITL) evolution strategies leading to their successful nation-wide FTTC/xDSL broadband deployment, then discussed the events and considerations leading to the recent BPON-based FTTH trials and broad-band services of 100 Mb/s to the user and practical deployment considerations. It also described the research and development activities on WDM PON for the next-generation FTTH optical access networks, to meet the ultimate broadband services needs in Korea.

(5) Jeff Fishburn of DynamicCity Metronet Advisors for Project UTOPIA, Utah, USA, contributed the 5th chapter on 'Broadband Fiber-to-the-Home Technologies, Strategies and Deployment Plan in Open Service Provider Networks: Project UTOPIA.' In Chapter 5, the concept and realization of an open service provider network was described in detail, which formed the basis of the design and implementation of the well-known multi-municipality community network in the State of Utah. This network, the Utah Telecommunications OPen Infrastructure Agency (UTOPIA) network, currently interconnects 14 cities over Utah. The operation model, the guiding principles, the access technology options and platforms, the outside plant and architectural considerations, and the network operation and management issues, were discussed in detail.

(6) Jeroen Wellen of Lucent, Bell Labs Advanced Technologies, The Netherlands, wrote the 6th chapter, 'High-Speed FTTH Technologies in an Open Access Platform – The European MUSE Project.' As an objective, the Multi Service Access Everywhere (MUSE) project targets first-mile solutions that are capable of providing 80 % of the European end-users with 100 Mb/s by 2010. In Chapter 6 there are in-depth discussions on the vision and missions of this well-known European Project MUSE, the future broadband services needs, the optical fiber access topologies, the comparison of PON and P2P architectures, and their cost factors. The need for Terabit Routers in the central

offices was discussed, and the new AsPON (asymmetric P2P/P2MP PON) was proposed and discussed with its relative merits. These and the various design and deployment considerations are the subjects of interest in the project MUSE, a representative broadband optical access project with the goal of making the European 'Broadband for All' vision a reality.

(7) Frank Effenberger of Motorola Telecom Access Solutions, Andover, Massachusetts, USA, contributed Chapter 7, on 'Residential Broadband PON Systems.' This chapter briefly reviewed the history of PON at British Telecom and FSAN, ATM-PON trials at NTT and Bell South, the evolution of various PON architectures including B-PON architecture for triple-play, and examined the system technologies and tradeoffs in B-PON system design and practical deployment considerations for FTTH and FTTP, including economic comparisons, the RF video engineering options, and ONU powering issues, etc.

(8) Dave Payne, Russell Davey, David Faulkner and Steve Hornung of British Telecom (BT), Martlesham Heath Ipswich UK, wrote Chapter 8 on 'Optical Networks for the Broadband Future.' Chapter 8 started with a brief review of the history of optical fiber access including the early PON work at BT, and the subsequent FSAN consortium activities, leading to several PON standards including A-PON, B-PON, and G-PON and with interoperability considerations. The chapter then discussed the services drivers for FTTH, such as Internet, digital imaging and HD video, digital CPEs, etc. as well as the three possible future broadband service scenarios leading to high-bandwidth demands in the near future. The chapter then also proposed a new long-reach PON access architecture, with the use of high-power EDFAs, as a possible lower cost future access architecture which enables large increases in bandwidth, geographical range, and optical splits compared to a conventional PON where only passive splitters but no EDFAs are used.

(9) James O. Farmer of Wave7 Optics, Inc., Atlanta, Georgia, USA, contributed Chapter 9 'An Evolutionary Fiber-to-the-Home Network and System Technologies – Migration from HFC to FTTH Networks.' Jim, an expert from the Cable TV industry, reviewed the state of Cable TV's hybrid fiber coax (HFC) networks in North America, and compared the HFC and FTTH architectures and system technologies for delivering broadcast video and IPTV, data and voice services in both architectures. He concluded that with an proper evolutionary path and deployment strategy, 'FTTH systems are really complementary to HFC systems and not competitive with them.' The chapter concluded that Cable TV operators can continue to operate their HFC plant while adding FTTH for business or new-build residential applications in an evolutionary manner, noting that FTTH systems offer superior quality and reliability, and operational costs are a small fraction of what they are with HFC. This will certainly be of great interest to Cable MSOs in their strategic planning and positioning in the evolution from their current HFC network to a future with FTTH for the ultimate broadband optical access.

(10) Jianli Wang and Zishen Zhao of Wuhan Research Institute of Posts & Telecom, Wuhan, China, wrote the 10th chapter on 'FTTH Systems, Strategies and Deployment Plans in China.' In this chapter, the current status of broadband access in China was reviewed, including the rapid services growth and broadband service requirements, the key players, and existing access technologies. The chapter indicated that the drivers for

FTTH/FTTO include increased bandwidth demand (IPVOD, HDTV, etc.), reduced system cost, and increasing competition. The target bandwidth need is estimated to be about 60 Mbps in an average family in China, assuming widespread HDTV penetration. The chapter then discussed China's FTTH initiative, and the FTTH trials in Wuhan and Chengdu, as well as other FTTH trials being planned in Beijing, Shanghai, Hangzhou, etc. The chapter also introduced major FTTH players and key FTTH products of EPON and P2P systems. The market size for FTTH is expected to be huge in China. Assuming 100 million FTTH subscribers within 5 years, it is estimated that China's average annual FTTH market, in terms of FTTH systems and equipments devices, components, accessories, deployment, services, etc., is between $30–40 billion USD. The chapter concluded with a brief but insightful summary of the deployment strategies to target these huge future FTTH market opportunities.

(11) Ton Koonen and Anthony Ng'oma from the COBRA Institute-Technical University of Eindhoven, The Netherlands, contributed Chapter 11, on 'Integrated Broadband Optical Fiber/Wireless LAN Access Networks.' In this chapter, a special topic of research interest, radio-over-fiber in an integrated fiber/wireless LAN system, is discussed. Radio-over-fiber techniques enable simplified antenna stations and centralization of radio signal processing functions, and thus may reduce the costs of broadband wireless communication networks. Optical frequency conversion techniques such as heterodyning and optical frequency multiplying (OFM) relax the requirements on the fiber link bandwidth. With the OFM technique, microwave signals can be delivered to simplified Radio Access Points through single-mode as well as multimode fiber networks; this enables the integration with fixed services such as Gigabit Ethernet in in-house multimode fiber data networks.

(12) Chinlon Lin of the Lightwave Communications Laboratory, Center for Advanced Research in Photonics, Chinese University of Hong Kong, wrote the final chapter, Chapter 12, on 'Broadband Optical Access, FTTH and Home Networks – the Broadband Future.' In this final chapter, the current status of FTTH and broadband optical access deployment trends were briefly summarized. A few examples of worldwide broadband competition in several leading countries were described to illustrate the strong competitions the established carriers and service providers are facing on a daily basis. An example of such competition between the key broadband players in Hong Kong, one of the leading telecom and broadband services cities in the world, was also illustrated, especially regarding Cable TV and the new IPTV services. The chapter also covered the topic of emerging broadband home networks. The current research trend of WDM PON and related topic (such as colorless ONUs) for the next-generation optical access network was then discussed. Finally, the vision of the Broadband Future and its potential impacts on the human civilization in the next century was also briefly discussed.

Overall, these 12 chapters should collectively provide a very good cross-sectional view of the FTTH and broadband optical access in network architectures, systems technologies, and deployment strategies up to the end of year 2005. However, one should note that broadband is a very fast changing area. It can be seen that currently there is a great momentum for FTTH and broadband optical access in several leading broadband countries, but full FTTH deployment worldwide will take many years. On the other hand, we are witnessing very strong competitions emerging and transforming the telecom landscape – many Cable TV companies offering VoIP telephone services over their HFC networks, Telcos offering IP TV, and VOD services over their broadband access infrastructure, and new comers like Yahoo!BB in Japan and Hong Kong BB using FTTH with high-speed, 1000 Mb/s or Gb Ethernet services competing against both Cable TV companies and traditional Telcos. In fact, because of the wide availability of broadband access to IP networks and Internet, many non-telecom Internet industry companies can also become competitive broadband IP service providers if they wish to, as many of them are already offering IP video streaming services over the Internet. The year of 2006 could be interesting and unique, as a critical year of large-scale FTTH deployment and worldwide broadband advances. Indeed year 2006 could be called the year of FTTH and the year of HDTV. With the expected popularization of large-screen HDTV and high-capacity HD DVD (and Blu-ray DVD) in 2006, and perhaps also the bandwidth-demanding IP HD video streaming VOD over the Internet, and HD video-based 'Blog,' etc. the great transformation into a true Broadband Future has just begun.

ACKNOWLEDGEMENTS

I thank Dr Charles K. Kao, now retired in Hong Kong from his last post of Vice-Chancellor of the Chinese University of Hong Kong, for being an encouragement to those in the field of optical fiber communications, a field he pioneered 40 years ago.

I also wish to thank Prof. Nick Holonyak, Jr. of the University of Illinois, Champaign-Urbana, and Prof. T. Ken Gustafson of the University of California, Berkeley, whose encouragement and kind help in my graduate student years were important to me.

I would also like to thank my dear wife, Helen (Huanhuan) Chou, for being loving, caring, and considerate, as well as being very patient with me. Her warm support at home has been essential to me and has made my writing and editing tasks much more enjoyable and more effective.

The editorial staff of John Wiley & Sons in Chichester, UK, have been most helpful. I wish to express my appreciation to Joanna Tootill, Richard Davies, and Birgit Gruber, for their assistance, suggestions, and patience.

Chinlon Lin

Holmdel, NJ, USA, and Shatin, Hong Kong
January 2006

Acronyms

AAA	Authentication Administration and Accounting
ADSL	Asymmetric DSL
ADSL	Asynchronous Digital Subscriber Line
ADSL 2$^+$	Asynchronous Digital Subscriber Line 2$^+$
AES	Advanced Encryption Standard
AGN	Aggregation Node
AN	Access Node
AON	Active Optical Network
APC	Angled Polished Connector
APON	ATM Passive Optical Network
ARPU	Average Revenue Per User
AsPON	Asymmetrical Passive Optical Network
ATM	Asynchronous Transfer Mode
AWG	Arrayed Waveguide Grating
BAN	Body Area Network
BFW	Broadband Fixed Wireless
Bidi	Bidirectional (transceiver, transmission)
B-ONT	Broadband Optical Network Terminal
B-PON	Broadband Passive Optical Network
Cable MSO	Cable (TV) Multi-System Operator
CAPEX	Capital Expenditures
CATV	Cable TV or Community Antenna TV
CLEC	Competitive Local Exchange Carrier
CM	Cable Modem
CMTS	Cable Modem Termination System
CNNIC	China Network Information Center
CO	Central Office
CO	Communication Operator
CPE	Customer Premises Equipment
CTS	Common Technical Specifications
CWDM	Coarse-Wavelength-Division-Multiplexing
DA	Distribution Area
DBA	Dynamic Bandwidth Assignment
DDN	Digital Data Network
DLNA	Digital Living Network Association
DOCSIS	Data-Over-Cable Service Interface Specifications
DSL	Digital Subscriber Line
DSLAM	Digital Subscriber Line Access Multiplexer
DVD	Digital Video Disk

DWDM   Dense-Wavelength-Division-Multiplexing
E1   European equivalent of DS1
EDFA   Erbium-Doped Fiber Amplifier
EML   Element Management Layer
EMS   Element Management System
EN   Edge Node
EPON   Ethernet Passive Optical Network
EU   European community
FA   Feeder Area
FBT   Fused Biconical Taper
FCC   Federal Communications Commission
FD   Feeder
FDM   Frequency-Division-Multiplexing
FITH   Fiber In the Home
FITL   Fiber In the Loop
FM   Frequency Modulation or First Mile
FSAN   Full Services Acces Networks (Consortium)
FTTB   Fiber To The Building
FTTC   Fiber To The Curb
FTTCab   Fibre To The Cabinet
FTTH   Fiber To The Home
FTTN   Fiber To The Neighbourhood
FTTN   Fiber To The Node
FTTO   Fiber To The Office
FTTP   Fiber To The Premises
FTTR   Fiber To The Riser
FTTx   Fiber To The x
FTTx   Fibre To The [building/curb/home/node, etc.]
FWA   Fixed Wireless Access
GbE   Gigabit Ethernet
GE   Gigabit Ethernet
GEM   G-PON Encapsulation Method
GEPON   Gigabit Ethernet Passive Optical Network
GigE   Gigabit Ethernet
GPON   Gigabit Passive Optical Network
G-PON   Gigabit Passive Optical Network
G-PON   Gigabit-capable Passive Optical Network
HAN   Human Area Network
HDTV   High-Definition TV
HE   Headend
HFC   Hybrid Fiber/Coax (network architecture)
HSI   High Speed Internet access
IAD   Integrated Access Device
IDATE   Institut de l'Audiovisuel et des Télécommunications en Europe
IEEE   Institute of Electrical and Electronics Engineers
IGMP   Internet Group Management Protocol

ILEC	Incumbent Local Exchange Carrier
IP	Internet Protocol
IPTV	IP Television
ISDN	Integrated Service Digital Network
ISP	Internet Service Provider
IT	Information Technology
ITU	International Telecommunications Union
ITU-T	International Telecommunications Union – Telecommunication Sector
L2	Layer 2
L3	Layer 3
LAN	Local Area Network
LCD	Liquid Crystal Display
MAC	Media Access Control
MDF	Main Distribution Frame
MDU	Multi-Dwelling Unit
MII	Ministry of Information Industry
MM	Multimode (fiber)
MoCA	Multimedia over Coax Alliance
MPEG	Motion Pictures Expert Group
MPOE	Minimum Point of Entry
MTU	Multi-Tenant Unit
MUPBED	Multi-Partner European Testbeds for Research Networking
MUSE	Multiservice Access Everywhere
NCTA	National Cable Telecom Association
NGN	Next Generation Network
NID	Network Interface Device
NML	Network Management Layer
NOBEL	Next Generation Optical Network for Broadband in Europe
NOC	Network Operations Centre
NT	Network Termination
OADM	Optical Add-Drop Multiplexer
OAM	Operation Adminisration Management
OAN	Optical Access Network Working Group
ODN	Optical Distribution Network
OECD	Organization for Economic Cooperation and Development
OLT	Optical Line Termination
OMCI	ONT Management and Control Interface
ONT	Optical Network Terminal
ONU	Optical Network Unit
OPEX	OPerational EXpenditures
OTDR	Optical Time-Domain Reflectometer
OXC	Optical Cross-Connect
P2MP	Point-to-Multipoint
P2P	Point-to-Point
PAN	Personal Area Network
PBX	Private Branch Exchange

PC	Personal Computer
PFP	Primary Flexibility Point
PLC	Power-Line Communications
PLC	Planar Lightwave Circuit
PMD	Physical Media Dependent
PMD	Polarization-Mode-Dispersion
PON	Passive Optical Network
POTS	Plain Old Telephone Service
PSTN	Public Switched Telephone Network
PTS	Post och Telestyrelsen (the national post and telecom agency)
PTT	Post, Telegraph and Telephone administration
PWLAN	Public Wireless Local Area Network
QAM	Quadrature Amplitude Modulation
QoS	Quality of Service
RBOC	Regional Bell Operating Companies
R&D	Research and Development
RFP	Request for Proposal
RG	Residential Gateway
RGW	Residential Gateway
RN	Remote Node
RTP	Real-Time Protocol
SAI	Serving Area Interface
SBU	Small Business Unit
SCM	Sub-Carrier Multiplexing
SCTE	Society of Cable Telecom Engineers
SDH	Synchronous Digital Hierarchy
SDSL	Symmetric Digital Subscriber Line
SDTV	Standard-Definition Television
SDV	Switched Digital Video
SFD	Start Frame Delimiter
SFU	Single Family Unit
SM	Single-Mode (fiber)
SOA	Semiconductor Optical Amplifier
SP	Service Provider
STB	Set-Top Box
T1	Digital transmission system operating at 1.544 Mbps
TC	Transport Convergence
T-CONT	Transmission CONTainers
TDM	Time-Division-Multiplexing
TE	Techno-Economic
TFF	Thin-Film Filter
UDP	User Datagram Protocol
UPC	Ultra-Polished Connector
UPS	Uninterruptable Power Supply
USD	US Dollars
VC	Virtual Circuit or Virtual Connection

VCSEL	Vertical-Cavity Surface-Emitting Laser
VDSL	Very-high bit-rate Digital Subscriber Line
VGW	Voice GateWay
VLAN	Vurtual Local Area Network
VOD	Video On Demand
VoD	Video on Demand
VoIP	Voice over IP
VOIP	Voice Over IP
V-OLT	Video Optical Line Termination
WDM	Wavelength-Division-Multiplexing
WDM-PON	WDM-based Passive Optical Network
WEPON	Wavelength Ethernet Passive Optical Network
WiFi	Wireless Fidelity (IEEE 802.11)
WiMAX	World Interoperability for Microwave Access (IEEE 802.16)
WISP	Wireless Internet service providers
WLAN	Wireless Local Area Network
XC	Cross-Connect
xDSL	x Digital Subscriber Line

Contributors List

Hiromichi Shinohara and Tetsuya Manabe
NTT Access Network Service Systems Labs, Ibaraki, Japan

Eugene Edmon, Kent Mccammon, Renee Estes and Julie Lorentzen
SBC Communications, San Ramon, California, USA

Claus Popp Larsen, Örjan Mattsson and Gunnar Jacobsen
Acreo, Stockholm, Sweden

Kilho Song and Yong-Kyung Lee
Korea Telecom, Seoul, Korea

Jeff Fishburn
DynamicCity Metronet Advisors, Project UTOPIA, Utah, USA

Jeroen Wellen
Lucent, Bell Labs Advanced Technologies, The Netherlands

Frank Effenberger
Motorola Telecom Access Solutions, Andover, Massachusetts, USA

Dave Payne, Russell Davey, David Faulkner and Steve Hornung
BT, Adastral Park, Martlesham Heath, Ipswich, UK

James O. Farmer
Wave7 Optics Inc., Atlanta, Georgia, USA

Jianli Wang and Zishen Zhao
Wuhan Research Institute of Posts & Telecom, Wuhan, China

Ton Koonen and Anthony Ng'oma
COBRA Institute-Technical University of Eindhoven, The Netherlands

Chinlon Lin
Center for Advanced Research in Photonics, Chinese University of Hong Kong, Hong Kong, China

1

Broadband Optical Access Technologies and FTTH Deployment in NTT

Hiromichi Shinohara and Tetsuya Manabe

NTT Access Network Service Systems Labs, Ibaraki, Japan

1.1 INTRODUCTION

The NTT access network encompasses NTT central offices and customer terminals and networks, and consists of transmission devices for individual customers in each NTT central office, transmission lines, and corresponding devices on the customers' premises. Unlike trunk relay lines between NTT central offices, access networks are not designed to carry concentrated communication traffic. Since cost reduction strategies such as multiplexing and task splitting cannot be employed, it is important that the component parts are as inexpensive as possible.

Providing an access network for ordinary customers that covers the length and breadth of the nation represents an enormous undertaking. For instance, the process of upgrading facilities to alleviate congestion in telephone services necessitated a substantial 20-year investment program that started in the 1950s, when NTT was known as the Nippon Telegraph and Telephone Public Corporation. Given the scale of the task, it is especially important to develop simple, well-designed technologies and systems for the construction, operation, and maintenance of the access network.

The transmission characteristics of the access network have a direct bearing on service provision. Equipment and system designs must strike an appropriate balance between the need to keep costs down at the initial stage and the need for the scalability and flexibility to accommodate the future diversification and modernization of the services.

Thus, the access network is governed by a wide range of conditions and a different level of requirements. Much is expected from NTT's optical technology research program, which began with the trunk network. In particular, it is hoped that the optical

Broadband Optical Access Networks and Fiber-to-the-Home: Systems Technologies and Deployment Strategies
Chinlon Lin   © 2006 John Wiley & Sons, Ltd

technology will provide NTT with greater flexibility to respond to the very rapid changes in the marketplace that followed privatization in 1985. These changes have included the introduction of competition, which has forced prices down; the shift towards mobile phones at the expense of landline services, which has affected equipment investment; and the paradigm shift brought about by the advent of Internet and broadband services.

The technologies involved in access networks are many and varied. They include optical transmission systems and optical cable and associated hardware, optical wiring based on demand and environmental considerations, and construction, cable laying, and maintenance systems. These technologies have evolved in line with changes in demand patterns and the socio-economic environment. Similarly, NTT's equipment investment and service development strategies are both influenced by technological progress.

This chapter provides a brief overview of the development of optical access technology in Japan, before moving on to a discussion of modern broadband services. Finally, it will examine the current state of optical access systems used in broadband services, and consider the use of access network technology in the broadband services of the future.

1.2   HISTORY OF OPTICAL TECHNOLOGY IN JAPAN

1.2.1   THE FIRST RESEARCH ON SUBSCRIBER OPTICAL TRANSMISSION SYSTEMS

Research into optical transmission technology for subscriber systems and existing access systems began in 1977, and research into trunk relay optical transmission technology began shortly thereafter. Trunk relay optical transmission research was geared towards such objectives as multiplexing, long-distance transmission, and relay-free system design. On the other hand, the goals of optical transmission technology for access systems were to achieve overall economic targets, design outdoor facilities whose simplicity of construction and operation made them suitable for mass production, and develop services that fully utilize the benefits of optical access networks.

1.2.2   FROM MULTI-MODE FIBER TO SINGLE-MODE FIBER

Multi-mode fiber was originally used as subscriber optical fiber. In 1988, NTT decided to introduce single-mode fiber, which had originally been developed for trunk networks, into the optical access network as well. This decision followed research into the best way to introduce the fiber into the subscriber network [1].

The decision to go with the potentially more costly single-mode fiber in the subscriber system, where cost considerations were paramount, was indeed a bold one, especially considering that most other countries were considering multi-mode fiber for subscriber systems. However, given the low-loss characteristics of single-mode fiber and its suitability for broadband applications, it was definitely the correct decision for the future. The introduction of single-mode fiber was made possible through the development of new technology in a number of areas. In particular, these developments related to high-precision manufacturing (using a process known as total synthesis), the more economical production of long-wavelength light sources, photoreceptors, and fiber interconnection and installation techniques that enabled optical fiber cable to be laid more efficiently. There were also significant economies of scale to be derived from the standardization of fiber with the trunk network.

The focus of research subsequently shifted to more advanced areas, including simple, low-cost wiring configurations such as multi-core cables, thinner wires and multi-core optical fiber connectors; low-maintenance, remote-operated gas-less systems; and fast, low-cost multi-core alternatives.

1.2.3 DEVELOPMENT OF CT/RT SYSTEM

Research into the use of optical technology in ordinary subscriber telephone services began in the late 1980s with the Central Terminal/Remote Terminal (CT/RT) system, which is employed in the section of the network equivalent to a feeder line. CT/RT delivers economies of scale through subscriber multiplexing, and allows systems to be scaled to suit line capacity. It also permits the diversification of services into areas such as Integrated Services Digital Networks (ISDN) and dedicated service support. As the results of research and development began to translate into cost reductions, it became possible to design smaller RT cabinets suitable for installation along roadways, on telegraph poles and inside buildings. The first indoor model, installed at the Chiyoda pilot plant in Tokyo in February 1993, featured 1000-strand multi-core optical fiber cable. The CT/RT system, which was later named FALCON, was deployed nationwide in Japan as a typical optical access system.

1.2.4 MOVING TOWARDS FTTH

Fiber-To-The-Home, or FTTH, cannot be achieved until we develop new forms of technology capable of both delivering substantial cost reductions across the board and facilitating a seamless transition from metal to optical cable systems. FTTH will require advances in a range of areas from hardware, such as cables and transmission system components, to nonhardware items, such as network design, construction, operation, and ongoing maintenance. Following the successful development of CT/RT, NTT continued research and development with the goal of realizing FTTH.

The Passive Double Star (PDS), which makes it possible for NTT central offices and multiple customers to share cables, was considered particularly promising with respect to cost reduction. The system consists entirely of passive components, and so unlike the CT/RT system it does not need an electricity supply (or the associated space requirements). Furthermore, the same cost-sharing benefits as the CT/RT system are complemented by impressive cost reductions in terms of the components used at the RT end. Another reason for the appeal of PDS was the network topology, which provides excellent compatibility with distribution services such as CATV.

A number of two-way communication systems were developed, including Synchronous Transfer Mode PDS (STM-PDS) [2,3], Asynchronous Transfer Mode PDS (ATM-PDS) [4–7], and Sub-Carrier Multiplexing PDS (SCM-PDS) [8,9].

1.2.5   OPTICAL SYSTEMS AT METAL-WIRE COSTS

In 1994, NTT announced its intention to accelerate the development of optical fiber access networks and provide an upgraded communication infrastructure for future broadband expansion. To achieve the objectives set out in the published statements, it would be necessary to achieve further economies with FTTH technology. To this end, in 1995 NTT embarked on research into cost reduction initiatives, particularly in relation to distribution systems and user-end optical distribution line systems. When the research program began, the cost per subscriber of optical access was some six to eight times greater than that for access via conventional metal wire; by 1997–1998, this had been reduced to approximately double the cost, and by 2000 the costs were the same. The core technological developments that enabled NTT to achieve its cost-reduction objectives were based on five key concepts:

(1) new approach to network configurations;
(2) low-cost optical PDS;
(3) low-cost optical fiber cables and associated technology;
(4) simplicity of installation and maintenance; and
(5) optical operation systems.

The newly developed technology first appeared in the optical access system, commonly known as the π system [10], which essentially involved grafting an optical access network onto an existing metal-wire telephone infrastructure. The optical network was brought as close as possible to subscriber homes (usually to the nearest telephone pole or the outer wall of an apartment building). It was terminated at a π-ONU device carrying the traffic of 10–20 subscribers, which was then linked to the existing metal cables. In July 1996, NTT announced its plan to deploy the π system to upgrade the access network.

1.2.6   ACCESS NETWORK OPTICAL UPGRADING PROGRAM

The construction of an optical access network requires a huge investment. It is therefore necessary to prepare a proper equipment investment program that takes account of service supply and demand considerations. In addition to upgrading metal-wire equipment to optical fiber, efforts were made to introduce technology developed for new services tailored to meet unknown demand levels and business requirements. The combination of the π system and optical fiber cable at the same cost as metal cable meant that optical access systems could also be extended to ordinary user districts where the upgrading of aging metal cable equipment was required.

1.3   TRENDS IN BROADBAND SERVICES

1.3.1   GROWTH OF BROADBAND SERVICES IN JAPAN

We first describe the trend of broadband services in Japan. Figure 1.1 shows the recent growth in the number of broadband users in Japan. We have three categories of broadband communications. The dominant broadband medium is the ADSL service with more than 13.6 million users. CATV comes next, but its growth is slackening off, and has become

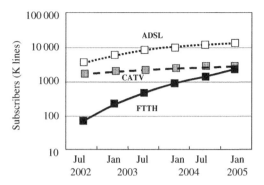

Figure 1.1 Increase in demand for high-speed Internet access. *Source*: Ministry of Internal Affairs and Communications, JAPAN.

nearly flat. The last is FTTH, which is growing rapidly and the number of users has exceeded 2.8 million and is still growing. Figure 1.2 shows the quarterly increment in new broadband access users. The number of new FTTH users surpassed the number of new ADSL for the first time in the first quarter of 2005. This is the Japanese network environment for which we are now developing technologies.

Another issue with the FTTH service is so-called 'regional disparities.' Although ADSL broadband services are steadily spreading throughout Japan, FTTH (which brings optical fiber directly to the home) is still confined to Tokyo and Osaka and their immediate surroundings, with little penetration into regional cities. Research and development programs will also need to focus on the issue of eliminating regional disparities with respect to broadband services.

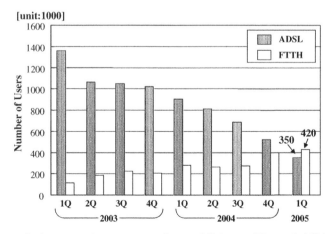

Figure 1.2 Quarterly increment in new users. *Source*: Ministry of Internal Affairs and Communications, JAPAN.

1.3.2   VISION FOR A NEW OPTICAL GENERATION

1.3.2.1   Broadband Leading to the World of Resonant Communication

In November 2002, NTT released a statement entitled 'Vision for a new optical generation: Broadband leading to the world of resonant communication,' describing its outlook for the ubiquitous broadband era of around 2008 based on optical technology [11].

In the optical fiber 'resonant communication environment' envisaged by NTT for the year 2008, video and related technology will provide more natural, real-time communication than has been possible with narrowband, thereby achieving the ultimate aim of communication, which is to demolish the barriers of time and distance. Breaking down the time barrier translates into more free time for people, while breaking down the distance barrier will allow individuals and corporations to expand their spheres of activity. This in turn promotes human activity on a global scale, transcending national and regional boundaries as well as industries and generations, and enables available knowledge and wisdom to be shared among individuals, companies, and borderless business operations. In this way, individuals and corporations will be able to access knowledge from around the world in what is termed the 'cooperative creation of knowledge.' This will revolutionize both the way we live and the way we do business.

The announced vision has been adopted as a common theme for the entire NTT Group, which is working to develop advanced business models by stimulating demand for optical services and creating business solutions. At the same time, partnerships with both domestic and overseas businesses in a variety of different fields are helping to promote and develop the resonant communication environment in industry.

1.4   OPTICAL ACCESS TECHNOLOGY BEHIND BROADBAND SERVICES

In this section, we describe the access network technology that is currently used to provide broadband services. The key requirements of the access network are economy, performance, and ease of use for both installation technicians and customers. Ease of use has a direct bearing on cost, since a well-designed network is easier and hence less costly to set up and maintain. And lower network setup and operating costs mean that services can be provided in both rural and metropolitan areas, thereby reducing regional disparities. Below, we discuss some of the technologies that have been developed with these objectives in mind.

1.4.1   OPTICAL ACCESS TECHNOLOGY FOR CURRENT BROADBAND SERVICES

1.4.1.1   Optical Access Systems

Figure 1.3 shows the configuration of an optical access network. Optical technology provides an economical way to transmit large volumes of data. Thus, an optical fiber system extends from NTT substations to points near customers' homes, such as footpaths and telephone poles. At the telephone pole or its equivalent, optical signals are converted into electrical signals, which are then fed down metal cables to individual households.

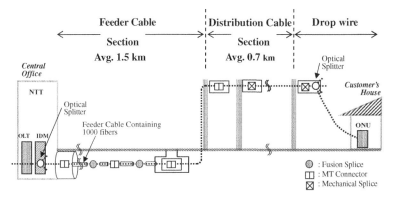

Figure 1.3 Overview of NTT's optical access network.

With FTTH, optical fiber is connected right through to customers' homes to provide faster broadband services. The FTTH system shown in the diagram employs the PON configuration, where a signal from one device at a substation is used by many customers. The signal is distributed to different customers via an optical coupler, which is a simple low-cost device made from glass. There are substantial economic benefits when a single substation device can service multiple customers.

NTT has finished developing a carrier-grade Gigabit EPON system for commercial deployment, as shown in Figure 1.4. The GE-PON system [12,13] is one example of an optical access system based on FTTH in a PON configuration. It provides a transmission capacity of 1 Gb/s through optical fiber at the substation and has an Ethernet interface. Key technologies provided by the GE-PON system include encryption and burst reception, both vital components of the PON setup, and bandwidth control and delay suppression, which are used to maximize system performance. NTT is also involved in the development of key carrier-grade operation functions.

Figure 1.5 is a diagram showing encryption technology. Encryption technology is employed during transmission from a substation to ONU devices in customers' homes. In a PON system, an optical star coupler is used to split the optical signal from the substation to the ONU devices. Since the entire signal reaches all ONU devices, a different form of encryption is used for each ONU device. This ensures that each ONU reads only signals intended for that customer.

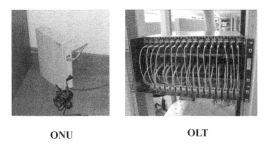

ONU　　　　　　　　　　OLT

Figure 1.4 Photographs of commercial gigabit EPON (GE-PON) system.

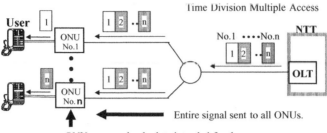

Figure 1.5   Downstream signal encryption.

Figure 1.6 is an illustration of burst transmission technology, which is employed during the transmission of an upstream signal from an ONU to an OLT. One function of burst technology is to prevent collisions between signals sent from customers' homes. A single receiver at the OLT accepts signals from multiple ONU devices. Since different signals cannot be received simultaneously, times for transmitting signals to the OLT are allocated to the ONU devices, and collisions are avoided by processing these transmissions in a set order.

Also, the different lengths of optical fiber between the coupler and the various ONU devices generate differences in path loss. This means that although each ONU device generates optical signals at the same strength, by the time the signals reach the OLT the power levels will vary somewhat. The receiver must be capable of processing signals of various strengths in rapid succession. This is achieved through the use of burst transmission technology.

Now let us look at some of the technologies used to maximize the performance of the PON system. The first is bandwidth control. Figure 1.7 shows the dynamic bandwidth allocation. Different users require different bandwidths at different times. Sometimes they need to transfer large amounts of data, while at other times they may need to transfer only a little. Rather than allocating the same bandwidth to every user, the system makes its 'best

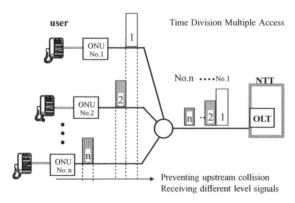

Figure 1.6   Receiving burst signals from different ONUs.

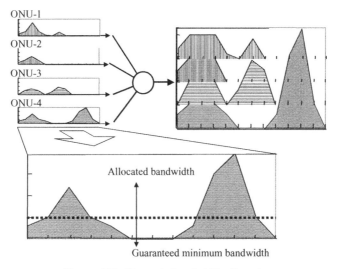

Figure 1.7 Dynamic bandwidth allocation.

effort' to redirect unused bandwidth to users who need it the most. In this way, all users are guaranteed a minimum level of bandwidth, and are also provided with additional bandwidth when required, which is allocated economically on a best-effort basis.

The priority control, shown in Figure 1.8, is another form of technology used to maximize the PON system performance. Video and audio data must be transmitted in real time. GE-PON gives priority to the regular transmission of signals associated with such services, and transmits other signals as time permits. Information classified as 'minimum delay class' is given priority, with regular transmissions at shorter cycles, while information classified as 'normal delay class' has lower priority. These functions are controlled equitably on a best-effort basis in order to provide users with an appropriate environment in which to enjoy audio and video services.

The technique of wavelength division multiplexing (WDM) significantly increases the transmission capacity of optical fiber by enabling signals with different wavelengths to be transmitted together along the same fiber. The ITU-T has established standards for the WDM of up- and downstream signals and video signals in each optical fiber in an optical access line. In accordance with these standards, the upstream line from the ONU to the substation

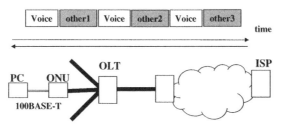

Figure 1.8 Low-delay transmission class applied (for example) to voice signals.

uses the 1.3 μm wavelength band, while the downstream line from the substation to the ONU uses the 1.48–1.5 μm wavelength band. For video signals, the standards specify up- and downstream lines that use different wavelengths in the 1.55 μm band. In this way, a single optical fiber can be used to carry upstream, downstream, and video signals, using prism-like optical elements (coarse wavelength division multiplexing or CWDM elements) to split and then recombine the various wavelengths.

NTT has developed the STM-PON, B-PON, and GE-PON systems to meet the increasing transmission capacity requirements. In light of the anticipated need for even greater capacity in the future, NTT continues to work on the development of new systems designed to harness the broadband capabilities of optical fiber.

1.4.1.2 Optical Access Installation

In order to provide FTTH services to customers at affordable prices, it is necessary to reduce both the cost of optical access systems and the expenses associated with the installation of optical fiber and associated hardware by simplifying the installation of optical drop cables, optical cabinets, and indoor wiring. NTT is working on developments in a number of areas designed to improve the installation process.

One example is bend-resistant optical fiber. Conventional optical cable requires a minimum bending radius of 30 mm during installation to avoid damage. NTT has developed an optical cable that can be bent at 15 mm without fear of damage. Figure 1.9 shows photographs of conventional and newly developed bend-insensitive indoor optical fiber cables. Not only is the cable easier to work with, it has a cleaner look since the bent sections are less prone to swelling. This improvement will have a significant impact on the installation of optical cable in ordinary homes. Furthermore, the introduction of a curled form of optical fiber with a low bending loss makes the cable as easy to handle as metal wire cable such as telephone receiver cords. Curled cable will greatly facilitate the ONU connection procedure shown in Figure 1.10 [14].

The installation of the optical cabinet used for the optical drop cable has been made easier. Figure 1.11 shows photographs of conventional and newly developed cabinets [15]. It is no longer necessary to provide a length of fiber with its sheath removed at the fiber termination. A length of exposed optical fiber must be provided at the end of the drop cable for connection to the indoor optical cord via a mechanical splice. The optical fiber can be connected directly to the drop cable using a special connector, and this can be accomplished

Bending radius: 15 mm   **Bending radius: 30 mm**

Figure 1.9 Photographs of newly developed bend-insensitive (left) and conventional (right) indoor optical fiber cables.

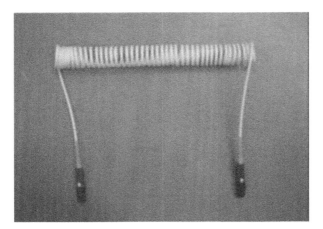

Figure 1.10   Photograph of flexible curled cable.

on site during the wiring process, eliminating the need for exposed optical fiber cord. Meanwhile, the bend-resistant fiber cord has reduced the required length from 30 to 15 mm. As a result, the volume of the new optical cabinet is around one-third that of the conventional cabinet.

Conventional drop cable contains metal wire as a reinforcing tension member, and this needs to be grounded to protect against, for example, lightning. The new cable uses non-conducting Fiber Reinforced Plastic (FRP) as the reinforcing tension member. Grounding is no longer required and this reduces the time needed for the installation of lead-in wires by around 20 %.

1.4.1.3   Wireless Access

In some cases it is not easy to service the demand for broadband access. Laying new optical fiber throughout existing apartment buildings, for instance, can present difficulties. WIPAS [16], a wide-area FWA system, was developed as a means of providing broadband services in such situations. Figure 1.12 shows the WIPAS system. Access is provided from a small wireless base station mounted on a telephone pole, which transmits wirelessly to outdoor

Newly developed cabinet.   Conventional cabinet.

Figure 1.11   Photographs of newly developed (left) and conventional (right) optical connector cabinets.

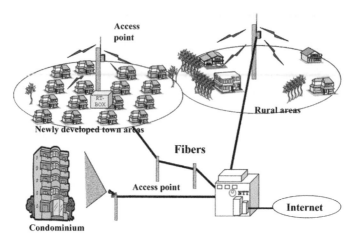

Figure 1.12   Wireless IP access system (WIPAS).

terminals usually installed on the verandas or rooftops of user households. WIPAS provides an inexpensive way to deliver broadband services tailored to local conditions in areas lacking a fully developed optical fiber network. In this way, wireless technology can be used to supplement FTTH in certain circumstances and provide a quick and inexpensive alternative source of broadband services.

1.4.2   BROADBAND ACCESS NETWORK TECHNOLOGY IN THE FUTURE

This section discusses the future of access technology. The target of the Medium-Term Management Strategy of the NTT Group is to provide optical access to 30 million customers in Japan by 2010. To achieve this target, it will be necessary to deliver further improvements in cost, performance, and ease of handling. We shall now examine these parameters with respect to optical access systems, optical access installation, and wireless systems.

1.4.2.1   Optical Access Systems

The primary development objective for optical access systems is performance enhancement, particularly with respect to cost, handling, and capacity. In the preceding section, we discussed optical systems that use different wavelengths for the up- and downstream communication lines. It should be possible to boost the cost-effectiveness and performance further by utilizing the broad optical wavelength domain of optical transmission routes via techniques such as WDM, which increases transmission capacity by transmitting multiple wavelengths through a single fiber. In terms of ease of use, systems can be designed to utilize existing networks and facilities. This approach minimizes installation costs and makes it easier to introduce and add new services. In addition, the introduction of common specifications for optical interface modules promotes component sharing for improved economy and user-friendly design.

There are two ways to harness the broad optical wavelength domain of optical fiber: wavelength division multiplexing, WDM, as described above, and code division multiple access, CDMA, which is already used for mobile telephone systems. We begin by looking at the general features of WDM in access systems.

The only extra initial investment required for WDM, in addition to the optical transceiver and transmission paths used in conventional systems, is the CWDM (coarse WDM) unit. There are two types of CWDM unit: one which splits light from the reception side into multiple wavelengths, and the other which combines optical signals of different wavelengths from the transmission side. Normally, the aim is to keep initial investment to a minimum, since there is rarely any demand for high-volume data transmission capacity at the outset. Ideally, the system design should be able to accommodate future capacity expansion in line with demand growth without great expense. WDM provides this feature. In most cases it is possible to expand capacity simply by adding more optical transceivers; there is no need to lay additional optical fiber. Similarly, if the original design capacity of 100 Mb/s is insufficient, it can be upgraded to 1 Gb/s, for instance, simply by installing a new 1 Gb/s transceiver. The changeover can be performed without affecting systems operating at other wavelengths.

The same advantages apply equally in the case of access systems, making the future expansion of system capacity an inexpensive proposition. WDM can be used in access networks as an economical means of boosting transmission capacity by adding different wavelength transceivers, but it also promises economic benefits in other areas. For instance, if demand for Internet services is accompanied by demand for video and other content delivery services, these new services can be accommodated on new wavelengths over the existing optical fiber. It is also possible to allocate separate wavelengths to users who require greater transmission capacity. In this way, WDM represents an economical way to utilize existing networks, allowing the flexibility to upgrade performance and functionality by increasing capacity and providing new services.

Thus, WDM makes maximum use of existing networks and equipment, and provides an economical means of delivering advanced services simply by upgrading terminals as required. The underlying system concept presupposes compatibility with existing networks and equipment. Next, we consider economic and ease of handling issues as regards WDM equipment and components.

Conventional WDM devices consist of integrated sets of electrical and optical components for each separate wavelength, gathered together on a board. With this approach, the specifications for each wavelength are slightly different. Fabrication costs are higher, since it is necessary to produce a large number of slightly different devices to make up a single unit.

In contrast, with the new devices the interfaces between the optical transceiver, electrical processor, and CWDM are common to all manufacturers, and are designed as interchangeable modules. This approach allows the mass production of fewer components for less total cost. Furthermore, the optical transceivers are integrated into the body of the unit, making the entire relay both smaller and cheaper to produce. The same principles also apply to the CWDM module [17–19].

NTT is using this approach to develop WDM optical modules designed to reduce the cost of WDM systems. The optical transceiver modules used to emit and receive light at various wavelengths and the CWDM modules are designed to be readily interchangeable and thus allow greater flexibility in terms of system configuration. For instance, it is possible to design a point-to-point transmission system, as used for relay transmission, consisting of the light-to-electric conversion of the user signal, an electrical processor, electric-to-light conversion, and an optical fiber transmission path. A wavelength converter can be configured with similar ease. The cascade format, used extensively in LAN design, can be configured to generate a number of different wavelengths (normally at the server), with successive repetitions of the reception, transmission, splitting, and recombining of different wavelengths for each user.

Further into the future, we can envisage the use of CDMA technology in optical communication systems. The advantages of CDMA are already well known in relation to mobile telephone systems. First, the code division format offers excellent resistance to signal interference, or in the case of optical communication, to light interference. Thus, if a customer inadvertently introduces a wavelength into the PON network that acts as interference, this will have only minimal impact on other customers. As with wireless communication, CDMA is a more efficient use of the available wavelengths (spectrum) than WDM. This technology could potentially prove vital in the event of a shortage of optical fiber bandwidth, as it will help us to make better use of the available wavelengths. Also, allocating codes rather than separate wavelengths to individual users allows more flexibility in planning user deployment.

In this way, the next-generation GE-PON, in the form of systems such as WDM that are designed to utilize the broadband capabilities of optical fiber, will deliver economic and performance benefits and be easier to handle.

1.4.2.2   Optical Access Installation

We now consider optical access installation technology. By creating small holes in the fiber, it is possible to change the fiber properties [20,21]. This technology enables to develop fiber with a very small bending radius and substantially reduced bending losses. The configuration and bending loss characteristics of this fiber, which is known as Hole-Assisted Fiber, is shown in Figures 1.13 and 1.1.4, respectively. Because such a fiber could easily be bent with a radius of less than 15 mm, fiber cords containing such fiber could be laid quickly indoors in any required configuration without compromising aesthetic requirements.

1.4.2.3   Wireless Access

Finally, we briefly discuss the wireless access technology used to complement optical access services. Wireless systems on the so-called 'last one mile' segment are capable of

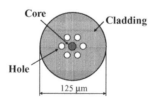

Figure 1.13   Configuration of hole-assisted fiber.

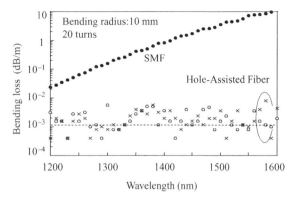

Figure 1.14 Bending loss characteristics of hole-assisted fibers.

transmission speeds of 50 Mb/s, which is about half that of optical drop fiber operating at 100 Mb/s. The aim is therefore to increase this speed to that of optical fiber. The deployment of additional base station antennas must not be too expensive, since this will occasionally be needed to ensure quality of service. In addition, it will of course be necessary to reduce the overall costs. Future development of wireless access technology is geared towards these objectives.

1.5 CONCLUSION

In this chapter, we briefly reviewed the history of optical communication systems in NTT, Japan. In terms of broadband services, the current focus of research and development is on servicing a range of user demands and reducing regional disparities in service levels. We discussed optical access technology for broadband services in the form of the GE-PON system, installation technology, and wireless access technology, and analyzed NTT's research and development with respect to the broadband environment. Finally, we looked at the development of network access technologies for future broadband services.

Such factors as the increasing interest in high-volume applications including Internet video distribution services and large data file exchange, the expectations of guaranteed services to complement existing best effort services, and the ongoing diversification of the network environment emphasize the need for the further development and enhancement of transmission services. Access networks represent the entrance through which customers gain access to such services. As such, more effort must be poured into the research and development of optical access networks in line with the advent of ever more advanced broadband services tailored to the market demand.

REFERENCES

1. Shimada S, Hashimoto K, Okada K. Fiber optic subscriber loop systems for integrated services: the strategy for introducing fibers into the subscriber network. *J Lightwave Tech*, 1987; **LT-05**: 1667–1675.

2. Fujimoto Y, Ohtaka A, Yamaki K, Miki N. STM Shared Access System for High speed IP Communications. Proc. European Conf. on NOC 2000, June 2000, pp. 110–117.

3. Yoshino M, *et al*. DBA Function for Broadband Passive Optical Network Systems. Proc. OHAN/FSAN 2001, April 2001, pp. 3.1-1–3.1-8.

4. Shibata Y, Okada K. Development of FTTH/B B-PON System in NTT. OHAN/FSAN Workshop 2001 F7, April 2001.

5. Ford B. Development of Fiber to the Home. OHAN/FSAN Workshop 2001 F5, April 2001.

6. Laurette M, Abiven J, Durel S. France Telecom FTTH/O Developments and Plans. OHAN/FSAN Workshop 2001 F6, April 2001.

7. Ueda H, Okada K, Ford B, Mahony G, Hornung S, Faulkner D, Abiven J, Durel S, Ballart R, Erickson J. Deployment status and common technical specifications for a B-PON System. *IEEE Com Mag*, 2001.

8. Way W I. Subcarrier multiplexed lightwave system design considerations for subscriber loop applications. *J Lightwave Tech* 1989; **7**(11): 1806–1818.

9. Kikushima K, Yoneda E. Erbium-doped fiber amplifiers for AM-FDM video distribution systems. *IEICE Trans* 1991; **E74**(7): 2042–2048.

10. Ikeda K, *et al*. Development of New Optical Access System. 8th International Workshop on Optical/Hybrid Access Networks, Atlanta, 1997.

11. Vision for a New Optical Generation – Broadband Leading to the World of Resonant Communication –, NTT News Release, 11/25/2002, (http://www.ntt.co.jp/news/indexe.html)

12. Ochiai K, Tatsuta T, Tanaka T, Yoshihara O, Oota N, Miki N. Development of a GE-PON (Gigabit Ethernet – Passive Optical Network) System, NTT Technical Review, Vol. 3, No. 5, May 2005, pp. 51–56.

13. Abrams M, Becker PC, Fujimoto Y, O'Byrne V, Piehler D. FTTP Deployments in the United States and Japan – Equipment Choices and Service Provider Imperatives. *IEEE/OSA J Lightwave Tech*, Jan. 2005; **23**(1): pp. 236–246.

14. Hiramatsu K, Kurashima T, Araki E, Tomita S. Development of Optical Fiber Curl Cord. Proc of Comm Conf IEICE, B-10–3, pp. 374. March 2004.

15. Aoyama H, Tanaka H, Hoshino Y, Oda Y. Optical Wiring Technology for Home Networks for a Service-ready and Low-cost FTTH Service. NTT Technical Review, Vol. 3, No. 4, April 2005, pp. 33–37.

16. Nidaira K, Shirouzu T, Baba M, Inoue K. Wireless IP Access System for Broadband Access Services. Int Conf Comm IEEE, Vol. 6 WC15–1, June 2004, pp. 3434–3438.

17. Small Form Factor Pluggable (SFP) Multi Source Agreement, INF-8074, http://www.schelto.com/SFP/

18. Kimura H, Yoshida T, Kumozaki K. Compact PLC-based Optical Transceiver Module with Automatic Tunable Filter for Multi-rate Applications. IEE Electron Lett 2003; **39**(18): 1319–1321.

19. Yoshida T, Kimura S, Kimura H, Kumozaki K, Imai T. New 156M/2.5Gbit/s Multi-rate SFP Transceiver with Automatic Sensitivity Switching. Proc 10th OptoElectron Comm Conf (OECC 2005), July 2005, pp. 204–205.

20. Tajima K, Nakajima K, Kurokawa K, Yoshizawa N, Ohashi M. Low-loss Photonic Crystal Fibers. Proc Optical Fiber Comm Conf (OFC 2002), ThS3, 2002; pp. 523–524.

21. Nakajima K, Hogari K, Zhou J, Tajima K, Sankawa I. Hole-assisted fiber design for small bending and splice losses. IEEE Photon Tech Lett 2003; **15**(12): 1737–1739.

2

Today's Broadband Fiber Access Technologies and Deployment Considerations at SBC

Eugene Edmon, Kent Mccammon, Renee Estes, Julie Lorentzen
SBC Communications, San Ramon, California, USA

2.1 INTRODUCTION

There is no doubt that the age of Fiber to the Home (FTTH) has arrived. The fits and starts of the past are over. In North America, every major Telecom provider and many smaller service providers are proceeding with deployments or trials of fiber technologies, predominately Passive Optical Networks (PONs). Most providers in Asia are also deploying or are extremely interested in doing so. Much of this activity can be attributed to achieving installation cost points that are competitive with alternative copper-based products that can deliver at least voice and data. The key to the realization of FTTH is the standardization of PON, allowing component vendors and system vendors to focus on a single solution.

Favorable regulatory policies are also helping. In the United States, the Federal Communications Commission recently clarified rulings about FTTH to the effect that no unbundling will be required for new deployments. This decision reflects the competitive environment for new home developments where builders are seeking providers of full service networks with advanced capabilities.

In this chapter we review the background of the extensive work we have done in SBC on the path to current deployment plans. The overall effort started with substantial investment in standards work with the FSAN group of companies. Then the opportunity arose to use PON technology as specified by ITU-T to meet the challenge of providing an FTTH network at a new building site, Mission Bay, in San Francisco, CA.

After the success of the Mission Bay network, SBC moved forward, with Verizon and BellSouth, to release a joint RFP for FTTH that specified compliance with ITU-T PON standards. This joint effort was successful in giving direction to the industry about the desired product standards.

Broadband Optical Access Networks and Fiber-to-the-Home: Systems Technologies and Deployment Strategies
Chinlon Lin © 2006 John Wiley & Sons, Ltd

While SBC is planning extensive FTTH deployment, the full story of fiber deployment also includes extensive use of Fiber to the Node (FTTN). Use of the dual platforms is expected to provide the most cost-effective strategy for delivering a 'triple play' of services, including voice, high-speed Internet access (HSI), and video. In this chapter we review the technologies of both fiber solutions.

In the following material we will discuss the FTTH and FTTN architectures, describe some of the key fundamentals of PON technology, review our learnings from current deployment efforts, show how the FTTH will complement our FTTN plans, and describe the general solutions we are working on for home networking.

2.2 FIBER-TO-THE-NEIGHBORHOOD (FTTX) ARCHITECTURE

Figure 2.1(a) provides a high-level illustration of the SBC strategy. For greenfield areas, we will deploy FTTH. In brownfield/overbuild areas, we will deploy a FTTN platform that utilizes VDSL in the last mile. Both networks will support switched digital video (SDV) employing IP as the end-to-end protocol.

2.2.1 FTTH ACCESS ARCHITECTURE

Figure 2.1(b) provides a high-level illustration of the FTTH architecture that will be deployed in SBC. It is an integrated platform capable of providing telephony, data, and video services to residential areas, which may include a mix of single-family homes/units (SFUs), multi-dwelling units (MDUs), small business offices/units (SBUs), and multi-tenant offices/units (MTUs). The system contains seven basic building blocks:

- The Optical Network Terminations (ONTs), which interface the system to customers' home telephony, data, and video networks.

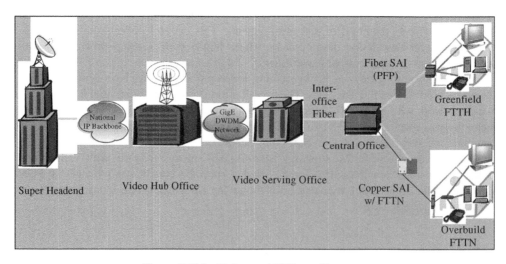

Figure 2.1(a) End-to-end FTTx architecture.

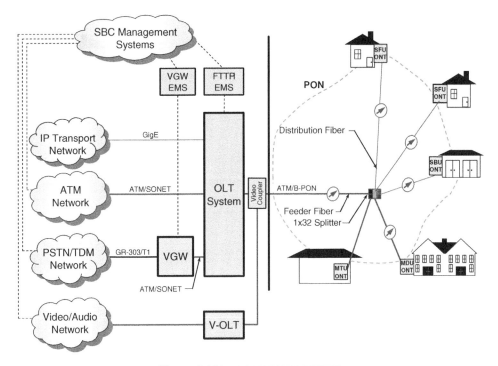

Figure 2.1(b) FTTH access network.

- The Optical Line Termination System (OLT), which manages the ONTs, aggregates/cross-connects voice and data traffic from multiple PONs/services, and interfaces the system to core transmission networks.
- The Voice Gateway (VGW), which interfaces the system to the legacy PSTN/TDM network.
- The Video OLT (V-OLT), which receives and amplifies/regenerates video signals from a video headend and inserts local video signals. (As described below, SBC has no plans to deploy this element.)
- The Element Management Systems (EMSs), which interface the different network elements to SBC's core operations network(s).
- The ATM network, which aggregates/switches ATM traffic from multiple core networks to the OLT(s).
- The Passive Optical Network (PON) or Optical Distribution Network (ODN), which connects the ONTs to the OLT and provides the optical paths over which they communicate.

Currently, the FTTH architecture is based on the ITU-T B-PON access network, which is standardized in the G.983 series of recommendations. Eventually, it will migrate to the ITU-T G-PON network standardized in the G.984 series of recommendations.

The B-PON network is an ATM-based, integrated platform capable of providing telephony, data, and video services to residential and small business customers over a single

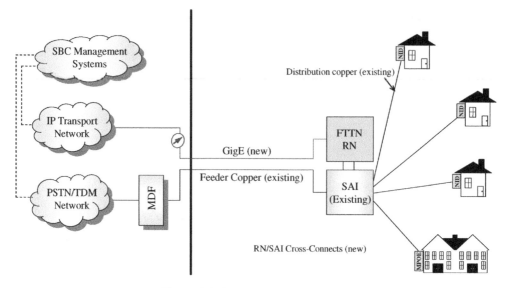

Figure 2.1(c)   FTTN access network.

fiber. One feature of this network is an overlay wavelength that can be used to provide conventional video services. While this is a compelling feature, it will not be implemented in SBC because of our desire to have a common product suite and transport network for both FTTH and FTTN. Instead, video over FTTH will be based on the SDV IPTV format and will be carried over the B-PON 'basic' bands.

2.2.2   FTTN ACCESS ARCHITECTURE

Figure 2.1(c) provides a high-level illustration of the FTTN architecture that will be deployed. It is an integrated platform capable of providing telephony, data, and video services to residential customers. The access network has basically one key new network element: the FTTN Remote Node (RN). Broadband transport/services are provided to this element from/to the Central Office (CO) by Gigabit Ethernet (GigE) fiber; these are then cross-connected to existing twisted-pair copper in the Serving Access Interface (SAI), and are transported to/from the customer using Ethernet-based VDSL.

2.3   ITU-T PON STANDARDS

The Full Service Access Network (FSAN) group, which consists of 22 operators and approximately 30 vendors from around the world, has been highly instrumental in the development and ongoing enhancements of the PON standards. Two of the FSAN Working Groups provided the foundation and much detailed input on B-PON and G-PON to ITU-T. The Optical Access Network (OAN) Working Group provided input to ITU-T Question 2/Study Group 15, under which the G.983 and G.984 series were developed; the Operations and Maintenance (OAM) Working Group detailed specifications to ITU-T Question 14/ Study Group 4, under which the Q.834 series was developed.

2.3.1 ITU-T G.983 B-PON STANDARDS SERIES

Standards pertaining to B-PON have been developed and published through two ITU-T Recommendation series: ITU-T G.983 and ITU-T Q.834. The G.983 series began with standardization of the physical and transmission convergence layers, and of the ONT management and control interface (OMCI). Later, standards support was added for an overlay wavelength, dynamic bandwidth assignment (DBA), survivability, increased line rates, enhanced security, and enhanced ONT management. The Q.834 series of recommendations pertains to management of B-PON networks. Table 2.1 provides a listing of key features of the G.983 and Q.834 Recommendations.

Table 2.1 Key features of ITU-T G.983.x and Q.834.x Recommendations.

ITU-T Recommendation		Key features
G.983.1	Broadband optical access systems based on passive optical networks (PON)	• Provides specifications for 155/155 Mbps and 622/155 Mbps systems with 20 km reach
G.983.1 Amendment 1	G.983.1 Amendment 1	• Extends G.983.1 to support 622/622 Mbps systems
G.983.1 Amendment 2	G.983.1 Amendment 2	• Extends G.983.1 to support 1.244 Gbps downstream and to support AES security option
G.983.2 (2002)	ONT management and control interface specification for ATM PON	• Includes ONT management/ control support for: voice (POTS); data (MAC bridged LAN); and video (card/port functions)
G.983.3	A broadband optical access system with increased service capability by wavelength allocation	• Enhanced wavelength plan to allow WDM expansion including adding video broadcast service • OLT downstream wavelength of 1480–1500 nm specification allowing wavelength expansion over B-PON
G.983.4	A broadband optical access system with increased service capability using dynamic bandwidth assignment	• Improves upstream bandwidth utilization • Allows flexible provisioning of bandwidth
G.983.5	A broadband optical access system with enhanced survivability	• Addresses protection of G.983.1 systems
G.983.6	OMCI specifications for B-PON systems with protection features	• Enhancements to G.983.2 to support protected B-PON systems

(Continued)

Table 2.1 (*Continued*)

ITU-T Recommendation		Key features
G.983.7	OMCI specification for DBA B-PON system	• Specifies enhancements to G.983.2 to support DBA-capable ONTs
G.983.8	OMCI support for IP, ISDN, video, VLAN tagging, VC cross-connections, and other select functions	• Includes enhancements to G.983.2 to support IP, ISDN, video, VLAN tagging, and VC-cross-connections
G.983.9	OMCI support for wireless Local Area Network interfaces	• Enhancements to G.983.2 to support wireless LAN interfaces
G.983.10	OMCI support for Digital Subscriber Line interfaces	• Enhancements to G.983.2 to support DSL interfaces
Q.834.1	ATM-PON requirements and managed entities for the network element and network views	• Describes B-PON information model, focusing on NML/EML interface
Q.834.2	ATM-PON requirements and managed entities for the network view	• Describes B-PON information model, focusing on NML/EML interface
Q.834.3	A UML description for management interface requirements for B-PONs	• Defines part of the management aspects for network resources
Q.834.4	A CORBA interface specification for B-PONs based on UML interface requirements	• Defines CORBA interface for B-PON

2.3.2 *ITU-T G.984 G-PON FOR HIGHER SPEEDS*

SBC began deployment with a standards-based B-PON access network. While B-PON meets SBC's current needs for PON, G-PON (based on the ITU-T G.984 Recommendation series) is seen as the best direction for continued full service networks supporting IP video. Table 2.2 gives a brief overview of the G.984 Recommendations. Use of the G-PON Encapsulation Method (GEM) protocol will allow for highly efficient delivery of Ethernet packets over G-PON.

GEM utilizes flexible frame sizes to transport data and also allows frame fragmentation. Using GEM, a header is applied to each data frame or frame fragment that is destined for or coming from a user. This header provides information including the length of the attached frame fragment in order to support delineation of the user data frames and a traffic identifier used to support traffic multiplexing on the PON. When an Ethernet packet is mapped into a GEM frame, the Preamble, and Start Frame Delimiter (SFD) bytes are stripped off and no Inter-Packet Gap is needed. This, combined with GEM's flexible frame size and support for frame fragmentation, allows for efficient delivery of Ethernet-based traffic over the PON.

In addition, the G-PON protocol allows for support of native TDM over GEM along with Ethernet packets. TDM services may also be supported on G-PON via a circuit emulation

Table 2.2   Key features of ITU-T G.984.x Recommendations.

ITU-T Recommendation		Key features
G.984.1	Gigabit-capable Passive Optical Networks (G-PON): General characteristics	• Provides summary of fundamental characteristics • Line rates ranging from 1.244/0.155 to 2.488/2.488 Gbps • 20 km max physical reach, 60 km max logical reach
G.984.2	Gigabit-capable Passive Optical Networks (G-PON): Physical Media-Dependent (PMD) layer specification	• Provides PMD layer specifications

G.984.3	Gigabit-capable Passive Optical Networks (G-PON): Transmission convergence layer specification	• Provides TC layer specifications
		  • ATM-mode, GEM-mode, or dual mode operation
		  • AES security
		  • Pointer-based media access control

G.984.4	Gigabit-capable Passive Optical Networks (G-PON): ONT management and control interface specification	• Provides ONT management/control support

approach. Support for both Ethernet and TDM on a common access system is a powerful combination to expand the suite of full-service network applications for G-PON. Enhancements to G-PON are a near-term active area of work in the FSAN OAN Working Group, with the intent to finalize in 2005 for possible 2006 deployments.

Another key aspect of G-PON is enhanced security and privacy protection using the Advanced Encryption Standard (AES). Similar to our B-PON deployment, our network will require the optical reach and hardened ONT options that are supported by the G-PON Recommendations. FSAN and ITU-T continue with enhancements to the G-PON standards to meet evolving requirements of worldwide operators.

2.3.3   THE ROLE OF STANDARDS IN INTEROPERABILITY

A goal of service providers, and a key factor in widespread deployment, is to establish equipment interoperability that will allow a multi-vendor supply environment. Today, the OAN group is actively working on issues pertaining to interoperability of B-PON equipment in a multi-vendor environment and has organized a series of interoperability efforts.

In March 2004, multi-vendor B-PON interoperability was demonstrated during conformance testing that included the TC layer, optical levels, and OMCI. Following this, in June 2004, an interoperability demonstration showing Ethernet service level interconnectivity among ITU-T compliant B-PON systems was exhibited by FSAN members during the ITU-T All Star Network Access workshop. Multi-vendor voice interoperability over B-PON systems was demonstrated in September 2004; four OLT vendors, eight ONT vendors, and one test vendor participated in this event. The series of interoperability events is described on

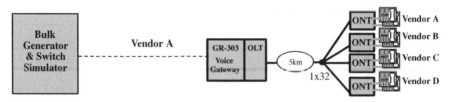

Figure 2.2 Configuration for voice interoperability demonstration.

the FSAN website at http://www.fsanweb.org/news.asp and http://www.fsanweb.org/presentations/page310.asp.

Figure 2.2 depicts the multi-vendor configuration for the voice interoperability event. FSAN continues to develop interoperability among OLT and ONT vendors at all layers, including the service level. The strong support for these interoperability efforts from both the operator and vendor communities serves as an indicator of the interest within the industry in developing and deploying standards-compliant B-PON systems capable of interoperating in a multi-vendor environment.

The operators within the OAN group are working on a document called the Common Technical Specifications (CTS) for B-PON systems. This document includes specifications from the physical layer up to the services layer, and is intended to provide additional benefit to the industry in developing systems beyond the protocol and physical layer of the G.983/G.984 Recommendations. The development of common specifications worldwide can build volume and lower costs for fiber access systems as well as provide additional structure to direct future interoperability efforts.

Along with its contributions towards further enhancements to B-PON and G-PON specifications and interoperability, FSAN continues to be a vibrant group working on the future use of fiber in access networks. SBC has realized great benefits from the availability of FSAN compliant access systems. The mechanism to enhance and maintain in both FSAN and ITU-T is vital to keep the system expandable to new services in a standards-based implementation with the level of specification necessary for interoperability. SBC has continuously contributed to the FSAN work activities since 1997 and will continue to work in FSAN to develop next-generation access systems.

2.4 PON TECHNOLOGY BACKGROUND

In this section, we review some of the key technology features of PONs that make them so attractive for FTTH.

2.4.1 UPSTREAM BANDWIDTH ASSIGNMENT

A key feature of PON is the aspect of shared bandwidth, which raises the question of how individual users will be allocated time/bandwidth on the network. Downstream allocation is relatively straightforward because there is one transmitter and bandwidth is broadcast to all ONTs on the PON. In the upstream direction, however, a problem of access control arises with the multiple upstream transmitters. PON solves this problem with grants from the headend controller to each ONT. Grant timing is communicated in downstream messages to all ONTs, which inform the ONTs of their time slots.

Utilization of the upstream bandwidth on the PON can be improved through the implementation of Dynamic Bandwidth Assignment (DBA), which was introduced in ITU-T Recommendation G.983.4. With DBA, the OLT monitors the upstream bandwidth requirements of the ONTs and adjusts how it distributes grants accordingly.

G.983.4 introduced the concept of Transmission CONTainers (T-CONTs), each of which can aggregate one or more physical queues into a logical buffer. When DBA is employed, grants are associated with individual T-CONTs. Each T-CONT has bandwidth-related parameters associated with it that are used in the grant assignment process. Four categories of bandwidth are identified for DBA – fixed, assured, nonassured, and best-effort (listed from highest to lowest priority in terms of granting). Five T-CONT types are defined with different combinations of these bandwidth categories. Each ONT can support one or more T-CONTs; the specific T-CONT type or combination of T-CONT types on a given ONT is tailored to support the quality of service (QoS) requirements of the traffic flows on the ONT (G.983.4 provides a guide indicating which QoS categories are supported by which T-CONT types). For example, T-CONT type 5 is the most flexible type, accommodating all four bandwidth categories, and a single type 5 T-CONT on an ONT can be used to accommodate multiple traffic flows with a variety of QoS.

There are two 'flavors' of implementing DBA – idle cell monitoring and status reporting. In idle cell monitoring, the OLT monitors how many idle cells are being sent from each T-CONT. In status reporting, the ONTs send reports to the OLT regarding the queue status/length of each T-CONT. The OLT then adjusts the allocation of grants based on the information it obtains regarding the T-CONTs. Particularly for scenarios where heavy utilization of the PON is found, it is expected that status reporting provides some advantages over idle cell monitoring in aspects such as cell delay. As such, deployment scenarios with heavy utilization involving MDUs and small businesses, for example, would be expected to benefit from implementing the status-reporting method of DBA.

2.4.2 RANGING

The physical distance between the OLT and the ONTs on the PON varies, which means that signals require different times to get to and from the different ONTs. A technique called ranging is used to adjust the timing between each ONT and the OLT. The ranging protocol in ITU-T Recommendation G.983.1 allows placement of an ONT anywhere within a 20 km distance from the OLT, providing flexibility in ONT placement in the ODN. To initiate ranging, the OLT sends a specific grant to the ONTs to trigger the ranging process and opens up a window during which it can receive ranging information from the ONTs. Upon receipt of this grant, an ONT sends a ranging cell back to the OLT. Based on the elapsed time between when the OLT sends the ranging grant and when it receives a ranging cell from an ONT, the OLT can determine the appropriate equalization delay to assign to that ONT.

2.4.3 SPLITTERS

The splitter can be considered a defining feature of PON, since it is the key technology that allows the access network to be electrically passive. A major cost advantage of PON is the reduced fiber requirements versus a point-to-point architecture with fiber direct from the CO to each home. This cost reduction is achieved using the splitter to take one fiber from the CO and serve up to 32 homes in the SBC network.

Significant improvements in splitter technology have occurred in the last 4 years, including improvement/advances in optical performance, reliability, and cost per port. These advances contributed to the selection of the PON topology for FTTH at SBC. Today, the performance of splitters has reduced excess loss to 1–1.5 dB above the ideal loss of the device and nonuniformity to less than 2 dB over a wide wavelength range and wide temperature range while achieving satisfactory cost per port.

Advances in fabrication and packaging technology for passive fiber splitters were driven by market demand for increased optical performance in CATV fiber distribution and optical networking applications. Reducing splitter excess insertion loss and uniformity of loss variation across all ports focused supplier investment in large port size (1×16, 1×32) devices using the planar lightwave circuit (PLC) technology. PLC fabrication involves creating optical waveguides in a planar substrate such as silica to form a splitting function. SBC has selected the 1×32 size predominantly with the 1×16 size a second option when additional fiber reach is required.

PON deployments in Japan were increasing and creating a larger market for splitters for PON applications. Industry leading suppliers provided improvements in reliability assurance programs to meet the requirements for splitters placed in the outside plant environment where temperature and humidity are not controlled.

SBC evaluated the performance of splitters fabricated using the fused biconical taper (FBT) process and the PLC process starting in 2000. The FBT fabrication process involves drawing two or more optical fibers together under heat and pressure to achieve the appropriate coupling ratio. Splitters with larger sizing are made by joining multiple 1×2 devices in a cascading fashion and providing a larger package size than PLC devices. We review the assessment of splitter optical performance and reliability collected since 2000 in advance of our early FTTH deployments, trials, and planned rollouts of FTTH.

2.4.3.1 Splitter Performance

SBC splitter requirements for loss and uniformity span the three wavelength bandpass regions designated for ITU-T G.983.3 B-PON systems, including a WDM overlay option for video signaling. The three bandpass regions have center wavelengths of 1310, 1490, and 1555 nm. The splitter optical performance is dependent on the splitter fabrication technology. Figure 2.3 illustrates the optical performance of two different 1×32 devices, one fabricated with a PLC process and the other with a FBT process. The results illustrate the variation of loss over the bandpass of interest for PON systems.

Each line represents the loss from the single input to one of the 32 output ports of the device. While the FBT device loss shown here does provide low loss windows centered on the commonly used 1310 and 1550 nm bands, the PLC device is more uniform over the bands and up to 1640 nm. A fiber access network infrastructure with a uniform loss across a large wavelength range simplifies the optical test and acceptance of the fiber network. Evolution strategies to additional wavelength bands in the future are simplified by the selection of PLC-based splitters with loss that has very low dependence on the wavelength.

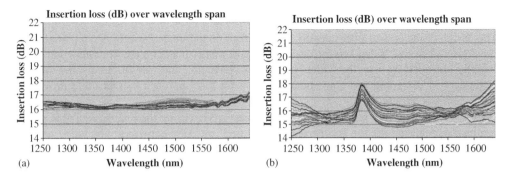

Figure 2.3   Loss versus wavelength comparison of 1×32 splitters made with (a) PLC device and (b) FBT device.

A PON installation with nonuniform loss and wider variance in loss with wavelength complicates planning due to uncertainty in the loss for any new wavelengths being added to the PON in the future. SBC has found selection of PLC to be advantageous as we plan for future WDM expansion on our PON deployments.

2.4.3.2   Splitter Reliability

Splitter devices placed within the SBC footprint require reliability under environmental extremes ranging from the elevated summer heat and humidity in Southern Texas to the low winter temperatures in Northern Michigan. Based on a comprehensive review of environmental and mechanical testing results from several PLC providers, we found that early issues of reliability with certain PLC devices were no longer a fundamental concern. Reliability results from PLC suppliers verified the availability of splitters with the required robustness for placement in uncontrolled environments.

2.4.3.3   Splitter Conclusions

Splitter evaluations have provided SBC with reliable and cost-effective devices achieving excellent uniformity and low loss over the contiguous bandpass from 1260 to above 1600 nm. Splitters fabricated with PLC technology are the superior choice over splitters made with FBT technology, and PLC-fabricated splitters were selected for our 2002 construction of our first B-PON deployment in Mission Bay and SBC continues to deploy only PLC splitters in B-PON deployments.

2.5   THE SBC FTTH NETWORK

Key characteristics contributing to the success of the SBC FTTH network (Figure 2.1(b)) are the triple play of services transported by the network, detailed design of the optical fiber/ distribution network, and the availability of a family of ONTs optimized for different applications.

2.5.1 THE OPTICAL FIBER/DISTRIBUTION NETWORK

The design of the optical fiber network is dependent on the transmission system planned for the desired services. Video service design can have a big impact on the fiber network. SBC initially intended to use the video overlay wavelength, and we discuss some of the challenges of designing and constructing a PON for that approach so that other providers can potentially benefit.

In 2001, our direction was to implement a video service using the readily available video headend equipment and Set-Top Boxes (STBs) used in CATV networks. The video service system transmits a multi-channel signal with a mixture of both analog and digital modulated RF carriers. The transmission of analog video would allow a video service to be provided without a digital STB at the televisions in the residence.

The video signal is broadcast downstream on the PON on a separate wavelength band compliant to the ITU-T G.983.3 specification. Support for analog video over a PON network with 32 splits and sufficient optical reach requires systems supporting the Class B optics specified in G.983.3 to provide an optical budget of 25 dB. To achieve maximum reach and contain cost of the analog video transport equipment, the passive optical network had to be built with products and methods for loss control not required of digital transmission systems commonly being deployed in other SBC fiber networks. Analog modulated RF carriers for video transmission were well known in the CATV industry to require the control of optical loss and optical reflection. Papers published in the early 1990's detailed the issues as the emerging Hybrid Fiber Coax (HFC) networks were being designed and evaluated.[1] Extending analog video to a PON with significantly greater optical budget than HFC networks was needed. SBC developed fiber design and construction guidelines for a passive optical network capable of supporting analog video transmission.

The design and construction methods to support analog video over a passive optical network require consideration of optical loss control, loss variation control, and reflection control. We report on the successful analog video service trial delivered over the SBC deployment in San Francisco using fiber products and methods enabling analog video transmission over a PON.

2.5.1.1 Loss Control

Operators deploying passive optical networks must consider fiber products and construction methods that lower the fiber network losses and provide sufficient reach. Fiber products with reduced optical loss include the following: lower loss optical splitters, low loss fiber cable, lower loss fusion splicing rather than mechanical splicing, and low loss fiber connectorization products. Construction methods to reduce optical loss include minimizing the use of fiber connectors, enhanced training to clean and inspect fiber endfaces for lower connector mating loss, and fiber management practices to reduce cabling loss from excessive bending in closures and cabinets.

Testing end-to-end loss within the required range of the PON system is a necessary verification of the fiber network before service activation. Three loss control guidelines

---

[1] Multiple-reflection-induced intensity noise studies in a lightwave system for multichannel am-VSB television signal distribution, Way, Lin, *IEEE Photonics Technology Letters*, Vol. 2, No. 5, May 1990, pp. 360–362.

promoted to meet the requirements are the following: splicing using fusion techniques only, greater attention to connector cleaning and inspecting, and the specification of lower loss splitters.

SBC studies have found fiber reach to be insufficient for active sites for trials and planned deployments using 32-way splitters. The reach limitations occur even with greater attention to additional loss control measures undertaken for PON when compared to point-to-point fiber systems used. B-PON reach with 32-way splitters becomes limited at distances exceeding 10 km and well short of the 20 km reach available by the ranging protocol in B-PON systems. Surveys of new housing developments have found 20 % of potential FTTH locations to be in the range of 10–20 km. Extending the reach to support these longer loops from the CO has become a significant issue in the use of FTTH to new housing developments. New developments in the SBC footprint are typically found in the undeveloped regions of cities which are further from existing central offices in the older part of a city. Improvements in the optical budget from advances in optical devices have occurred to provide on the order of 1–2 dB in recent years. SBC has specified Class B optics with a 25 dB maximum budget with enhancements to as high as 28 dB.

However, further advances in budget cannot be expected without cost impacts. SBC expects an ongoing requirement for extending reach, and several extended reach alternatives are considered. The two design approaches for extended reach include using 16-way splitters with lower splitter loss and applying budget for greater fiber reach or placing remote OLT cabinets. SBC plans to place remote OLTs due to the higher cost penalties from a lowered split to 16-way for the larger number of homes in new builds in our region. Implementation of loss control measures continues to a significant issue to support the deployment of PON throughout the region.

2.5.1.2 Loss Variation Control

Loss variation control is a unique requirement for RF-video fiber transmission with analog modulated signaling. Signal levels arriving at all ONTs sharing the PON must be within the dynamic range of the RF video optical receiver to provide adequate video quality. Figure 2.4 illustrates the key methods for the passive optical network to support analog video. The design of the distribution area fiber network to minimize loss variation includes the following: using optical splitters with enhanced uniformity over all outputs, minimizing the number of optical connectors that can each contribute to higher loss variation, and limiting a PON to a single distribution area thereby limiting the differential length to an area which is commonly 1 km but not more than 2 km in SBC. The optical loss testing and verification after fiber construction, prior to services delivery, will ensure that an analog video signal can arrive to each ONT on a PON.

Several methods, including the method to test for loss and install optical attenuators at specific points to lower loss variation in the network during construction, were detailed in a paper provided at the NFOEC in 2003.[2] The combination of analog-friendly PON design and construction guidelines, deployment of low uniformity splitters, amplifiers with low variation in power, and video receivers with wider dynamic range provide a viable approach to delivering analog video signals in the operating range to each ONT.

---

[2] F. J. Effenberger, K. McCammon, D. Cleary, Analog Video and PON Optical Loss Variations, NFOEC'03, p. 589–595, 2003.

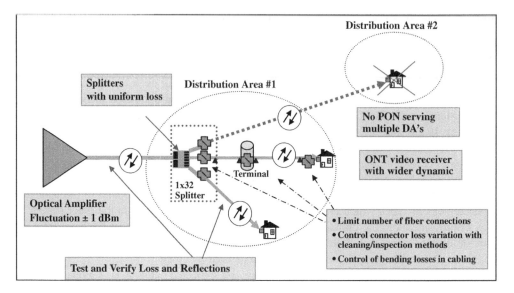

Figure 2.4 Methods to minimize loss variations in passive optical networks.

2.5.1.3 Reflection Control

Methods to control reflection in fiber joints are readily available by using fusion splicing exclusively and angled end-face fiber connectors. The target for reflectance of any fiber joint in the fiber network to support analog video was −50 dB to eliminate elevated noise from multi-path interference that degrades the viewing quality of analog video systems. Mechanical splicing can produce elevated optical reflections worse than −50 dB that occur at construction or degrade during environmental exposure in the outside plant. Angled fiber connectors have highly repeatable reflectance of < −50 dB with very low possibility of degradation even during exposure to hostile environments. In practice, angled fiber connectors' failure mechanism is elevated loss and not elevated optical reflection due to contaminated endfaces or separation and loss of contact.

Operators who choose to provide analog video service over a PON network can select fusion splicing and angled connectors to control optical reflection, eliminate one potential source of video picture quality degradation, and simplify optical layer troubleshooting and live optical testing on a working PON fiber system. Use of mechanical splicing and/or nonangled fiber connectors for an analog video service delivery over a PON will impose a greater need to test with OTDR to verify the control of reflection.

2.5.1.4 Optical Fiber Network Results at Mission Bay Deployment

The SBC Mission Bay deployment was the first deployment of PON at SBC (Mission Bay described further below). The Mission Bay fiber network was constructed using the optical design and construction methods developed by SBC to ensure an analog video transmission capability. Fiber connectors were used only at the CO location and the ONT connection inside the living unit at the ONT location. No connectors were deployed at the splitter location or at the building entrance location in the high-rise building. The design used only fusion splicing and only angled fiber connectors (SC/APC) to minimize optical fiber reflections from the video headend to the serving office and to the residence. End-to-end loss was tested and recorded from the CO cable termination to each ONT location. The range of losses measured was 19.9 to 17.1 dB over a total fiber distance of 2.2 km. The variation in loss was 2.8 dB for the first building constructed with FTTH in the Mission Bay deployment. The low variation was achieved by using high-performance splitters with low loss variation between the 32 output ports, by limited use of fiber connectors, use of core alignment splicers, troubleshooting measures to locate and repair excessive losses, and the attention to fiber cleaning including inspection of fiber endfaces. The network losses were tested and verified, and repairs done in advance of any services applied to the fiber network. The superior results for loss and loss variation were obtained with a skilled fiber construction crew and additional attention to transmission to analog video. The use of angled connectors and fusion splicing minimized the concerns over multiple optical reflections as verified by low optical return loss and back reflectance measurements taken at Mission Bay. After voice and data services were operational, the video wavelength was inserted into the working fiber network at the central office using a previously installed WDM coupler. No voice and data services were impacted during the insertion of the 1550 nm video overlay signal and video service activation. Video service quality for the analog signals was measured and verified to meet the analog and digital video service quality requirements. No adjustment to the optical network to adjust the optical level reaching the ONT optical receiver was required to be in the operating range of the video receiver. Our experience in Mission Bay showed that proper control of loss, loss variation, and reflection on the fiber network can successfully deliver analog video services over a passive optical network.

2.5.1.5 Evolving Optical Design at SBC

Since the Mission Bay deployment, SBC has given greater attention to lowering construction costs of FTTH. The Mission Bay successes in fiber network loss control and loss variation control were largely achieved due to the use of construction crews with previous experience in fiber handling, and the extra troubleshooting time to find and repair excessive losses in Mission Bay. SBC is investigating methods and products with improvements in fiber handling to allow a reduction in the construction costs for FTTH. A key construction cost driver for SBC is the cost of fusion splicing in the distribution area. An approach to lowering splicing cost was the introduction of additional fiber connectors in the FTTH trials following our first deployment in Mission Bay. SBC has trialed a fiber cross-connect cabinet where the PON splitters are placed.

The cabinet, which is pictured in Figure 2.5, is called a Primary Flexibility Point (PFP). The PFP serves as a single point for multiple splitters and serves a typical distribution area of 200–400 residences. In comparison, some operators are using separated 1×4 and 1×8 splitters, which form a logical 1×32 total splitting ratio. The PFP allows for higher utilization of the splitter and attached CO electronics since each new residence taking the PON services can be sequentially added using the fiber jumper flexibility in the PFP. Each of the 32 PON splitter outputs can then be dedicated and filled for the first 32 customers taking the service in the distribution area.

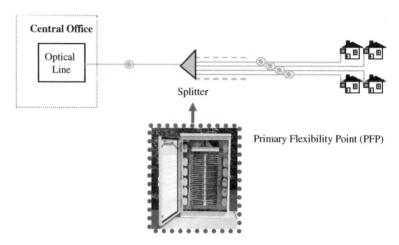

Figure 2.5   The primary flexibility point (PFP) concept used in SBC trials in 2004.

In this design, troubleshooting is enhanced with fiber connector access for test equipment insertion which can be necessary to locate faults towards the subscriber from the PFP location. The PFP concept also has the advantage of allowing simplified replacement of splitters at one centralized location.

SBC has concluded that the SC type connector is the superior connector type with the best reliability when exposed to testing consistent with placement in the hostile environments in the PFP and near the residence. Smaller form factor fiber connectors would be an advantage over the SC connector due to the smaller size PFP, but must have improvements in reliability. Future fiber connector improvements in physical size, reliability, and better immunity to airborne contaminants are needed to keep connectors a benefit and not a liability for network reliability. SBC now specifies the Ultra-Polish Connector (UPC) endface polish for FTTH deployments planning. The SC/APC connector with an angled endface is no longer an SBC requirement to eliminate fiber reflections from connector pairs which is consistent with the SBC removal of analog video as a requirement for delivery going forward.

2.5.1.6 Optical Network Summary

SBC developed design and construction guidelines for the optical distribution network for FTTH deployments that supported the transmission of analog-modulated video signaling. The results from the Mission Bay deployment showed an analog video service can be delivered successfully over a PON network with proper attention to the guidelines developed by SBC. For PON networks without RF-video signaling, the requirements for loss variation and reflection control are greatly relaxed, and loss control becomes the primary design concern. The relaxation of the optical network design and construction requirements without analog video will be leveraged by SBC to reduce the deployment costs for FTTH in the future.

Table 2.3 FTTH ONT types and characteristics.

| ONT/ONU type | Service interfaces | | | | Description |
	POTS	Data	Video	DS1	
Single-Family Unit (SFU)					
SFU:ONT (Triple Play)	4	1 Ethernet	1	–	Hardened. Maximum dimensions $13'' \times 13'' \times 6''$.
SFU-P-D:ONT	4	1 Ethernet	–	–	Locally powered by a separate 12-Vdc UPS, with min backup for 4-hr POTS & 1-hr data.
Multiple-Dwelling					
Unit (MDU) MDU:ONT	24	12 Ethernet and/or xDSL	1	–	Locally powered. Data to be supplied in modular units of 4.
Small Business Unit (SBU)					
SBU-V:ONT	8	1 Ethernet	1	2	Hardened. Locally powered by a separate 12-Vdc PSU. Number of Service Interfaces indicated are minimum values.
Multi-Tenant/Shared					
Business Unit MTU:ONT	24	8 Ethernet	1	4	Hardened. Locally powered. Number of Service Interfaces indicated are minimum values.

2.5.2 FTTH ONTs

The ONT is one of the highest cost components of the FTTP system because it is located at the customer end of the loop and is thus shared by the fewest customers. In addition, it determines to a great extent the type and quality of service available to the customer. A number of different ONT types for different applications could be specified, ranging from a single-family residential ONT that provides two voice and one data ports to a multi-tenant business ONT that provides multiple voice, data, and video service ports. It is unlikely that any company will deploy all the different types of ONT because inventory-management/volume-discount issues prescribe a smaller number of types. Table 2.3 provides a list and description of the ONTs that are planned for use in SBC.

2.5.2.1 SFU ONT

The SFU ONT is intended for use in residential applications with single family/detached and small (e.g., 2–4 unit) multi-dwelling/attached homes. It will provide, as a minimum, four POTS interfaces and one 10/100-bT Ethernet interface. (Current versions of the ONT also provide a coax interface intended for video over the B-PON overlay wavelength, but this will not be used/required in future designs.) The SFU ONT is environmentally hardened and will be installed on the outside of the home – replacing the current passive Network Interface Device (NID). Powering for the SFU ONT will be provided locally by an DC Uninterruptible Power Supply (UPS), which will be installed inside the customer's home or garage.

2.5.2.2 MDU ONT

The MDU ONT is intended for use in apartment complexes, condominiums, and townhouses that contain five[3] or more living units and house long-term residents. (In the future, these ONTs may also be used for short-term resident applications, such as university dormitories and hotels.) Each MDU ONT will be capable of serving 12 living units, and provide a minimum of 24 voice interfaces, 12 VDSL interfaces in modular units of 4, and 1 RF video interface with addressable tap. MTU ONTs may be installed in several different types of locations (e.g., inside a communication closet or terminal room, on the outside building wall, or in an exterior pedestal/enclosure), and hence these ONTs must be environmentally hardened. Powering for the MTU ONT could be provided either by a local DC UPS or by an existing −48 Vdc supply (e.g., in existing buildings).

2.5.2.3 B-ONT

Two types of B-ONTs are required for the FTTP system: the SBU ONT, which is intended to provide service to one business; and the MTU ONT, intended to provide service to four or more small businesses. The SBU-ONT is similar to the triple-play SFU ONT, in that it is intended to provide triple-play services for use by one small/home office. It will provide, as a minimum, eight POTS interfaces, one 10/100-bT Ethernet interface, and two DS1 interfaces. Like the SFU ONT, it will be environmentally hardened for installation on the outside of the office/home (replacing the current passive NID), and will be powered locally by a DC UPS, which will be installed inside the home/office. The MTU ONT is intended to serve four to eight small businesses, and will typically be located in a small business park or strip mall. It must provide, at a minimum, 24 POTS interfaces, 8 10/100-bT Ethernet interfaces, and 4 DS1 interfaces. Like the MDU ONTs, the MTU ONT may be installed in several different types of locations (e.g., communication closet, terminal room, exterior wall, or exterior pedesta), and hence must be environmentally hardened. Powering for the MTU ONT will be provided by either a local UPS or a −48 Vdc supply.

2.5.2.4 FTTH ONT Powering

While one of the advertised advantages of the PON architecture is a wholly passive plant, in actuality power must be provided to the ONT located at the customer end of the network. Power has long been considered the 'Achilles' heel' of fiber to the home because the same fiber that brings megabits per second of information to a customer separates that customer from the typical 'always on' power plant familiar from standard POTS service.

Over the years, many powering schemes have been proposed, tested, and deployed. In general, these can be categorized into 'centralized' and 'local' powering schemes (Figure 2.6).

In centralized powering schemes, power is provided to several/many ONTs from a central network site such as a CO, RT, or remote power node. The primary source is generally commercial AC, rectified and converted to DC at the site, and the backup source is typically batteries and engine generators. Both primary and backup power are transmitted to the ONT over metallic media, typically conventional twisted copper pair. Variations to the basic

---

[3] Buildings with fewer than five units will be served by SFU and/or SBU ONTS.

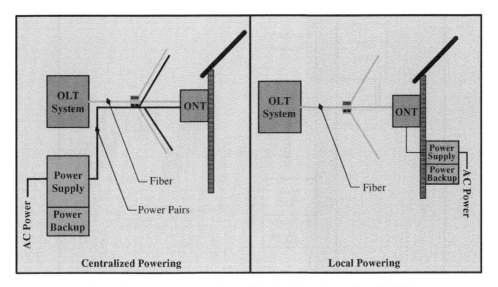

Figure 2.6 Comparison of centralized and local powering of ONT.

centralized power scheme that have been explored over the years include use of solar energy, wind energy, and fuel cells as the primary source, use of flywheel energy storage and various new/evolving battery technologies as the backup source(s), and even the use of the fiber as the power transport medium.[4]

In local powering, power is provided to an ONT from its own dedicated source, which is located near/at the ONT. The typical primary source is again commercial AC, rectified and converted to low-voltage DC by the source/supply, and the typical backup source is batteries. Variations to this basic scheme include use of solar energy and fuel cells as the primary source, and use of flywheel energy storage, mechanical power converters, and various new/ evolving battery technologies as the backup source(s).

2.5.2.4.1 Recommended Powering Architecture

Centralized powering can provide high power reliability and can be easier to maintain and operate than local powering. However, it experiences much high power loss (e.g., transmission loss and multiple power conversions), presents a single point of failure, and mitigates/ removes many of the reliability, maintainability, and operational advantages of an all-passive optical network. Because of this, SBC chose a local powering architecture for its FTTH deployments. This architecture is illustrated in Figure 2.7 for the SFU ONT.

As indicated in the figure, the ONT will be powered from a DC UPS, which can be located as much as 100 feet away from the ONT. The UPS will provide low-voltage DC power to the ONT; obtain primary input power from a commercial 120 VAC power connection in the

---

[4] In fiber powering, high-energy light from a high-power laser in the CO is pumped into and transported by the distribution fiber. At the customer/far end, this light is converted to electric power by a photo diode. Studies have indicated that about 1 W of power can be transported in this way. High cost, safety issues, and the relatively low amount of power have generally precluded this power scheme for FTTH.

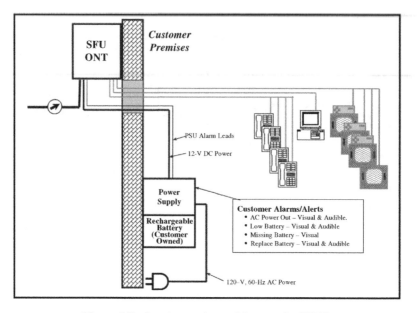

Figure 2.7 Local powering architecture for FTTH.

customer's premises; and obtain secondary/backup input power from a rechargeable battery located within the UPS housing. Key features of this powering scheme include:

- The power supply and batteries will be located inside the customer's home in a more weather-controlled environment to enhance battery capacity and life (SBC is currently investigating a hardened power supply to facilitate installation).
- During AC power outages, a power-down scheme is used to disable nonessential services. This will promote longer life for voice services as the backup battery will last longer.
- The UPS will alert the customer of various powering events (i.e., an AC power outage and a missing, failed, and discharged battery) to help ensure uninterruptible powering of the ONT.

2.5.3 SBC'S MISSION BAY TRIAL

Mission Bay is a 4-billion dollar redevelopment project in San Francisco, CA that will convert over 300 acres of former landfill (from the 1906 earthquake/ fire) and rail-yards into a virtual 'city in a city.' The development is located south of downtown and is equal in size to San Francisco's entire downtown business district.

At completion (expected to take 10–20 years), Mission Bay will include 6000 residential housing units, 6 million square feet of office/life science/technology commercial space, a new University of California research campus, 800 000 square feet of retail space, a 500-room hotel, 49 acres of public open space/parks, a new public school, and new fire and police stations.

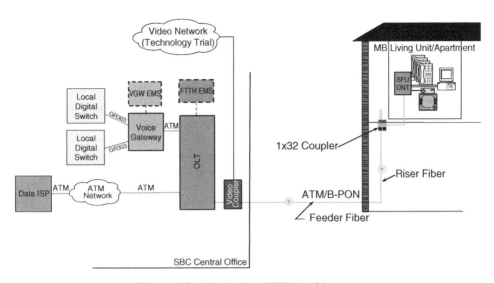

Figure 2.8 Mission Bay FTTH Architecture.

The project was spearheaded by the Catellus Development Corporation, who envisioned an innovative, state-of-the-art community supported by a 'broadband technology infrastructure that will provide homes with voice, video, and data' (Catellus).

To provide the broadband infrastructure, Catellus issued a competitive RFP to telecommunications carriers/providers in 1999.

Pacific Bell/SBC won the proposal, in large part because of its offer to provide a 'fiber-to-the-home/apartment' technology to the residential units. On the basis of the RFP, SBC and Catellus signed a comarketing relationship in January 2002, in which SBC is the preferred provider for all voice, high-speed internet access (HSI), and video services for Mission Bay.

Figure 2.8 shows the FTTH system that was presented to Catellus and is now deployed in Mission Bay. It is the same as the generic B-PON FTTH architecture discussed in previous sections, except that the 1×32 splitter and the ONT are located inside the apartment/condominium building. Key elements of the system are:

- The OLT is an Alcatel 7340 P-OLT (Packet OLT) system, which supports up to 36 PON interfaces and up to 1052 SFU-ONTs.
- All ONTs used to date are the Alcatel 7340 H-ONT, which supports up to four separate POTS interfaces, one 10/100baseT Ethernet interface, and one RF-video interface. In the future, MDU-ONTs may be used in some installations; these ONTs will serve up to 12 apartments/condominiums and provide POTS, VDSL, and RF video interfaces.
- The Voice Gateway is the General Bandwidth G6 Packet Telephony Platform, which supports up to 3360 simultaneous calls and 26 880 ONTs.
- The 1×32 splitters are housed inside the buildings in a cabinet provided by Tyco. The splitters are made by NEL of Japan. A single fiber is routed from the splitter cabinet to each living unit using fusion splicing and a single SC/APC connector in each living unit to connect to the ONT.

Residential voice and data service over B-PON in Mission Bay launched in April 2003. Since then, growth in these services has followed the Mission Bay build and occupancy rates. As of October 2004, there were about 500 voice lines on B-PON in Mission Bay, with a penetration rate of over 80 % for the HSI access service.

Video services in Mission Bay are not currently offered over the B-PON system. However, SBC performed a limited technology trial of these services using the video overlay wavelength from June 2003 through July 2004. In the trial, video services were provided only to customers in one 32-unit condominium. Services included analog and digital video, interactive and broadcast video (over 300 broadcast channels were offered including local, commercial-free digital music, Pay-per-View movies and sports, and digital premium and multiplex channels), and Standard-Definition (SD) and High-Definition (HD) video. The trial was very successful and demonstrated/verified the vast potential of the B-PON video enhancement band.

2.6   SBC FIBER TO THE NODE (FTTN) NETWORK

The SBC plan will make FTTN the dominant triple-play network in terms of homes served. This is due to economic evaluation favoring FTTN and the convergence of several technologies allowing the support of video. These technologies include advanced video compression, standard VDSL, and carrier class Gigabit Ethernet.

The fiber feed for FTTN is Gigabit Ethernet. Gigabit Ethernet meets the bandwidth requirements for feeding video to up to 200 homes, allowing a lower cost network for video distribution. At the node, there is a VDSL DSLAM that handles switching of all the video and other services to DSL ports that supply VDSL to the home at distances up to about 5000 feet on a single pair.

The DSLAM implements IGMP processing to allow replication of channels to homes for a single video stream on the GigE link. On the VDSL link, packet mode is also used, so ATM has been eliminated from the system. This lowers cost and eliminates unused overhead.

The other technology advancement that makes the whole solution viable is advanced video coding, reducing the total bandwidth of an HDTV channel eventually to perhaps 6 Mbps. This allows the support of four channels, with one or even two HDTV streams, over the VDSL link with a bandwidth between 20 and 25 Mbps.

The planned deployment of FTTN did end up having a significant impact on the SBC FTTH solution. FTTN dictated the use of Switched Digital Video (SDV) in that architecture due to the approximate 20–25 Mbps of bandwidth available. In order to have a single video solution, it was determined that the best option was to also use SDV on PON as well. Thus, initial plans to use the video overlay wavelength were abandoned in favor of SDV.

2.7   THE HOME NETWORK

The final stage in the delivery of services for both FTTH and FTTN is the home network. A major goal of all this high-speed networking of course is the delivery of triple-play services, including a full complement of entertainment video. When that video arrives at the house as high-speed data, some new solutions are called for. However, we desire to use standards and industry trends as much as possible.

An advantage of more or less simultaneous implementation of FTTH and FTTN is that similarities in home networking can be optimized. Thus, the two solutions have the same design once the respective physical layer is terminated. The full solution set for a FTTH subscriber who has video service and high-speed data is to deliver all the traffic out of the 100 Mbps Ethernet port on the ONT and run it over CAT5 to the Residential Gateway (RG). The RG then routes the traffic either to a STB or to a PC on the home LAN.

The solution for FTTN is exactly the same after the data is delivered to the RG. The only difference is that FTTN of course will have VDSL as the physical layer input to the RG and this is carried from the telephone interface into the house via CAT3 or COAX.

The demands of distribution from the RG are challenging. Communication to multiple PC locations may be required as well as multiple TVs. SBC plans to provide service for up to four TVs and the bandwidth needed for video is high. A key element is to minimize cost by re-using existing inside wire if at all possible, so this eliminates approaches like running all new CAT5. Unfortunately, no existing wireless scheme works well enough for video and a wired solution is a must.

For video distribution we will use the technique for Ethernet over COAX promoted by the Multimedia over COAX Alliance (MoCA). This supports the bandwidth required, and in many cases, the wiring to the TV location is already in place. Of course, if the customer desires the TV at a new place, some wiring may have to be done.

For data distribution the best bet seems to be a combination of 802.11 wireless and HPNA, which reuses existing telephone twisted pair.

The home architecture will further support VoIP, allowing full conversion of all services to IP. VoIP traffic will be given the highest priority for both downstream and upstream handling.

2.8   MOTIVATING THE NEW NETWORK – IPTV

A fundamental goal for building these new network capabilities is to give consumers new options in video entertainment delivery, in particular a full offering of digital entertainment TV carried in IP packets throughout the network. We review here some of the additional basic network features that support this exciting new way of video delivery.

Multi-casting is a key feature the network must support, even as planning allows for substantial migration to video-on-demand (VOD) as customers expand desires to view what they want when they want. To conserve bandwidth for basic TV service, the network should carry only a single channel as far as possible from the acquisition point to the subscriber. To support this, each node in the path needs to be multi-cast enabled. This includes at least four points in a typical case: the home router, the Access Node, the first aggregation switch, and the first router. This would allow two or more TVs in the home to be watching the same channel and only one version of that channel appears on the SBC network.

With the multi-casting approach one of the main concerns is sizing each component for the required number of multi-cast streams. Special consideration may be needed when this number exceeds 4–8 per home, which can easily be the case when supporting multiple TVs.

Managing quality of service (QoS) for video is also key. The network is multi-service of course, carrying voice and data as well as video. The video needs to get priority treatment over the data and in such a way as to maintain a very high-quality viewing experience by the customer. This QoS can be supported with appropriate Ethernet tagging resulting in high-priority treatment and excellent loss, jitter, and delay results. Multiple queues for video in the Access Nodes for handling normal video versus VOD will also allow better guarantees for the most watched programs.

The separation of services via VLAN tags is also important to overall service control. Options for VLAN assignment include per service and per customer. In per service VLAN tagging for example, IPTV would have one VLAN assignment and VoIP a different one but all traffic of a given type to the home would have the same assignment.

More complicated models also allow per service VLANs to be mapped to per subscriber assignments at different layers in the network.

Finally, to enable IPTV an appropriate set-top box (STB) is required. IP STBs have emerged on the market, but the selected one must run the appropriate middleware and applications for the service as well as handling the selected video compression coding. For example, STBs able to support HDTV with MPEG-4 are early stage at this time though now becoming available.

SUMMARY

In this chapter we have reviewed the experience and plans for extending fiber in SBC to provide customers with choices for advanced services. It is truly an exciting time for SBC and the industry as a whole, with the ultimate payoff being great new options for customers. Customers will be served by FTTH or FTTN and be offered the same or similar services. These networks will meet the demands of customers for more advanced television, including HDTV and VOD, as well as growing desires for higher speed data and IP services.

GENERAL REFERENCES

Ballart R, Getting inside SBC's OSP, *OSP Magazine*, August 2005.
Carey E, SBC Project Lightspeed, *Alcatel Telecommunications Review*, 2Q2005.
Edmon E, *et al*. Full service access networks at SBC, *Proceedings of OHAN 2002*, June 2002.
Erickson J, *et al*. FTTH architectures and systems – SBC requirements and deployment plans. *Proceedings of OHAN 2001*, April 2001..
Estes R, *et al*. Powering options for fiber in the loop, *Proceedings of ISSLS 1993*, September 1993.
ITU-T Recommendation G.983.1. *Broadband optical access systems based on passive optical networks (PON)*, October 1998.
ITU-T Recommendation G.983.2. *The ONT management and control interface specification, for B-PON*, June 2002.
ITU-T Recommendation G.983.3. *A broadband optical access system with increased service capability by wavelength allocation*, March 2001.
ITU-T Recommendation G.983.4. *A broadband optical access system with increased service capability using dynamic bandwidth assignment*, November 2001.
ITU-T Recommendation G.983.5. *A broadband optical access system with enhanced survivability*, January 2002.
ITU-T Recommendation G.983.6. *ONT management and control interface specifications for B-PON system with protection features*, June 2002.
ITU-T Recommendation G.983.7. *ONT management and control interface specification for dynamic bandwidth assignment (DBA) B-PON system*, November 2001.
ITU-T Recommendation G.983.8. *B-PON OMCI support for IP, ISDN, Video, VLAN tagging, VC Cross-connections, and other select functions*, March 2003.
ITU-T Recommendation G.983.8. *B-PON ONT management and control interface (OMCI) support for wireless Local Area Network interfaces*, June 2004.
ITU-T Recommendation G.983.10. *B-PON OMCI support for Digital Subscriber Line Interfaces, June 2004.*

ITU-T Recommendation G.984.1. *Gigabit-capable Passive Optical Networks (GPON): General characteristics,* March 2003.

ITU-T Recommendation G.984.2. *Gigabit-capable Passive Optical Networks (GPON): Physical Media Dependent (PMD) layer specification,* March 2003.

ITU-T Recommendation G.984.3. *Gigabit-capable Passive Optical Networks (G-PON): Transmission convergence layer specification,* February 2004.

ITU-T Recommendation G.984.4. *Gigabit-capable Passive Optical Networks (G-PON): ONT management and control interface specification,* June 2004.

Mestric R, *et al.* Optimizing the network architecture for Triple Play, *Alcatel Telecommunications Review,* 3Q2005.

SBC Confirms Project Lightspeed IPTV Field Trial, *http://www.tvover.net/SBC + Confirms + Project + Lightspeed + IPTV + Field + Trial.aspx,* 3 November 2005.

3

FTTH: The Swedish Perspective

Claus Popp Larsen, ørjan Mattsson and Gunnar Jacobsen
Acreo, Stockholm, Sweden

3.1 INTRODUCTION

This chapter presents the Swedish perspective on broadband access with special focus on fibre-to-the-home, FTTH. Sweden was an early adopter of both Internet and fibre-based access with more FTTH connections than any other European country, so there is a large accumulated knowledge base on both how to design, build and operate these networks. Almost 20 % of all broadband connections in Sweden are of FTTH-type, which is the highest share in the world [1–3]. A single Swedish consensus view on broadband and FTTH does not exists. Instead, this chapter will describe the Swedish broadband access market and try to capture some general trends and driving forces – technically, historically and politically. Also, some of the leading actors, operators as well as vendors, are described in more detail in order to exemplify the different perspectives (Figure 3.1).

Although the perspective varies with the different actors, there is a common view on some issues, describing the Swedish perspective on broadband and FTTH. Below are some highlights about Sweden and broadband access – all of these will be further elaborated in this chapter:

- Sweden has the largest share of households connected with FTTH in the world.
- The requirements from triple play services (data, voice and TV over IP) are shaping the broadband networks.
- The large number of municipal networks (stadsnät) running so-called *open networks* and the operator B2 with *vertical integration* are driving the FTTH deployment in Sweden.
- Sweden has the largest share of 2 Mb/s connections and above in Europe with some of the lowest prices.
- There is a strong belief in fibre access as the only really future-proof solution.
- Point-to-point (P2P) Ethernet is to date the exclusively dominating FTTH technology as opposed to PONs.

Broadband Optical Access Networks and Fiber-to-the-Home: Systems Technologies and Deployment Strategies
Chinlon Lin   © 2006 John Wiley & Sons, Ltd

Figure 3.1   Sweden has a population of 9 mio living on 411 000 sq km – an area slightly larger than California. Most people live in the densely populated areas in the south.

3.2   CONTENTS

The vibrant area of broadband access and in particular FTTH is currently experiencing an immense development across the world with regards to technologies, business models, national strategies, etc. The intention of this chapter is to give a snapshot of the situation in Swedish by using statistics primarily from the end of 2004 and with current events up to mid-2005 – in just a few years from now the situation may be very different. Special Swedish characteristics are discussed in more detail, and an attempt to explain the historical background for certain technological and political directions are given for some of these characteristics.

The chapter is structured as follows. In the section *Definitions* the key terms 'broadband', 'FTTH', 'muni net' and 'residential area network' are defined. *Background for the Swedish FTTH boom* shortly describes the historical and political background for the present situation. In *The Swedish broadband market today* Sweden is compared with other countries with regards to broadband penetration and FTTH deployment. Further, the national broadband market is analysed with regards to market shares, technologies, etc., and the collaboration between industry and academia is described. The section *Open networks versus vertical integration* explains the concept and consequences of deploying so-called open or operator neutral networks. Also business cases and the incentives to build such open networks are outlined. This is compared with the more traditional vertical integration of the whole value chain. *Access network technologies* briefly compares PON with point-to-point Ethernet. This is followed by a more detailed description of the role of IP and Ethernet equipment in such access networks. The section *Drivers, services and trends for the future broadband networks* intends to capture what drives the development and further deployment of FTTH in the short to medium term. Finally, *Swedish FTTH players* describe the following companies in slightly more detail: The access system suppliers Ericsson and PacketFront, the operator TeliaSonera, the estate company Svenska Bostäder and the fibre infrastructure owner Stokab.

3.3   DEFINITIONS

3.3.1   BROADBAND DEFINITION

There is no generally accepted definition of broadband, but here follow some attempts. The International Telecommunication Union (ITU) recommendation I.113 defines broadband as a transmission capacity that is faster than primary rate ISDN, at 1.5–2 Mbit/s. OECD defines broadband as a technology providing downstream speed in excess of 256 Kb/s (and upstream access speed in excess of 128 Kb/s), while the US Federal Communications Commission (FCC) has defined broadband as a connection of at least 200 Kb/s in one direction.

We will simply use the following definition:

> Broadband is here defined as a high-speed Internet connectivity service that has the potential to always be connected

Exact definitions change with time and vary considerably depending on who is asked: The IT-technician, the politician, the advanced user, the marketing director, etc. Also, the definitions may differ widely among different nations.

3.3.2   FTTH DEFINITION

We are using the definition from [1]:

> 'We refer to FTTH as all fibre connections to the building, whether private houses or apartment blocks and, in the latter case, assume that the distribution of these connections is carried out via xDSL, Ethernet or wireless technology'

In Sweden the distribution is almost always carried out via Ethernet. The term FTTH is therefore used here in the same sense as FTTP (fibre to the premises) and FTTB (fibre to the building). FTTN/FTTC (fibre to the node/curb) is *not* covered by FTTH.

3.3.3   MUNI NET DEFINITION

A muni net or municipality network ('stadsnät' in Swedish) is a local or regional network operated or alternatively initiated and supported (or otherwise subsidized) by a municipality and which interconnects the local authorities and provides broadband access to households and offices. Observe that this is not a strict definition as muni nets are still a relatively new phenomenon. Muni nets are also sometimes called city networks, urban networks, community networks or something similar. Muni nets embrace the access network and to some extent the services, and they partly have the objective to support the local community. Swedish muni nets are characterised by being open or operator neutral. Open networks are described later in the chapter.

3.3.4 RESIDENTIAL AREA NETWORK, DEFINITION

A residential area network covers a neighbourhood, for instance one or several multi-dwelling units. Usually the IT infrastructure in a residential area network is controlled by the telephone operator, the internet service provider (ISP) and/or the cable TV operator. However, the past few years many Swedish real estate companies have routinely installed a fibre infrastructure in their new projects, and rather than handing over the control or ownership to a *single* service provider they run the network much like a muni net in the sense that they open it to *several* service providers. Such an open residential area network is almost always of the FTTH type. From a technical and operational point of view the open residential area networks are quite similar to muni nets but in a smaller scale, and the business models and the incentives to build them are different.

3.4 BACKGROUND FOR THE SWEDISH FTTH BOOM

Sweden has a strong tradition in IT and telecommunications. The country was an early user and a leader in fibre optics by the end of 80s and the beginning of the 90s very much depending on efforts made by Ericsson and the PTT Telia (became TeliaSonera in 2002 after merging with the Finnish incumbent) in cooperation with university research. Sweden was early in using PCs at home and has today one of the highest PC penetrations in the world. In mobile communication Sweden was one of the early adopters alongside the other Nordic countries, and Ericsson together with Finnish Nokia are today the leading suppliers of mobile systems.

During the 90s the government took a number of steps to deregulate the telecommunications market, and Sweden is today one of the most deregulated countries in the world with the market monitored by the governmental authority PTS[1] [4]. The previous monopoly of Telia was broken and a large number of new competing telecom operators appeared, challenging the incumbent. Some of these owned their own fibre, some used dark fibre, and some were virtual operators. The (publicly owned) utility companies and the railway company (Banverket) all had impressive fibre installations, so there was never any shortage of fibre in the core networks.

By the end of last millennium Sweden was covered with fibre, and the Swedes were experienced IT users and in all aspects exemplary consumers of new telecom and datacom equipment. But still remained the issue of how to reach the households and be able to offer them high-speed Internet access.

Preceded by a major public debate the government issued in 2000 a bill 'An Information Society for all' including a 1BUSD support program for the municipalities to implement broadband. This initiated a substantial rollout of fibre in the municipalities (Figure 3.2). The political ambitions were manifold; some of the ideas behind were that this would kick-start high volume rollout of broadband connections and that all Swedes would get a high-speed Internet access. This in turn would stimulate local industry and the Swedish broadband equipment suppliers, and hopefully it would reduce the

---

[1]The National Post and Telecom Agency (PTS or Post & Telestyrelsen) is the governmental authority that monitors and regulates the electronic communications (telecommunications, IT and radio) and postal sectors.

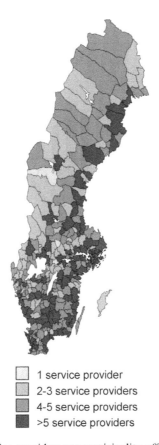

1 service provider
2-3 service providers
4-5 service providers
>5 service providers

Figure 3.2 The number of service providers per municipality offering broadband connections [5].

migration from rural areas to the big cities. The intention was also to offer public services online, which would benefit democracy and reduce public administration simultaneously. By that time many of the municipalities already operated their own network to connect the local authorities in a low-cost fashion, so they did not start from scratch. The economical compensation was so favourable that the majority of Sweden's municipalities accepted, and by the end of 2004 almost all of Sweden's 290 municipalities had got broadband networks or had at least started to build an IT infrastructure. Virtually all of the muni nets support the local authorities and the majority have also private households connected. EU's structural funds have also contributed to broadband build-out.

3.5 THE SWEDISH BROADBAND MARKET TODAY

This section gives a broad overview of the Swedish broadband market – both in comparison with the rest of the world as well as internally. The comparisons are based on statistics from OECD [2], IDATE [1] and the Swedish PTS [3,5].

percent

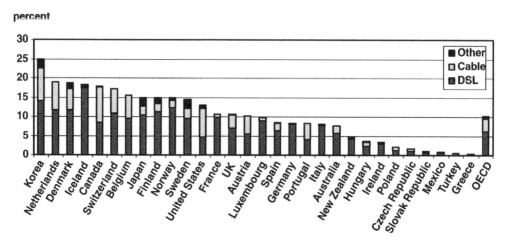

Figure 3.3   Broadband subscribers per 100 inhabitants, by technology, December 2004 [2].

3.5.1   BROADBAND PENETRATION COMPARED TO THE OECD

According to OECD (statistics from the end of 2004) Sweden has the highest FTTH penetration in the world with 2.5 % of the inhabitants. This is closely followed by South Korea and Japan with 2.3 % and 2.2 %, respectively. No other countries are close to these rates. Despite Sweden's record high share of FTTH connections, the country is far from leading the broadband penetration league. In Figure 3.3 the number of broadband subscribers per 100 inhabitants for different countries and different technologies is compared. The technologies are split into DSL (using the telephone infrastructure), cable modem (using the cable TV infrastructure) and other technologies. The latter include fibre optics, local area network (LAN – usually based on Ethernet), satellite, powerline communication (PLC) and fixed wireless access (FWA). In Sweden the category 'other' is predominantly FTTH technology that is also LAN based. Also in South Korea and Japan 'other' is almost equal to FTTx while in other countries 'other' may include very little FTTx [1].

Compared to the other OECD countries Sweden is characterised by not only a high FTTH penetration but also a low share of cable modems. In total, Sweden is number 11 with just less than 15 % broadband subscribers per 100 inhabitants, but with a growth rate slightly below the countries in front.

It should be noted that other reports and analyses may show different results with regards to FTTH penetration. This is mainly due to diverging definitions of FTTH. Also observe that due to the rapid development in the field with many national and local initiatives, the 2005 statistics may be quite different from the present one. For instance, there are massive investments in FTTH in the Netherlands, which will certainly show in the OECD statistics for 2005.

A historical fact: In 1999 about 50 % of the Swedish broadband households were connected by FTTH, thanks to an early start by the operator B2 (Bredbandsbolaget, acquired by Norwegian Telenor in mid-2005).

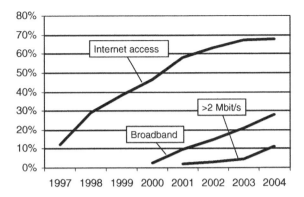

Figure 3.4 Fraction of Swedish households with Internet access, December 2004 [5].

3.5.2   THE BROADBAND MARKET AND ACCESS TECHNOLOGIES IN SWEDEN

The situation in Sweden today is characterised by a fierce competition in the broadband marketplace which has led to a number of consolidations in 2004. Almost 70 % of Swedish households had Internet connectivity by the end of 2004. Twenty-eight percent of these, corresponding to 1.2 million subscribers, had broadband access (compared with 63 % in South Korea [1]), which is a 35 % increase compared to 1 year earlier and there are no signs of saturation. See Figure 3.4 [3] and observe that it measures the *household* penetration whereas Figure 3.3 measures the *inhabitant* penetration. Forty percent of all Swedish broadband subscribers had a connection with a bandwidth of 2 Mb/s or above by the end of 2004, and this share is rapidly growing. No other European country has such a high share of 2 Mb/s connections or above.

It is not trivial for the consumer to compare monthly fees for broadband connections as they also may depend on installation fees, subscription time and bundling with other services. Therefore, the Swedish Consumer Agency and the National Post and Telecom Agency support a public website with detailed pricing information for broadband connections, Telepriskollen (literally 'Tele-price-check' [6]).

By October 2005, a total of 25 different ISPs across Sweden with 57 different products offered bandwidths of exactly 10 Mb/s, downstream. Products with higher or lower downstream bandwidths were excluded here. Fifty one of the 57 products were FTTH based while the rest were based on DSL, cable and radio. The monthly fee varied between 10 and 66 € with an average of 27 €, the installation fee varied between 0 and 416 € with an average of 49 €, while the subscription time varied between 0 and 12 months with an average of 4 months. The fees are among the lowest in Europe, and during 2005 they have constantly decreased due to the very strong competition.

Sweden had by the end of 2004 150 ISPs offering Internet access [3] and the vast majority of these offered broadband connections. Most of these are local or regional, but there are around 20 national providers. The largest broadband provider is TeliaSonera with a market share of 39 % by the end of 2004; see Figure 3.5 [3]. TeliaSonera offers predominantly DSL (mostly ADSL). They own most of the copper access network, but must by law offer it to its

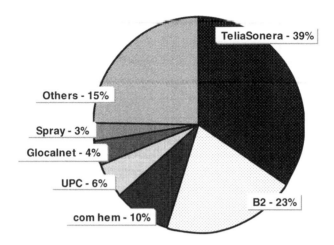

Figure 3.5 Figure shows the market shares of the Swedish broadband providers [5].

competitors. The second largest broadband provider is B2 with 23 % of private households. They operate the incomparably biggest FTTH network in Sweden but also offer DSL. The dominating cable operator Com Hem is on third place with 10 % of the private broadband market. In the enterprise sector TeliaSonera, Song Network (acquired by Danish TDC in 2004) and Telenor are the major players.

The split between the different access technologies are shown in Figure 3.6. DSL has a market share of just over 50 % but as can be seen, Sweden has a high percentage of FTTx, slightly less than 20 %, based on P2P Ethernet technology.

In conclusion, the following situation characterizes the Swedish broadband market:

- Sweden has the highest share of FTTH in the world.
- ...but is only number 11 in broadband penetration world wide.
- Sweden has few cable connections compared to other countries.

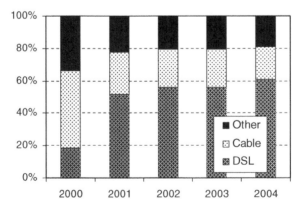

Figure 3.6 Market shares for different access types.

- Sweden has the highest share of high-end broadband access in Europe.
- Swedish prices for high-end broadband access are among the lowest in Europe.
- Sweden has extreme competition on the service provider side.
- Sweden also has very high competition on the network provider side.

3.5.3   EQUIPMENT SUPPLIERS

The Swedish broadband industry is rather fragmented. During the post-bubble consolidation years in the beginning of the millennium, Ericsson reduced their R&D activities and product offerings in fixed networks but is now recovering. A lot of start-up companies around optical networking were created as spin-offs from Ericsson and TeliaSonera. These include Qeyton (which was acquired by Cisco), NetInsight, Transmode (which merged with Lumentis in mid-2005) and Wavium making infrastructure based on optical transmission and switching. The same thing happened in the access area with companies like PacketFront (access equipment, will be described later) and 42 Networks (FTTH customer premises equipment), and many of these are today successful companies acting worldwide.

The introduction of triple play (Internet access, telephony and TV in one package) has opened up a market for companies developing IP-based set top boxes (such as Kreatel and I3Micro), residential gateways (RGW – the equipment in the home terminating the access connection) and also large number of service- and content-oriented companies have been started.

3.5.4   THE SWEDISH BROADBAND INDUSTRY

From hardware to operators and service developers, some 350 companies create the value chain of Sweden's broadband industry in 2004:

- Residential broadband access: 119 companies.
- Local infrastructure: 102 companies.
- Business broadband access: 72 companies.
- Equipment makers: 41 companies.
- Network integration: 25 companies.
- Network rollout: 21 companies.
- National infrastructure: 11 companies.
- Software: 11 companies.
- Broadband services: 8 companies.
- Hotspots/WISP: 7 companies.

Observe that the sum of the above is larger than 350 which is because several companies are present in more than one area [7]. The whole market is being consolidated through mergers and acquisitions, and since the compilation of this list some companies no longer exist while new have appeared.

3.5.5   COLLABORATION BETWEEN INDUSTRY AND ACADEMIA

The move to an IP-based broadband network that is a convergence between Internet, telecommunication and broadcasting creates of course a number of challenges for the R&D

community. How to build a future-proof cost-efficient quality of service (QoS) broadband network meeting the requirements of existing and future services is a critical question. To verify the service and infrastructure requirements testbeds with real end users have been implemented across Sweden. The biggest testbed is the research institute Acreo's national broadband testbed with more than 15 vendors involved, more than 15 operators and more than 10 universities and a number of public authorities [8–10]. In the testbed, end-to-end performance of future services and functionality in new systems can be verified, different access technologies and network architectures can be compared and interoperability, scalability, manageability, etc. can be checked. The end users come from the muni nets Hudiksvall Stadsnät and Sollentuna Energi, both with large FTTH installations but also ADSL and ADSL2+. The FTTH installations are mainly P2P Ethernet, but also EPON (see later section) is tested for comparison.

The testbed is also used for pure research activities dealing with high-speed transmission systems, integrated optics, switching and tunability. Figure 3.7 shows the testbed structure.

The Acreo national broadband testbed has extensive collaboration with both national and international industrial partners and academic institutions, and some of the most important collaborations are the three European Union 6th Framework Program projects in the area 'Broadband for All'. These are MUSE [11] (working with multi-service access networks),

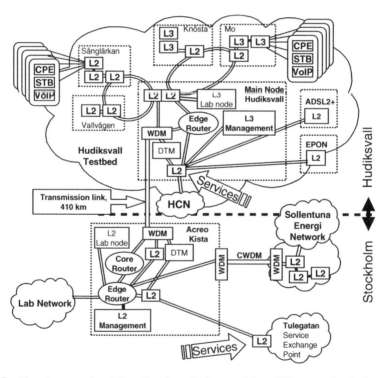

Figure 3.7 The Acreo national broadband testbed comprising different technologies, emerging services, new architectures. Partners come from operators, service providers, equipment vendors, academia, public authorities and muni nets, and everything is tested on real end users.

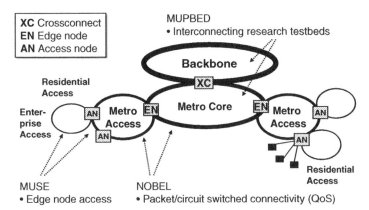

Figure 3.8 Acreo participates in European broadband projects covering access, metro, backbone networks and the interconnection between these.

NOBEL [9,12] (working with next generation optical networks for broadband) and MUPBED [13] (interconnecting testbeds across Europe). Acreo is the only research institution participating in all three projects (see Figure 3.8). The projects embrace participants from the leading European operators and equipment vendors including Swedish TeliaSonera and Ericsson.

3.6 OPEN NETWORKS VERSUS VERTICAL INTEGRATION

The field of broadband networks is heavily developing but still rather immature. There has been only limited consolidation with regards to the choice of technologies and business models. Concerning business models there are two fundamentally different 'schools', one in favour of vertical integration of services, operations, connectivity and infrastructure – the 'classical' network operator role – and another one supporting the so-called open network concept.

3.6.1 THE OPEN NETWORK

An integral part of the Swedish IT politics has been to separate the roles of a network operator and a service provider. It is envisioned that such open networks will benefit the end user, the industry and eventually the society:

- The end-user can freely select the service from a given service provider that offer him the most attractive conditions, and he can combine different services from different providers.
- The service provider gets a chance to reach many users without having to own or operate infrastructure or active equipment.
- Competition will stimulate growth and enable public services to be more easily offered to the end-users which will benefit the whole society.

A network where:
- the **network operator** and the **service providers** are separated
- the relationship between an **end-user** and a **service provider** depends on mutual agreements
- the **network operator** is not involved in that relationship beyond the connectivity service
- all **end-users** can choose a service from all **service providers** over the common infrastructure operated by the **network operator**

is called an *open network*.

Open networks are also called operator or service neutral, multi-operator network, etc. Most muni nets in Sweden are built as open networks, however with slightly different business models and different degrees of openness. Also many new residential area networks are built as open networks. An open network puts more stringent requirements on interfaces and standards and/or practices – both from a technically and a business model point of view – because the different functions are separated between different organisations. The open network model is generating interest abroad and has inspired, for example, installations in the Netherlands and the UTOPIA project in the US [14].

3.6.2   FUNCTIONS WHEN OPERATING MUNI NETS

The main functional layers in operating a network are seen in Table 3.1. The table exemplifies different implementations of muni nets in Sweden. In an open network a neutral part is appointed the role as communication operator, CO, by the infrastructure owner – typically a municipality or an estate owner. The role of the CO is to open the network to several service providers and TV suppliers in order to assure competition and with the intention to reduce prices and shorten the subscription period.

Typically the CO owns the active equipment in the network – This includes routers, switches and transmission equipment.

In a vertically integrated network the different functions are managed by one organisation, the vertical operator. However, also here the owner of the passive infrastructure could be a municipality or an estate owner. Observe that there are hybrids between open networks and vertical integration, and even within open networks as depicted in Table 3.1. For instance, and not indicated in the table, in some cases the owner of the active equipment and the connectivity operator can be separated because the municipality or estate owner owns the

Table 3.1   The functional layers in operating a muni net. Typical implementations of an open network and vertical integration are shown.

	Open network	Vertical integration	
Operation of service	Several ISPs and SPs	Vertical operator	
Owner of active equipment and operation of connectivity	Communication operator	Vertical operator	
Owner of passive infrastructure	Municipality/ estate owner	Municipality/ estateowner	Vertical operator

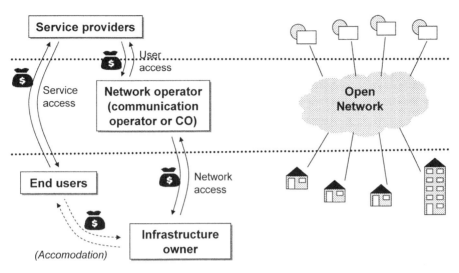

Figure 3.9   The relationships between the different parties of an open network are illustrated to the left. If the infrastructure owner is an estate owner, he may own the home of the end-user as shown in the lower left corner. The left figure also gives an idea of the business model in an open network. The physical connectivity of an open network is shown to the right.

active equipment. Also, for large networks there may be an extra layer between the service provider and the communication operator. This is the service broker that selects, makes agreements and maintains the relations with the service providers. Otherwise, the service broker role is part of the communication operator's responsibility. For muni nets the municipality often acts as communication operator, but the different roles may belong to different organisational units.

3.6.3   RELATIONSHIPS AND MONETARY FLOWS IN MUNI NETS

The relationships between end-users, service providers, the communication operator and the infrastructure owner are shown in Figure 3.9. The service providers (ISPs, TV-suppliers, etc.) pay the CO to get access to the network or rather the end-users. The end-users pay the service providers for the service, and the CO is *not* involved in this transaction (apart from providing the connectivity). The CO has been appointed by the owner of the passive infrastructure and pays for the rights of operating the network. In the case the infrastructure owner is an estate company, the end-user may live in an apartment owned by the estate company. Compare that with a vertically integrated network where the service provider and network operator and in some cases also the infrastructure owner are identical.

3.6.4   OPEN NETWORKS VERSUS VERTICAL INTEGRATION

The open network is commercially very important in Sweden and its political significance cannot be overestimated. The concept has led to strong competition among the service providers and is definitely a major contributor to Sweden's high-bandwidth connections and low prices compared to other countries.

The situation today is that the muni nets with open network models have materialised in the vast majority of the Swedish municipalities. They use different business models and network technologies, and their interfaces towards the service providers thus become different. This is not surprising as the deployments are relatively new and because open networks by nature require more interfaces than vertically integrated networks. However, in the next phase – starting now – it is of paramount importance that interfaces between network layers and towards service providers become standardised and more coordinated in order to reduce equipment costs and minimise operational expenses. The Swedish association of muni nets is working with these issues.

The true aficionados of open networks on one side and vertical integration on the other can provide many more pros and cons for the one solution versus the other. It is not the scope of this chapter to decide which have the better or more future-proof solution; it can just be observed that both models experience great success in Sweden these years. B2 representing vertical integration have 41 % of Sweden's FTTH installations (spring 2005 [5]), while the municipalities and residential areas with their open networks have the majority of the rest.

3.7   ACCESS NETWORK TECHNOLOGIES

In this section different fibre access technologies are discussed in the context of Swedish deployments. Point-to-point, P2P, Ethernet (also sometimes called switched Ethernet or active Ethernet) is being deployed much more than Ethernet passive optical network (EPON), and other types of PONs are almost absent in new installations. There is so far no clear consensus as whether to build Ethernet (Layer 2, L2) or IP (Layer 3, L3) based access networks, but 'pure' L2 networks are rarely seen. Some degree of L3 functionality is necessary as the network will otherwise be inadequate with respect to security and integrity. Customer premises equipment is beyond the scope of this description and will not be further discussed here. Nevertheless it should be mentioned that there is no consensus or trends as whether to install any kind of active equipment or just one or several Ethernet connections in the homes and offices – All solutions are tested at the moment.

3.7.1   PON VERSUS POINT-TO-POINT ETHERNET

Whereas DSL is broadband over the existing first mile copper infrastructure for telephony and cable operators also offer broadband services over their first mile coax, FTTx requires new installations. Either in a greenfield scenario where the fibres or ducts are wired together with other installations in a new residences or office buildings or in a brownfield scenario where the fibres or ducts are retrofit.

There are basically two kinds of FTTx networks; the passive optical network, PON, or P2P Ethernet. The PON is a point-to-multipoint network over a passive fibre plant comprising fibres and splitters/combiners, and with active equipment at the ends. P2P Ethernet uses switches instead of splitters/combiners (see Figure 3.10). PONs come in different flavours – More information is found at the PON Forums website [15]. For a more thorough description of P2P Ethernet, see [16]

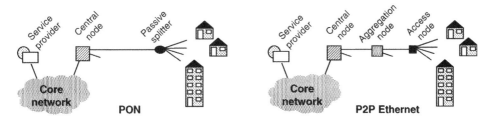

Figure 3.10 Two different FTTH technologies. In a PON (to the left) there are only passive splitters between the central node and the optical/electrical conversion (in the node/curb/home). In a typical Swedish muni net or residential area network P2P Ethernet (to the right) there are aggregation nodes (mostly L2 switches) and access nodes (L2 or L3 equipment) between the central node and the optical/electrical conversion (in the home).

While PONs are gaining popularity in North America and Asia (in particular EPON in Asia and BPON/GPON in North America), P2P Ethernet is more popular in Europe [1]. And indeed in Sweden the FTTH installations are completely dominated by P2P Ethernet with only a limited base of PONs being installed. This is the case for the incumbent TeliaSonera, for the leading FTTH operator B2 and for the majority of municipal networks. Part of the reason for Sweden's Ethernet preferences can probably be traced back to the aforementioned close collaboration between Ericsson, TeliaSonera and academia. In this relatively tight community the business models were often adapted to fit the technology rather than the other way round. FTTH has been on its way for many years now, and from a technical point of view P2P Ethernet has been regarded more interesting and potentially more future proof than PONs.

Without indulging in a technical discussion of the benefits and downsides of PON versus P2P Ethernet it should be observed that – regardless of the system architecture – for an area with no existing cable ducts, the cost of trenching associated with fibre infrastructure deployment will completely overshadow the access system cost.

3.7.2 L2 VERSUS L3 ACCESS ARCHITECTURES

The generic design of a P2P Ethernet-based FTTH access network is shown in Figure 3.10 to the right. Observe that no single terminology exists since the access network is a melting pot of different technologies. The central node (also called central office or head end) is connected to the core network and gets thereby access to the service providers. Towards the access network the central node is connected to an aggregation node (also called regional node or distribution node). The aggregation node is in turn connected to a number of access nodes which in turn are connected to customer premises equipment, CPE. Inside the house or apartment or office different end-user equipment (e.g. computer, TV, IP-telephone) is connected to the CPE.

Usually (but not always) the central node is an IP router connected to the core network. From the central node the IP traffic is directed to the individual CPEs. The intermediate equipment is layer 2 or layer 3 or a mix hereof. In Swedish muni nets and residential area networks, the aggregation node is most frequently an L2 switch while the access node is either an L2 switch with limited L3 functionality (explained below) or an L3 device. The services that are IP based are always carried encapsulated in Ethernet frames. For comparison, in traditional DSL the IP traffic is carried over ATM, and the access node would in this case be a DSLAM.

As indicated, many different solutions and mixes of L2 and L3 equipment are used in the Swedish FTTH implementations. Often some level of physical redundancy is assured through duplicating the central nodes and/or coupling the central nodes and the aggregation nodes in a meshed topology.

The choice between an L2- and L3-based access network is determined by a number of factors such as how future proof the solution is, security, cost, the availability of standardised interfaces, how much network supervision is required, etc. For instance, a large network needs constant supervision and a high degree of automation. Another example: A municipality may have a very limited budget and cannot afford to make the network too future proof.

Whereas an L3 network with IP-routers meets the basic requirements to an adequate access network, this is generally not the case with a pure L2 solution comprising Ethernet switches as security and integrity today cannot be properly guaranteed with a reasonably scalable design. This is however relatively easily solved by integrating limited L3 functionality (also called L3 awareness) in the switch design. And indeed, the leading access switch vendors have all integrated limited L3 functionality in their L2 equipment.

Without going into a more detailed technical discussion it should be noted that despite that Swedish muni net operators have implemented different broadband access networks the following points are common for the vast majority of networks supplying FTTH:

- P2P Ethernet is used.
- The services are IP based and transported between the nodes from the central node to the CPE on Ethernet.
- The central nodes are predominantly IP routers.
- The access nodes are either L3 devices or Ethernet switches with limited L3 functionality.
- The access nodes for apartments are placed in the basement of a multidwelling unit serving one or several other units. In case of several units all the apartments are connected with fibre. In case of only one unit the apartments are connected with either fibre or copper.
- The equipment terminating the optical signal is almost always placed *inside* the building, that is, in the house, in the basement of a multi-dwelling unit or in the apartment.

3.8   DRIVERS, SERVICES AND TRENDS FOR THE FUTURE BROADBAND NETWORKS

Sweden is now on the threshold to the next phase in broadband going from best effort network primarily offering Internet access to a converged multi-service network offering Internet, telephony and TV, triple play, all based on IP. During 2005 there will be an exciting infrastructure-based competition between the three largest broadband providers that are all offering triple play services (see Figure 3.5). B2 is offering triple play over FTTH, Com Hem is starting roll-out of IP telephony, enabling them to offer triple play over cable, and TeliaSonera have initiated IPTV over FTTH and DSL to complement their current offerings.

3.8.1   OPERATORS AND NETWORK OWNERS

FTTH providers are more or less born with a single infrastructure which they are now exploiting with new service offerings. Cable operators (Com Hem in Sweden) want their share of the future revenues and have started offering triple play services. However, whereas the customers receives a triple play package, the cable operator needs a parallel infrastructure to the TV distribution in order to offer telephony and data – Only the first mile copper is shared. Meanwhile the traditional telecom operators worldwide are discussing converged infrastructures, and some of them are starting to build such networks too. They predict savings in operational expenses as well as capital expenditure when procuring and operating a single network rather than multiple parallel networks as is typically the case today.

No matter what kind of operator, the big incentive to offer triple play services and constantly higher bandwidth is the anticipation of huge future revenues as well as reduced churn rates. All consumer tests show a demand for various TV and video services, particularly for the emerging high-definition TV (HDTV). And evidently there is a high acceptance to pay extra for both quantity and quality.

Whereas an IPTV channel requires around 5-10 Mb/s today, an HDTV channel coded in MPEG2 requires around 20 Mb/s, and 10–50 % better with MPEG4 coding. Observe, however, that some coding schemes can compress an HDTV channel even more, but at the expense of reduced quality. An average household would need at least two or three parallel TV channels apart from IP telephony and an Internet connection. So in order to support a high-revenue HDTV service the operators have to support a triple play package with a minimum connection of about 40–60 Mb/s.

As previously indicated, the large estate companies routinely install fibre in their new housing projects. The additional cost for this is comparably very low, and the valuation increase of the apartments or houses easily offsets the fibre investment. Moreover, the estate companies use the broadband connection to rationalise and optimise the operations of their buildings. At the same time and with a minimum of administration, the estate companies get extra revenues from the network operator that takes the responsibility for operating the network – as discussed in business case section.

3.8.2   AUTHORITIES

Sweden was early to use Internet and broadband and the government has strongly pushed the development for broadband access since the late 90s. We have however lost some momentum the last years due to unclear overall strategy, mixed messages, lack of visions, economical support that was 'frozen,' etc., but basically the government sees widespread broadband access as key to develop democracy, to maintain the welfare in Sweden and to increase international competitiveness.

Broadband access is in Sweden a cornerstone for implementing the so-called 24-h e-governance services. This is a vision about the citizens' possibilities to communicate with and use the services of the local or national authorities at all times. Towards the citizens, the 24-h e-governance services are initially about making information and forms electronically accessible, that is, designing user-friendly websites with the possibility to download forms. In the next step a higher degree of interactivity is envisioned.

The tax authority has got furthest of all authorities with all their forms available to download, interactive services such as 'Erik – the digital assistant' who can answer questions related to the income tax return, and above all the e-services. E-services are by the tax authority defined as web-based services that require a high-security digital signature. The perhaps most prominent of these services is the possibility to make changes in and approve income tax return forms – a service that got a major break-through during 2005 in Sweden. Apart from digitally signing of documents it is also possible to retrieve personal information about oneself from the tax database.

Common for the 24-h e-governance services and related projects are that they aim to create better, safer and more efficient systems. Ease of use and high penetration is crucial to obtain the most rationalisation benefits, and as such public e-services are a rare example of how economical concerns and development of democracy can coexist in a happy symbiosis. Other common factors for such services are that they require a high degree of security (as the digital signature) and that they today do not require a particularly high bandwidth. But for ease of use an always-on broadband connection is much preferred compared to a dial-up connection.

However, even by heavily subsidising the national broadband build-out, it is impossible to get close to 100 % penetration. At least for another generation there will be people with either no access to a computer and/or limited ability or motivation for using the Internet. Work is therefore done to make electronic services 'terminal independent' in the sense that they should be accessible from a mobile phone and a television set as well as the computer. The television has close to 100 % penetration, and tests are carried out in Svenska Bostäder's installations in Vällingby (see later section) where simple forms to the local social insurance office can be filled in using the remote control of the TV set.

The drivers towards broadband for municipalities are to some extent similar to the national situation, however in a smaller scale. Also here democracy and potential savings in administration for the 24-h governance play a big role. Apart from that, the broadband build-out has been heavily subsidised by the government, so the investment risks have been minimal. Many municipalities have had visions and clear ideas of what they wanted to obtain with broadband access, and the national build-out has generated such a momentum that even municipalities with no visions or ideas have jumped the wagon anyway because everyone else did. Correspondingly, whereas many municipalities have seen the prospects of attracting new citizens and small business to the region by deploying broadband, others have rather been driven by fear of losing citizens and businesses to other, better connected regions.

3.9   DESCRIPTION OF KEY SWEDISH FTTH PLAYERS

This section contains brief descriptions of some of the most influential Swedish FTTH and broadband access players: Ericsson where broadband access is just one of many business areas and PacketFront that has specialised in broadband networks. The incumbent TeliaSonera is of course one of the main players, with the largest broadband access market share. The estate owner Svenska Bostäder has driven the development for FTTH to apartments. And finally Stokab, which owns a massive fibre infrastructure in the capital Stockholm, facilitates brownfield FTTH deployments.

3.9.1 PACKETFRONT

PacketFront [17] was founded in 2001 by former employees from Bredbandsbolaget (B2) and Cisco, having vast experience from designing, deploying and operating the largest Ethernet broadband networks of the world. Based on this experience, PacketFront has designed its broadband solution for sustainable and cost-efficient broadband network operations.

PacketFront's offering is a combination of centralised software for automation and control *and* purpose-built access routers for advanced triple play broadband networks. The solution is designed to support multiple access technologies and multiple services delivered by multiple service providers – all distributed over one single IP infrastructure.

Included in the automation and control solution is BECS^{TM} – the control and provisioning system – which plays an important role in PacketFront's offering. BECS provides unique features for provisioning network elements, for service and policy management and for operations and support of the broadband network. BECS controls the triple play services with a possibility to manage each individual service to the end user, as opposed to management of the whole access line. This high granularity is important in order to allow differentiation of services and increase the average revenue per end user, ARPU.

Thanks to automation in the control of services and the access routers, deployment and operation of the network are performed cost-efficiently. PacketFront's 'access switching router' (ASR), is based on L3, and is designed to control services instead of access lines. The ASR is managed through BECS, and is offered for point-to-point FTTH access or as a broadband aggregation router for ADSL2+. The aggregation router also supports a multitude of other access technologies, for example, xDSL, PON, wireless, powerline communication or Ethernet over coax.

PacketFront has broadband networks deployed in Europe, Asia and North America with, as of October 2005, almost 500 000 ports sold (Figure 3.11).

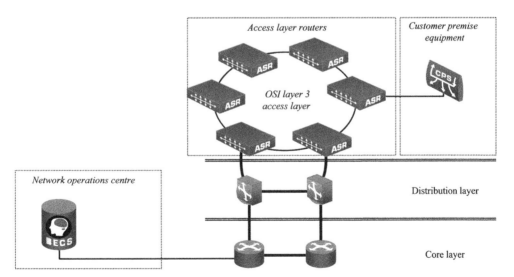

Figure 3.11 The key offerings in PacketFront's broadband solution are BECS, the control and provisioning system, and the ASR broadband routers. Reproduced by permission of PacketFront.

3.9.2   ERICSSON

Ericsson [18] offers a wide range of different Ethernet-based broadband access technologies
including DSL, WiMAX and fibre in different flavours. With all access technologies
integrated into one system solution it is possible to deploy a flexible solution adapted to
the specific requirements. Ericsson furthermore offers network support and services including operation and management for broadband access networks.

The fibre access platform is a highly scalable solution supporting any Ethernet-
based technology. The preferred fibre access solution is based on P2P Ethernet but also
GPON and EPON are supported. For a triple play-enabled access network, a control and
management architecture for service provisioning is included to ensure a proper QoS while
running services over a congested access network. Deploying a fibre access architecture is
more than just installing fibre, it is about providing an end-to-end architecture that allows
the end users to access any kind of services from any provider. Hence, much effort is spent
by Ericsson on having the access platform interact with upper layer architectures, such as IP
multimedia subsystem, IMS.

Ericsson is also participating in, and in many cases driving, work on fibre access and
architecture standardisation in various bodies such as IEEE, MEF, DSLF, TISPAN, etc., and
the company is a founding member of FTTH council Europe.

Once a fibre infrastructure has been installed it can be very costly to do upgrades.
Apart from broadband access equipment Ericsson also develop techniques that resolve
such problems. An attractive option that enables the network operator to grow and
upgrade the network infrastructure irrespectively of the topology chosen is the fibre
blowing technology. By installing low-friction microducts, in which fibres can be blown,
additional fibres can be added upon arrival of new users. This means that more
bandwidth, installation of new fibre types and connections to more users can be made
in a very simple and cost-effective way. With proper equipment it is possible to blow the
fibre in excess of 1000 m (see Figure 3.12).

3.9.3   TELIASONERA

The incumbent telecom operator TeliaSonera [19] evolved during the 90s from having the
monopoly on telecommunication (Telia in Sweden and Sonera in Finland) to being one of
several operators in competition. TeliaSonera has managed to maintain the largest broadband
access market share in Sweden, and since one of their main assets is the nation-wide copper
access network, DSL (primarily ADSL) is the dominant first mile technology, and will be
further developed. However, nowadays all TeliaSonera greenfield and rebuild customer
access deployments are fibre based. Either FTTH or FTTB architecture is chosen as
considered appropriate for each deployment. Economic prerequisites will often dictate
choice of technology and ultimately the decision will be made by the customer (e.g.
landlord).

As being the incumbent with a well-proven vertical integration of the value chain,
TeliaSonera is not primarily heading for the open network concept. Nevertheless, in order to
be well prepared to accommodate to new business models, TeliaSonera has engaged as a
communication operator in a number of FTTH installations, as in the case of Svenska
Bostäder as described below. In Malmö, the third largest city in Sweden, TeliaSonera is both

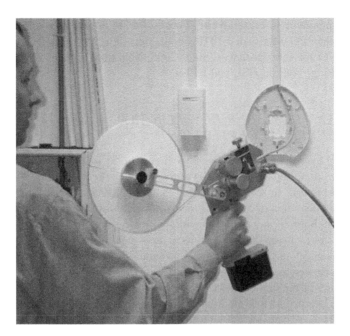

Figure 3.12 Ericsson delivers a wide range of broadband access products. Here is an example of a handheld, battery-powered fibre-blowing tool used in the Ribbonet® solution. (Courtesy of Ericsson). Reproduced by permission of Ericsson.

communication operator *and* ISP for a FTTH project encompassing more than 30 000 tenants.

In Sweden most inhabitants are concentrated to three densely populated urban regions, leaving large areas fairly sparsely populated. The figure below illustrates the local loop length in Sweden in TeliaSonera's copper plant for various population densities. The median distance to the local exchange is about 1.5 km (see Figure 3.13). Due to this loop distance

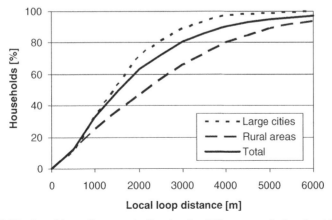

Figure 3.13 Local loop distances in Sweden for different population densities [20].

distribution, TeliaSonera is presently in favour of P2P fibre access to each residential customer rather than PON for their FTTH rollout, since P2P fibre architecture is considered more cost-effective for short distances. Since 2005 triple play services has been part of the TeliaSonera offering, both on DSL and FTTH.

3.9.4   SVENSKA BOSTÖDER IN VÖLLINGBY: A GREENFIELD DEPLOYMENT

The estate owner Svenska Bostäder (Swedish Homes) [21] is a good example of how new entrants can have large impact on FTTH broadband deployment – in this case by building a residential area network. In spring 2003, the community-owned Svenska Bostäder had its first FTTH deployment area of 5000 apartments in operation. The area, Vällingby just west of the capital Stockholm, is an urban area with multi-dwelling units, so the average last drop length is quite short, about 300 m. The local area network consists of 16 nodes, each connecting between 50 and 500 subscribers over multimode fibre.

Svenska Bostäder can as a community-owned real estate company apply long depreciation times which is one of the key enablers for the fibre deployment. The company chose fibre since it is future proof, has low environmental impact and improves the value of the real estates.

The FTTH network was an overbuild network over both the PSTN copper network offering ADSL and a cable TV network. In less than one year the FTTH market share had increased to about 40 % compared to the other broadband infrastructures. The take rate spring 2005 was 28 %, with a steady monthly increase of about one percentage point. The fibre network was built with 100 % initial penetration with fibre sockets installed in all apartments. When a subscription is activated only the media converter (converting the optical signal to an electrical signal) will have to be installed. The deployment cost was limited to about $900 USD per apartment, thanks to large duct availability.

The network is an open multi-provider network with TeliaSonera as the communication operator and Svenska Bostäder as network owner. The end-users can chose between five different ISPs and have access to local channels. IPTV was introduced during the autumn 2005.

With the positive outcome of this initial FTTH deployment, Svenska Bostäder has decided to extend the FTTH deployment to all their existing real estates (see below).

3.9.5   STOCKHOLM AND THE VÖLLINGBY MODEL: A BROWNFIELD DEPLOYMENT

In Stockholm city the fibre network is owned and operated by Stokab [22], a community-owned company. The historical reason for this has been twofold; to avoid uncoordinated digging in the city and to create a foundation for the future IT society. It is today a huge network comprising more than one million fibre kilometres – This corresponds to more than one kilometre per citizen in Stockholm. The network is unique in the respect that it supports both public and local administrations, big and small enterprises, operators and indirectly private households. It is an operator neutral fibre network which is used by more than 60 operators. It moreover connects the different local authorities as administrations, schools, hospitals in Stockholm, and for these purposes Stokab also acts as a communication operator.

With the positive outcome of its initial FTTH deployment in Vällingby – as described above – Svenska Bostäder has decided to extend the FTTH deployment to all their real estates, that is 45 000 apartments and 5000 business premises. The same decision has been taken by the two other city-owned real estate companies (Stockholmshem and Familjebostäder) which means that the plan is to connect more than 100 000 Stockholm apartments to broadband via FTTH by year 2009, all according to the open network concept. Three different communication operators have been selected. The plan is supported by the tenants association and a monthly rent increase by 6 USD to finance the real estate network is under negotiation. Similar plans are now worked out by the private real estate companies.

There are several business concepts for brownfield deployments, but in the city of Stockholm they often look something like this: The real estate owner installs ducts that reach every apartment in the building. Fibre is then blown through the ducts, and the tenants that have agreed so (either individually or through the tenants association) get a connection through a CPE. The access node (typically in the basement) is then connected to Stokab's fibre network that embraces a peering point to the service providers. The communications operator pays the real estate owner for access to the fibre and charges in turn the ISPs. The tenant gets a monthly rent increase from the estate owner and pays the ISP for services and Internet access. See also Figure 3.9.

Having access to Stokab's widespread operator neutral fibre network in the city is an enabling factor for these ambitious FTTH plans, for both publicly and privately owned residential area networks. The 'only' significant fibre installation is within the building connecting the apartments. Usually there is a fibre connection just outside the building.

3.10 SUMMARY

Traditionally, Sweden has had a very strong position in IT and telecommunications much thanks to combined efforts by Ericsson and the incumbent Telia in collaboration with academia. This includes fibre optics, mobile telephony, PC penetration, etc. Today Sweden has the highest penetration of FTTH closely followed by South Korea and Japan but well ahead of the rest of the world. This is mainly due to the operator Bredbandsbolaget (B2) that initiated national FTTH build-out in the late 90s and is still the operator with the incomparably highest FTTH market share.

Since the start of the century municipal networks subsidised by the government have appeared all over Sweden, many of these offering broadband connections to the inhabitants. Such muni nets are often built as open or operator neutral networks enabling competition at all layers of the network – except for the infrastructure. This is opposed to B2 and more traditional players that control the whole value chain. This has in turn induced competition between the muni nets and residential area networks representing open networks on one side against the traditional operators, B2 and the cable companies representing vertical integration on the other side. And it is probably this competition that has given Sweden the largest share of high-end broadband connections in Europe combined with low prices.

Another important driving force for the broadband build-out – which has directly led to the subsidisation of muni nets – is that the Swedish government sees widespread broadband access as key to develop democracy, to maintain the welfare in Sweden and to increase international competitiveness. The philosophy behind is to make the access networks as free and open to anyone as possible and to avoid monopolies and other constraints. This will lead to increased competition among the service providers which will eventually reduce prices so as many citizens as possible can get broadband access including access to the emerging electronic services offered by the authorities.

From a technical point of view Sweden has a different approach to FTTH than the US and Asia that are much focused on passive optical networks. Instead, and more in line with the rest of Europe, point-to-point Ethernet is completely dominating. There are two flavours of such P2P Ethernet access networks; they are either based on Layer 2 equipment (preferred by Ericsson) or Layer 3 equipment (advocated by PacketFront).

So what will the future bring for broadband Sweden? Despite its leadership in FTTH penetration Sweden is only number 11 in broadband penetration and the market share of FTTH compared with cable and DSL has been steadily decreasing for the past five years. However, rather than the FTTH deployment has weakened, the latter point is due to the fact that ADSL at the moment is growing faster than FTTH. Nevertheless, there is a widespread belief among most players on the market that fibre is the only really future-proof solution for broadband access. Finally, triple play with data, telephony and TV bundled together has got a breakthrough during 2005 with all three main infrastructures (B2 with FTTH, TeliaSonera with DSL and Com Hem with cable modems) competing along with some of the muni nets. A triple play package will need at least 20 Mbit/s to offer a data connection and two parallel TV channels of a decent quality – with HDTV the requirements double. The future is spelled FTTH.

ACKNOWLEDGEMENTS

The authors thank Hans Mickelsson from Ericsson, Astrid Lengdell and Magnus Olson from PacketFront, Caroline Olburs from TeliaSonera and Claes Engerstam from Svenska Bostäder for their valuable inputs to this chapter. Colleagues at the photonics group at Acreo AB are kindly acknowledged for helpful discussions and information.

REFERENCES

[1] IDATE, World Internet Access & Broadband Market, January 2005, @ www.idate.org
[2] OECD, OECD Broadband Statistics, December 2004, May 2005, @ www.oecd.org
[3] PTS, The Swedish Telecommunications Market 2004, PTS-ER-2005:34, July 2005, @ www.pts.se
[4] The National Post and Telecom Agency, PTS, www.pts.se
[5] PTS, Bredband i Sverige 2005 (in Swedish), PTS-ER-2005:24, July 2005, @ www.pts.se
[6] Telepriskollen, www.telepriskollen.se
[7] Invest in Sweden Agency (ISA), Broadband Sweden, September 2004, @ www.isa.se
[8] Olson M, Kauppinen T, Jacobsen G. The Swedish Broadband Testbed: Access related aspects. In Proceedings of 9th European Conference on Networks & Optical Communications, Eindhoven, The Netherlands, June 2004.
[9] Berntson A, Carlden H, Edwall G, Jacobsen G. Field trials and NOBEL related activities in the Swedish National Testbed. Proceedings of 9th European Conference on Networks & Optical Communications, Eindhoven, The Netherlands, June 2004.

[10]   Acreo, www.acreo.se

[11]   The MUSE project, www.ist-muse.org

[12]   The NOBEL project, www.ist-nobel.org

[13]   The MUPBED project, www.ist-mupbed.org

[14]   UTOPIA, www.utopianet.org

[15]   PON Forum, www.ponforum.org

[16]   Mickelsson H. Ethernet Based Point-to-Point Fibre Access Systems. *ECOC 2004*, paper Symposium Mo4.1.3, Stockholm, Sweden, September 2004.

[17]   PacketFront, www.packetfront.com

[18]   Ericsson, www.ericsson.com

[19]   TeliaSonera, www.teliasonera.com

[20]   IEEE contribution on loop lengths, grouper.ieee.org/groups/802/3/efm/public/jan02/mickelsson_1_0102.pdf

[21]   Svenska Bostäder, www.svebo.se

[22]   Stokab, www.stokab.se

4

Broadband Access Networks and Services in Korea

Kilho Song (and Yong-Kyung Lee ???),
Korea Telecom, Seoul, Korea

4.1 CHANGING ENVIRONMENTS AND FITL PLAN

In the early 90s, KT (former Korea Telecom) was facing an explosion of communication demands. More than half of its capital expenditures, totaling approximately 1 billion US dollars, were lavished on annual investments to lay copper cables, and CAPEX continued to be snowballing down the road. The Korean economy was expanding at a fast rate, and large buildings started to emerge day by day in the downtown of cities. Large buildings indicated that there were huge communication demands, and new copper cables had to be installed on the effected area. Although KT roll out copper cables based on 5-year demand predictions, the fast booming economy made it impossible to project the 5-year demand estimation. It was a yearly event that KT had to plan a new set of installations to meet the explosive communication demands. So, empty conduits were being filled rapidly and new civil engineering work took place frequently. However, it was extremely difficult to get the permission to dig the ground. Even if KT was able to get the approval, it was too expensive to lay new cables especially in the downtown areas.

To alleviate this problem, the adoption of loop carriers was proposed. By adopting loop carriers, KT reduced the demands for copper pairs as well as the possibility of requirements to lay new copper cables. Less cable installations meant less civil engineering work. As a consequence KT got to expect significant cost savings. Despite the continued controversy, the concept of Fiber Loop Carrier was defined.

So, KT established 'Fiber in the Loop (FITL) evolution strategy' in 1991. It comprised three steps, that is, FTTO, FTTC, and FTTH. KT also projected a plan to develop FTTx access systems, Fiber Loop Carrier (FLC) family, the concept of which is shown in Figure 4.1. FLC family largely falls into three types as their target areas: Fiber to the Office (FTTO) type for business areas, Fiber to the Curb (FTTC) type for densely populated residential areas like apartment complexes, and Fiber to the Home (FTTH) type for homes, which is the final destination of our plan.

Broadband Optical Access Networks and Fiber-to-the-Home: Systems Technologies and Deployment Strategies
Chinlon Lin © 2006 John Wiley & Sons, Ltd

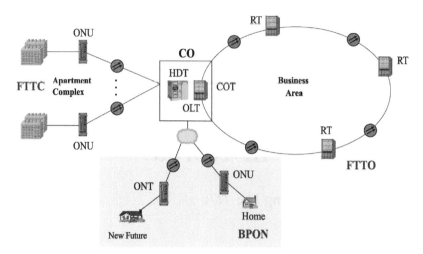

Figure 4.1 FLC family for FITL evolution strategy.

4.2 FLC-A AS THE FIRST MEMBER OF THE FITL

4.2.1 SERVICES CONSIDERED

The first version of fiber loop carriers was FLC-A. It was designed to provide the functionality of copper cable links without any additional service features. In other words, FLC-A was intended as a 'pair gain' system. Only existing services that had been provided by copper pairs were included. Various communication needs from large office buildings, such as simple black phone connections, enterprise PABX connections, and leased line connections with speeds up to 45 Mbps were met with FLC-A. Sometimes these buildings had Digital Subscriber Units (DSUs) for data communication, and they were accommodated in FLC-A.

4.2.2 HARDWARE CONFIGURATION

The FLC-A access network system consisted of Central Office Terminal (COT) and Remote Terminal (RT), and they are interconnected using high-speed optical links. The COT, which is located in a central office, was connected to such systems as PSTN, X.25 network, etc. The RT, which is installed in the customer's site, provides connections to black phones, ISDN, leased lines, pay phones, etc.

Both COT and RT are made of a common shelf and channel banks. The common shelf provides clock synchronization, SDH multiplexing/de-multiplexing, equipment management functions, and system power supply. A channel bank is equipped with a multitude of channel cards that provide specific functions such as black phone services, PABX connections, pay phone service, data services with speeds of 2.4–54 Kbps, and data services with speeds of n multiple of 64 Kbps. The total capacity of the system is STM-1 (155 Mbps) and can accommodate about 2000 voice channels. This capacity was determined to meet most buildings' demands with only one FLC system installation, while larger buildings could have multiple systems. Each RT was connected to COT with four optical fiber cores; two cores for one bidirectional connection while the other two were reserved for protection.

4.2.3 MANAGEMENT SYSTEM

The FLC-A system has a point-to-point network configuration between a COT and an RT. Inside the central office all COTs are interconnected via standard Ethernet 10baseT cables for Operation, Administration, Maintenance, and Provisioning (OAM&P) so that each COT and its corresponding RT may be managed from a central management system called FLC Management System (FMS).

KT has its own unique telephone line testing system, Subscriber Line Monitoring and Operating System (SLMOS). It is located at CO and monitors loop resistance, stray capacitance, etc. By monitoring such parameters, SLMOS can distinguish line faults from telephone hook-offs, thus eliminating unnecessary field trips for maintenance.

In order to utilize the end-to-end test capability of SLMOS even with the introduction of FLC into the loop, we incorporated into the FLC design a function that can test the loop portion between the RT and the customer terminal from CO locations. The FLC had features to specify a protocol to choose the loop to be checked, activate tests, and report test results back to SLMOS. This enabled subscriber lines connected using FLC-A to be serviced by SLMOS.

4.2.4 FLC-B AN FLC-A UPGRADE

FLC-A was successful. However, its full STM-1 capacity was too big to economically serve small-to-medium size buildings. And STM-1 capacity had to be shared between buildings. Therefore, a point-to-multipoint network configuration was required, and it resulted in a plan to develop FLC-B in 1995. FLC-B has point-to-multipoint configurability, that is, remote terminals might be connected to a COT through a bus, ring, or star configuration making use of SDH functionality. Also FLC-B has two new channel cards for ISDN BRI services. In addition, IDLC was implemented to eliminate requirements of channel banks at COT in case of connecting to public switches.

Although the loop carrier was generally deployed to connect a small number of residents far distant from suburban areas, KT has utilized it as an access and aggregation technology for providing services to a large pool of subscribers in metropolitan areas.

4.3 FLC-C

4.3.1 SERVICES CONSIDERED

FLC-A/B brought a tremendous success. FLC-A/B were produced with over 10 000 systems in total. Meanwhile, the competitors including operators abroad also used roughly 6700 FLC-A/B systems. Encouraged by its great success of FLC-A/B, KT landed on a plan to accelerate the development of FLC-C system. In 1996, KT started to develop FLC-C. At the early development stage of the system several critical issues concerning ONU installation environments were identified. Korea is relatively small by the size of its land but large in population. Metropolitan cities like Seoul and Busan are estimated to have approximately one-third to one-half of all their residents living in the area of apartment complexes. Moreover, one apartment complex amounts to at least dozens of buildings in number, and one building houses 50–300 households roughly. Therefore, we have chosen significantly larger Optical Network Units (ONU), where one ONU covers up to a capacity of 180 POTS basic plus 64 Switched Digital Video (SDV) optional subscribers and etc. By choosing such big ONU, KT had to abandon the grace of Passive Optical Network (PON), and had to choose active optical network configuration. This decision also led to some new challenges for KT, such as outdoor installation of bulky ONU cabinets, thermal management due to more power consumptions, more reliable powering, and longer battery backups.

4.3.2 HARDWARE CONFIGURATION

The Fiber Loop Carrier for Curb (FLC-C) system architecture is based on a Host Digital Terminal (HDT) corresponding to COT in FLC-A/B, and Optical Network Unit (ONU) comparable to RT in FLC-A/B. The HDT and ONU are interconnected using high-speed fiber optical links. Each HDT shelf has 16 STM-4c interfaces, so 16 ONUs may be connected to an HDT shelf. However, since each ONU is quite big, we usually use 1 : 1 protection. And eight ONUs can be connected to an HDT shelf in the case of 1 : 1 protection. Basically, ONU may have star configuration. However, some ONUs can be a master ONU and some ONUs can be linked to the master to form the double star configuration as shown in Figure 4.2. ONUs can be configured to be ring, star, double star, or a mixture of ring and star topologies.

The ONU interfaces to customer loops for providing POTS, ISDN, leased lines, and interactive video services. For the ONU interface of broadband services, Very High-Speed Digital Subscriber Line (VDSL) technology is employed so that existing conventional subscriber copper pairs are used as a transport medium without new installation. The HDT is connected to central office systems such as PSTN and ATM switches with standard interfaces of DS1E for POTS and STM-1 for video services, respectively. The optical link between the HDT and ONUs is based on STM-4c signals and carries Asynchronous Transfer Mode (ATM) cells. The configuration of the FLC-C system is shown in Figure 4.2.

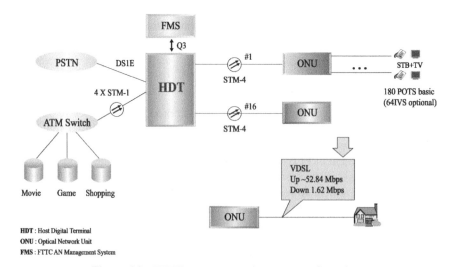

Figure 4.2 FTTC access network system configuration.

The ONU cabinet is installed outdoors in the vicinity of densely populated areas like apartment complexes. The cabinet is made of aluminum and equipped with telecommunications equipment, a power supply, and backup batteries, as shown in Figure 4.3. On the front door inside is attached a heat exchanger which is used to control the temperature inside the cabinet. The heat exchanger normally operates through AC power source, but operates through backup batteries during the failure of AC commercial power.

In installing ONU equipment outdoors, temperature and humidity are the two critical factors to be handled effectively. For most telecommunications equipment it is difficult to survive without using special electronic devices under a harsh environmental condition, for example, operating temperature of 60 °C or more. The thermal management problem of outdoor cabinets has been, therefore, a big issue from the very early development stage. The cabinet had to be designed to effectively remove internal heats generated from the dissipation of electronic load as well as heats from the sun.

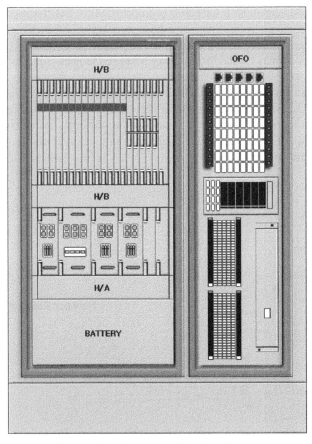

Figure 4.3 Outdoor ONU cabinet layout.

4.3.3 OPERATION SUPPORT SYSTEM

The FLC-C system has a star network configuration from an HDT to each ONU. In this configuration, one HDT shelf corresponds to a maximum of 16 ONUs as shown above in Figure 4.2. Inside the telephone office all HDTs are interconnected through an Ethernet Local Access Network (LAN) for Operation, Administration, Maintenance, and Provision (OAM&P) so that each HDT and related ONUs can be managed from a central management system, or FLC access network management system (FMS). The FMS operates as a manager and each COT carries out its task as an agent. It is accordingly possible that KT operators get the power and environment information of all ONUs from a central maintenance center called Power Operation Center (POC) which is connected to the FMS.

Through its standardized interfaces TMN gives a logical structure to exchange systems management information defined as Managed Objects (MOs) between management systems and networks equipment, and aims at interoperability, reusability, standardization, etc., between them. The exchange of management information is accomplished between a manager and an agent, which is shown well in Figure 4.4.

In the operation and maintenance system of FLC-C, Common Management Information Services/Common Management Information protocol (CMIS/CMIP) is used for the interface between a manager and an agent. MOs are used as a means of the exchange of management information between a manager and an agent. The definition of MOs is achieved through GDMO/ASN.1 (Guidelines for the Definition of Managed Objects/Abstract Syntax Notation One).

Figure 4.5 shows an example of GDMO template. MO Class templates are the most important ones in GDMO, which conceptually represent resources including the information of MOs. As shown in Figure 4.5, several items such as attributes, parameters, notifications, etc., are defined in MO Class templates and the common relationships among MOs are represented in Name Binding templates.

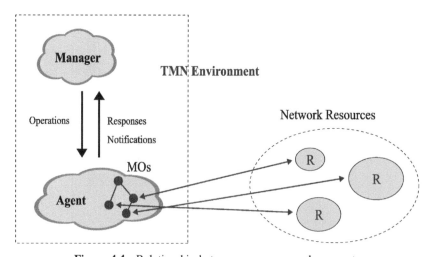

Figure 4.4 Relationship between a manager and an agent.

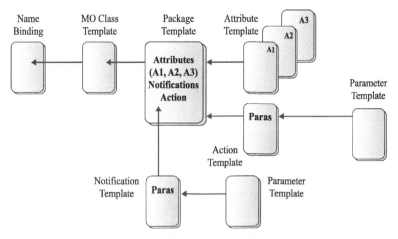

Figure 4.5   GDMO template.

Figure 4.6 and 4.7 show examples of FMS operator screen. The operators can handle the FLC-C systems through FMS remotely.

Once the operation and maintenance information from all HDTs and their ONUs reaches FMS, only the ONU power, and environment-related information is

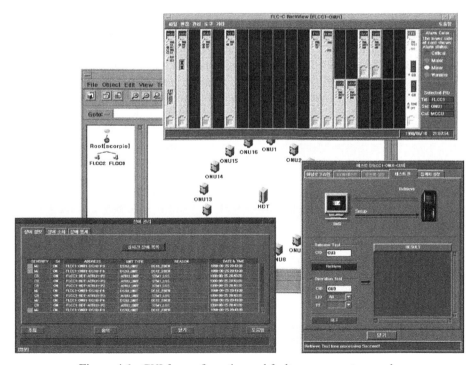

Figure 4.6   GUI for configuration and fault management example.

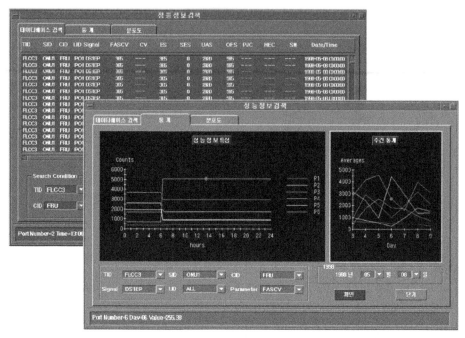

Figure 4.7   GUI example of performance management.

retransmitted to the POC after the FMS processes it. The interfaces to notice here are: (1) between the HDT (or ONU) telecommunications equipment and power supply, (2) between the FMS and HDTs, and (3) between the FMS and POC. Through the first interface which is internally specified in the form of Inter Processor Communication (IPC), the power and environment information is exchanged. Here, the environment information can be temperatures and humidity inside cabinets, door open, flood, etc. The second interface is based on TMN where Common Management Information Protocol (CMIP) is used. The last interface is X.25. The power monitoring terminal is used for debugging in process of development and for backup monitoring after development. Figure 4.8 shows schematically how the request and response messages are exchanged between the POC and ONU power system. The POC uses a X.25 protocol to connect the FMS for its remote operation and a CMIS/CMIP protocol is used between the FMS and ONU power system.

4.3.4   MULTI-VENDOR INTEROPERABILITY

It is required to take into account how to exchange the management information in a multi-vendor environment. ITU-T TMN standards recommends four factors to as inter-operability:

1. management platform
2. managed object modeling

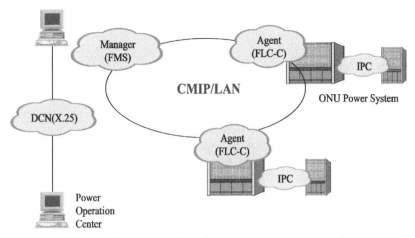

Figure 4.8 Configuration of power management network.

3. interface
4. management functions

First of all, the management platform is one of the key requirements to cut the overheads in constructing management systems. It helps to build a common view to implement management systems. Second, the managed object modeling is crucial to get the interoperability over access network systems among different vendors. For simultaneous provision of narrowband and broadband services, a tremendous amount of efforts are required to make relevant standards in the first place and to define consequent common managed objects for the new features of network elements (NEs) that are on the way. In addition, well-established managed objects modeling helps to provide smoother evolution. Third, the necessary interfaces should be defined to connect manager(s) and agent(s). Here, based upon the managed object model, Q3 interface is utilized to operate access network systems. And lastly, the management functions have to provide a common way of controlling NEs. Moreover, they should be designed to supply any requested information from upper layers. All of the recommendations above must be fully addressed and well established in a multivendor environment.

4.3.5 FLC-C EXPERIENCE

Although the development of FLC-C was able to reach nice completion, FLC-C did not achieve the successful result. It was too heavy to carry out POTS and the SDV service, which was expected as a killer application of FLC-C, so it never got started. Later on, despite the addition of internet access capability to FLC-C, only very limited number of FLC-C were deployed. The system was a beautiful build, but was too expensive to accommodate any new services. The failure of FLC-C was mainly doomed due to the failure of ATM. If there were any native ATM services at the time, FLC-C would have been of more use. ATM native services had never succeeded, and every new service had to be implemented on IP protocols rather than ATM, so FLC-C was not heading towards success.

4.4 BROADBAND ACCESS-XDSL

4.4.1 ENVIRONMENT

In the late 90s PC communication services emerged. It was required to have modems up to 56 Kbps to get connected, and people spent most of their time to use the telnet-like services, Hitel, Chollian, Nownuri, Unitel, Netsgo, Channeli, etc. Users got connected through modems, and mostly created and gathered in their own cyber-space communities. Bulletin Board Service (BBS) and FTP service were among the popular to share their interests and opinions. In 1999, more than 6 million PC communication users got logged on, and more than a million subscribers among them spent above $30 USD per month to get dial-up connections. Internet services were also expanding in a very fast pace.

4.4.2 INTERNET SERVICES IN KOREA

According to Korea Network Information Center (KRNIC), the number of high-speed Internet connections, based largely on digital subscriber lines (xDSL) and cable modems (CATV), surpassed 10 millions by the end of 2002 [1]. This figure corresponds to higher than 70 % of households and almost coincides with the 80 % penetration of personal computers. Out of these 10 million users, 6.1 million subscribed to xDSL services. KT alone covers 5.4 million xDSL subscribers, including 0.8 million users subscribed to very high-data-rate DSL (VDSL) services. The large number of xDSL subscriptions may be attributed to copper pairs preinstalled for POTS, which covers the entire country with no additional cable installation required. Cable modems over CATV networks is the second popular solution now with 3.7 million subscribers. These statistics are remarkable considering the current stagnation of economy in the telecommunications industry worldwide.

   The widespread availability of broadband Internet is now affecting every aspect of our daily life. A significant portion of offline activities have shifted to online networks. Nearly all transactions of banking, finance, tax payment, and stock trading are available at home, and Internet shopping has already grown to 10 % of the total retail market. The successful deployment of broadband Internet is the result of harmonization of network infrastructure, public policy, demographic profiles, and other factors. From the policy side, regulators strongly encourage controlled competition in the marketplace. This enables the normal user to get high-quality service at a low price. The service providers, in turn, are constantly striving to keep the cost low while increasing the quality of services. This has resulted in a wide range of service portfolios based not only on ADSL but also on other advanced technologies such as VDSL, Ethernet, and wireless LAN. KT has a broadband service portfolio under the umbrella brand Megapass. The service consists of several subcategories according to data speed and technology used, all with a monthly flat rate. Two-thirds of subscribers selected Megapass Lite rather than Megapass Premium because of its relatively high speed and reasonable price. On the other hand, Megapass Special and NESPOT users are fast growing because of the significantly higher speed and wireless access, respectively, of these two services. Another reason for this success is the high population density in the distinctive residential segments in Korea. Two distinctive segments are the apartment area and nonapartment area, which hereafter we call housing areas. The apartment area, where nearly 40 % of the total population lives, is especially suitable for providing economical fiber-based services. High population density and a distinctive residential environment

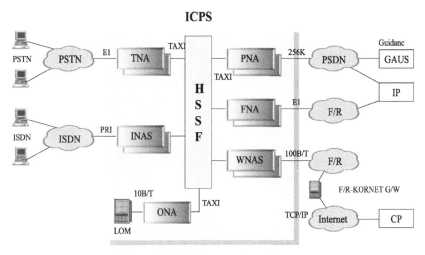

Figure 4.9   ICPS network architecture.

contribute to shortening the copper loop length. Accordingly, half of the subscribers are within 2 km from the central office (CO) and 80 % within 3 km. The average distance from CO to subscriber is around 2.5 km. These factors enabled cost-effective broadband deployment. KT, now serving almost 50 % of the Korean broadband market, has been in a pivotal position in the introduction and expansion of broadband services in Korea.

4.4.3   COMPETING TECHNOLOGIES

To cope with an excessive amount of demands for data communications towards the end of 90s, KT landed on a project based on public switches, Integrated Communication Process System (ICPS), and embarked offering 014×× services upon its completion. Figure 4.9 depicts its architectural layout and constituents including dial-up modem banks. The service provisioning processes are as follows. A customer dials up 014×× and public switches route the call to a corresponding modem bank of ICPS TNA. Followed by ICPS establishing a session to packet networks or frame relay networks, a data network connects the session through PNA to a specific host computer. This series of processes allow KT to bypass data traffic from trunks and makes use of mature packet backbone networks. However, modems can deliver only 56 Kbps and KT needed faster solutions.

In 1998 KT was hesitating between ISDN and ADSL. Although KT owned 40 % of the national hybrid-fiber-coax (HFC) networks, KT was not interested in cable modems as the future data communication solutions. KT was suffering from maintenance problems of the HFC network, and was very doubtful for uplink performance of cable modems due to the ingress noise. So, only ISDN and ADSL were the candidates for future broadband solutions.

KT had already spent plentiful resources in developing and deploying the ISDN, and it was very difficult to totally abandon ISDN investments. Peoples in R&D department pushed ADSL as their future solution; however, people from the financial department hesitated because of the cost to deploy ADSL. In 1998, equipment cost of ADSL was expected to be higher than $1200 USD per subscriber. It looked hopeless to an ADSL party to persuade financial people. Because it was a doubt that there are any applications or services that require high speed such as ADSL in comparison to ISDN, KT hesitated to deploy ADSL technology at that time.

However, everything changed suddenly. Thurunet begun the broadband access service in June of 1998 by using cable modems over Hybrid Fiber Coax (HFC) network, and Hanaro followed in February 1999 using Lucent's 'Any media' ADSL solution. There were already millions of subscribers who demand to be connected. Because both Thurunet and Hanaro adopted a flat rate of <$30 USD, the customers could have significant savings if they subscribe to the broadband access services. Higher speed and lower spending were perfect answers to customers, and broadband service proliferated. KT was alerted by these circumstances, ceased controversy, and changed his strategy quickly from deployment of ISDN to adoption of ADSL as main broadband solutions.

4.4.4 BROADBAND ACCESS NETWORK

The decrease of local and long-distance phone call traffics, mainly resulting from the introduction of cellular services, forced incumbent service providers to look for new growth engines beyond telephone services. KT started by introducing Internet access service, which was initially based on ISDN with a speed of 128 kbps. The service however proved insufficient for multimedia purposes and could not satisfy customer demands. Accordingly, ADSL was chosen as the broadband Internet technology, and intensive investment took place in 2000. ADSL is now a matured technology, and other enhanced services based on new technologies are emerging. Nevertheless, ADSL-based access service still occupies the largest portion of the current broadband access market.

Figure 4.10 shows two kinds of ADSL networks. The first one is ADSL directly from CO, the central office, called CO-ADSL. In this scheme, DSL access multiplexers (DSLAMs) are located in CO, and subscribers are directly connected to DSLAMs only through twisted pairs. In its early years, about 90 % of installed ADSL lines are COADSL, mainly provided for housing areas. COADSL, however, may not be effective for remote subscribers more than 3 km away from CO. About 20 % of all subscribers fall into this category, and they are the first candidates to be connected by fiber-based broadband access.

The second type of ADSL is based on fiber to the curb (FTTC). FTTC-ADSL mainly serves subscribers in apartment areas or, to avoid degradation of ADSL quality, in residential areas where the loop length is 3 km or more away from the CO. In this scheme, the optical line termination (OLT) is placed in CO, and the optical network unit (ONU) in apartment areas. So far, current access networks are based on asynchronous transfer mode (ATM) and ADSL. On the contrary, emerging broadband access networks are based on IP, Ethernet, and VDSL.

The successful deployment of an ADSL network depends heavily on the provisioning and maintenance capabilities of the service. For the provisioning side, shortening the delay between service request and actual provisioning becomes a key issue. On the maintenance side, timely and/or proactive service quality maintenance activities are crucial. For this purpose, KT has newly developed some operations support systems such as Telecom-Integration Management System (TIMS) and Access Network Support & Warranty for End Resources (ANSWERS) for successful OAM&P of ADSL networks. For example, when

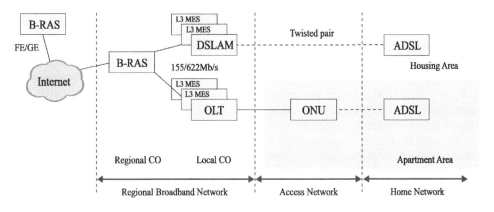

Figure 4.10   The ADSL access network of KT.

customer called for installation, the call receptionist put customer telephone number into the TIMS, then TIMS makes a query to Telephone line Operation and Management System (TOMS) to get the cable information. TOMS has the whole information about KT's access network. It has the geographic information of the conduits, cables, and so on. It also has the information of the wire gauge and the length of the cable. TOMS can give the cable length and the wire gauges of the usable copper pair between central office and the customer, and then TIMS can estimate the maximum speed of the ADSL connection. Therefore, TIMS can be used to determine whether KT can provide the broadband service or not. And if service is possible, TIMS can estimate the maximum speed of the service by inter-working with ADSL Transmission Line Analysis System (ATLAS) that provides service availability information and network quality. Before subscription, customers can get service availability information and network quality with the help of the ATLAS. KT called this function as 'preordering.' TIMS even has the configuration, performance, and fault management functions of the ADSL network.

An ANSWERS provides the customer care service and service assurance functions for maintaining xDSL service. It provides systematic line testing, problem localization and identification, management of trouble ticket, and dispatching outside field technician for problem handling at residential site.

For proper maintenance of the service, interoperability between devices from diverse vendors is crucial. Lack of interoperability can incur unnecessary cost when a device needs to be replaced. The customer's PC is another source of maintenance cost. Statistics show that a significant portion of customer requests for after-sale service come from malfunctions of customer PCs rather than problems related to Internet access functions. Since this incurs unnecessary cost, remote maintenance of customer PCs becomes an important approach for reducing service cost.

4.4.5   NETWORK ARCHITECTURES AND PROTOCOL

As a service provider network, KORea-telecom interNET (KORNET) provides Internet access services and contents. Now this network interconnects major nodes with 10 Gbps transmission lines. With the rapid increase of ADSL subscribers, user traffic is increasing rapidly as well and IP network gets more complicated. In order to provide better services to subscribers, KT increases network bandwidth and is considering to adopt Multi-Protocol Label Switching (MPLS) and IPv6 to improve routing structures, network security, and IP address deficiencies. Broadband Remote Access Server (B-RAS) is located in the Regional Operation Center, which has KORNET Point of present (POP). It concentrates DSLAMs from each CO, terminates ATM PVCs, and routes user traffic to KORNET.

DSL Forum recommended four architectures for access core networks: the transparent ATM core network architecture, the Layer Two Tunneling Protocol (L2TP) Access Aggregation (LAA) architecture, the PPP Terminated Aggregation (PTA) architecture, and Virtual Path Tunneling (VPT) architecture. KT selected PPP as a transport protocol to route IP traffic because this protocol can go with preexisting infrastructure for dial-up users. In addition, the protocol provides benefits of subscriber management, AAA functions, etc. To transport PPP protocol, two kinds of transport technology, PPPoA and PPPoE, are required according to CPE types. For core network access, we adopted the PTA architecture with PPPoE and PPPoA, since KT's ADSL subscribers have to get connected to KORNET in order to access Internet using ADSL and POP.

ATU-R comes in two types. The first one is a plug-in type and the other is a stand-alone type. As the plug-in type ATU-R, Network Interface Card (NIC) type with Peripheral Component Interconnect (PCI) interface is used. This type of ATU-R can provide IPoA (IP over ATM), PPPoA (PPP over ATM), and bridged mode by device drivers. As the stand-alone ATU-R, 10BaseT interface is used instead of ATM-25 because Ethernet card is much cheaper than that of ATM NIC and is prevalent. In case of NIC-type ATU-R, most of users can be supported with PPPoA architecture, but if users want to use PPPoE architecture, they should install a PPPoE driver for ATU-R and PPPoE software to their PC.

4.4.6   VDSL

Despite the widespread of broadband Internet services, the future success of broadband access businesses in Korea is not ensured. As the market becomes saturated, growth of new subscribers is leveling off, and broadband Internet service providers now face severe competitions. In terms of competitions, consumers have been asking for faster and faster speeds, and providers are always looking for the most cost-effective way to increase bandwidth. Termed post-ADSL, various newly emerging technologies are now being deployed to provide more advanced services in a cost-effective manner. These post-ADSL technologies include VDSL, Ethernet to the home, Metro Ethernet, and emerging wireless LAN.

In order to overcome sluggish growth, create new revenue sources utilizing existing copper, and maintain initiatives in the future broadband market, KT started to deploy VDSL in areas where churn is very heavy. Using VDSL means shortening the copper portion and moving the fiber closer to the neighborhood. This way we can expand our coverage and provide DSL in new areas. In July 2002, KT launched 13 Mbps symmetric

VDSL for the first time in Korea and also announced 25 Mbps asymmetric VDSL in the first half of 2003. Fifty megabit per second asymmetric VDSL service was launched in the second half of 2003.

4.4.7   SERVICES

The VDSL down and upstream bandwidth enables telcos to provide triple-play service of digital TV, video on demand, (VOD) and higher speed Internet and telephone service. Newly approved ITU-T Recommendation H.610 of the FS-VDSL Focus Group defines triple-play service using DSL technologies and makes it possible for telcos to provide video-centric full service [2]. To be successful as a triple-play operator, the network should support multimedia content and communication services that require bandwidth and quality management along the end-to-end path of the session. To exploit the most out of the increased bandwidth at the access network, and to avoid the quality problem associated with the backbone/peering traffic, the multimedia services are best provided at the edge of the core network. For multimedia services, KT has launched HomeMedia service that provides on-demand content service through a content delivery network comprising 10 regional server farms. For communication services, the quality control and management capability at the edge becomes crucial for delivering commercial-level services.

4.4.8   INTERFERENCES

While preparing for VDSL deployment in 2002, we found that the interference between VDSL and existing broadband access TLAN was severe. TLAN is one of the initial types of broadband Internet access that uses time-division duplexing LAN technology. As one of the early techniques, TLAN was based on proprietary technology and did not consider spectrum regulatory issues. Unfortunately, the spectrum of TLAN overlapped significantly with that of the newly deployed VDSL, and until it was fixed, it was the main cause of performance degradation. Another problem was the use of VDSL technology that was not standardized at deployment time. Mainly driven by the need to jumpstart the market, deployment of nonstandard VDSL was helpful in maintaining leadership in a fast-changing broadband market. In return, additional efforts were made to upgrade the nonstandard VDSL to prevent potential frequency interference with standard VDSL.

4.4.9   IP-VDSL

International Telecommunication Union – Telecommunication Standardization Sector (ITU-T) VDSL Recommendation G.993.1 specifies that the transfer mode of VDSL should be either ATM with transmission convergence (ATM-TC) or plesiochronous transfer mode with TC (PTM-TC) [3]. This means that while ADSL was based on ATM or synchronous transfer mode (STM), VDSL can use ATM or Ethernet. PTM-TC VDSL can be called simply Ethernet over VDSL. KT chose Ethernet as the next broadband protocol because of its wider availability, cost-effectiveness, IP conformance, and expected future-proofing. Thus, IP-VDSL was named in accordance with this viewpoint and represents Ethernet-based VDSL for an IP-converged broadband network.

4.4.10   MONUMENTAL SUCCESS OF THE XDSL

KT was the third to enter the broadband access service in Korea. However, KT planned massive deployment. It was possible because KT had budgetary advantage comparing to Thurunet and Hanaro. Also, KT had the advantage in the service area. Thurunet adopted cable modems, and because they did not own the HFC network by themselves, they had to negotiate with more than 30 individual cable TV operators for unbundling the HFC network. Hanaro had to install fiber cable first to deploy their fiber equipment. So, both Thurunet and Hanaro could deploy their system only in restricted regions. However, KT adopted copper base ADSL solution and virtually anywhere KT could begin the broadband service. Budgetary advantage, larger service regions, and good operation support systems such as TIMS, ANSWERS, TOMS, and ATLAS made KT win the broadband access market in Korea. KT is still the biggest broadband service provider in Korea and has more than 6 million broadband subscribers.

4.5   ETHERNET TO THE HOME AND WLAN

4.5.1   ENVIRONMENTS

Besides VDSL, another post-ADSL solution is Ethernet to the home. In Korea, UTP cables are already installed in some newly built apartments that enabled service providers to use pure Ethernet solutions to those residential areas. By using Ethernet technology directly to the home, we can connect at up to 100 Mbps line rate to customers while the speed of service is limited to 10 Mbps by provision. This service is characterized as Megapass Ntopia in Figure 4.11. Figure 4.12 shows IP-VDSL and Ethernet to the home for apartments. Both IP-VDSL and Ethernet to the home have the same network topology, except that the VDSL DSLAM is substituted by a fast Ethernet switch. Both of them use a Metro Ethernet network as the regional broadband network.

Megapass	*Main Technology*	*Max, speed (down/up, b/s)*	*Main Target Area*
Lite	ADSL	2M/640K	Housing areas
Premium	ADSL, VDSL	8M/640K	Housing areas
Special	VDSL	13M/13M 25M/3M 50M/7M	Apartment complexes
Ntopia	Ethernet	10M	Newly built apartment complexes
NESPOT	Wireless LAN	11M	Home, SOHO, hotspot (combined with wired access)

Figure 4.11   The umbrella brand of KT's broadband services.

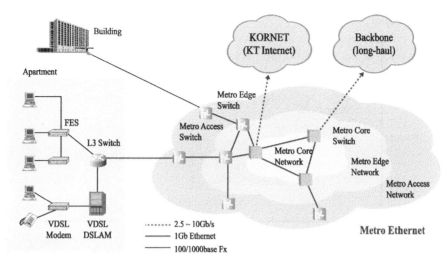

Figure 4.12   IP-VDSL, ethernet, and metro ethernet networks.

4.5.2   METRO ETHERNET

Figure 4.12 shows the overall architecture of IP-VDSL in accordance with the Metro Ethernet network. To overcome the limitation of loop length, the layer 3 switch and DSLAM are placed in the apartment complex. It is a simple but effective deployment because 40 % of the total population in Korea lives in apartment complexes where one apartment complex normally has more than 1000 households. The DSLAM terminates the VDSL and implements Ethernet switching, IGMP snooping, prioritized traffic handling, and so on. An L3 switch aggregates tens of DSLAMs and works as a gateway to the Internet. It implements L3 and L2 protocols including default routing, DHCP server/relay, multi-casting, and 802.1p/Q. The Metro access switch, Metro edge switch, and Metro core switch are all L3 Ethernet switches and hierarchically constitute the Metro Ethernet network. A Metro access switch is located in every local CO and aggregates tens of L3 switches via Gigabit Ethernet. The hierarchical position of ADSL DSLAM in Figure 4.10 corresponds to the Metro access switch in Figure 4.12. The Metro edge switch is located in the regional CO and aggregates several Metro access switches via Gigabit Ethernet. The Metro edge switch corresponds to B-RAS in Figure 4.10.

4.5.3   WIRELESS LAN

So far, we have only covered wired broadband access networks. Considering the increasing popularity of wireless information terminals, wireless LAN (WLAN) becomes another important post-ADSL solution for new growth. KT started public nationwide WLAN service in early 2002, and the customer base is about a half million users by the end of 2004. The service, called NESPOT, is currently based on the IEEE 802.11b standard, and it will evolve to 802.11a or 802.11 g for enhanced services. Figure 4.13 shows the position of WLAN in the service spectrum from wired to

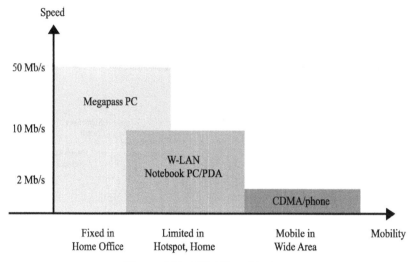

Figure 4.13   WLAN positioning.

mobile services. Wired broadband service such as ADSL/VDSL provides very high-speed Internet connection but does not support mobility. 3G CDMA service enables relatively low data rates, but supports mobility in wide areas. The position of WLAN is located between the two, and the service enables easy and high-speed Internet access through a notebook PC or PDA with restricted mobility in hotspot areas. Besides public WLAN, it is noteworthy that home WLAN service is very promising because Korea has a sufficient number of homes wired to high-speed Internet. With growing numbers of homes having multiple PCs and notebooks, the home WLAN market is emerging as a strong potential market combined with existing wired access. Using access points integrated in ADSL or VDSL modems, home WLAN provides a wireless home network that enables multi-PC connections while using one wired access line of ADSL/VDSL. The service is available to existing customers for a small additional charge. Current WLAN service has several drawbacks such as short coverage, limited number of hotspots, and radio frequency (RF) interference in the industrial, scientific, and medical (ISM) band. It also has to overcome problems such as ease of use, security, mobility, and network management [4,5]. However, with advances in wireless communication technology along with integration with other fixed and mobile technologies, the wireless broadband Internet market is believed to be another major market in the long term.

4.6   B-PON (BROADBAND PASSIVE OPTICAL NETWORKS)

4.6.1   ENVIRONMENTS

Although ADSL was very successful, the speed of the ADSL depends on the distance of the local loop. Figure 4.14 shows its network coverage roughly. And an ever-increasing speed

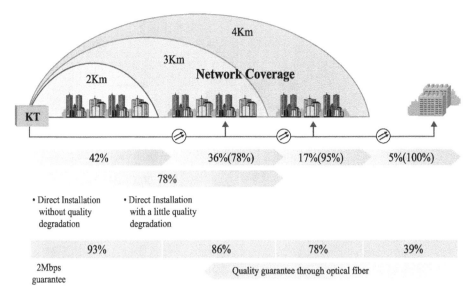

Figure 4.14 ADSL network coverage.

requirement makes KT engineers to remind the FITL plan once more. International standard organizations already were preparing PON solutions, and KT also was interested in B-PON system.

B-PON development was begun based on the Full Service Access Network (FSAN) standards. PON architecture seemed to be advantageous in maintenance and believed to reduce the operation cost. A major feature of the B-PON is that 32 customers can be concentrated on a single fiber to the central office using a simple passive optical splitter. Since this optical splitter requires no electrical power, it reduces a major cost and maintenance element. The B-PON supports a bit rate of 622 or 155 Mbps in the downstream and upstream direction, which is shared by the users through time division multiplexing. The multi-service capability and support of different QoS classes of ATM are well suited to efficiently deploy Ethernet data, video stream, as well as the legacy narrowband leased line services over a single full service access network infrastructure.

Within B-PON passive optical components are used in the optical distribution network (ODN) and at the optical interface connection of the termination equipment with the ODN. B-PON system consists of optical network unit (ONU) or optical network terminal (ONT) at the subscriber side and the optical line termination (OLT) at the network or CO side. B-PON system is based on ITU-T G.983.1 and can configure into various networks. OLT transmits the ATM cells to ONT/ONU after multiplexing the ATM-based broadband digital signal. The ODN takes care of the connections between the ONU/ONT and OLT. The ODN is splitting and combining the optical signal flows between OLT and ONU/ONT. The main components of the ODN are therefore the optical fiber and the optical couplers having, respectively, the transport and the splitting/combining functions.

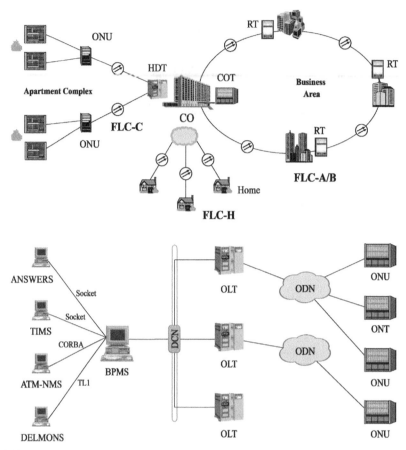

Figure 4.15 B-PON system and management system at KT network. (a) B-PON system configuration, (b) management system configuration for B-PON network.

4.6.2 MANAGEMENT

From the operator's point of view, management system should be easily operated without concerning the features of complex technology. Therefore, BPMS supports mirrored views that are simulating real Network Elements (NEs), network topology view, OLT/ONU/ONT rack view, rectifier view, etc. With these mirrored images, operator can look and feel the views as a real system. Figure 4.15(a) shows B-PON system configuration, and Figure 4.15(b) represents the management system configuration for B-PON in KT.

4.6.3 LEGACY OSS INTERFACES

B-PON management system comprises peripheral NMS interface, Management Modules, and OLT interface. Peripheral NMS interface takes part in cooperation with NMSs, which are TIMS, ANSWERS, and ATM-NMS at KT as follows.

System	Function	Protocol	Information type
ANSWERS	PSTN & ADSL warranty service	TCP/IP	Socket
TIMS	PSTN & ADSL installment and configuration	TCP/IP	Socket
ATM-NMS	ATM Link management	TCP/IP	CORBA
DELMONS	Leased Line management	TLI	Message

4.6.4 B-PON DEPLOYMENT

When KT planned B-PON, KT wanted to deploy the B-PON as the FTTC applications. In other words, each ONU serves multiple homes and targeted services include the POTS, internet access, VOD, etc. At the very early planning stage, FTTH seemed to be unrealistic. B-PON had advantages to the FLC-C in architectural aspects; however, if B-PON is used as FTTC system, the advantage of the passive optical network disappears because ONU must be fed electric power and equipped with battery backup system. And also ONU must be equipped with ATM to IP protocol conversion, IP encapsulation, and voice adaptation functions, and so on. And, since B-PON uses time division multiple access technology for uplink, B-PON implementation requires very complicated ranging algorithm and multiple access technology. Also burst mode receiver is needed. So, adopting B-PON as a FTTC solution resulted in abandoning the advantage of the PON configuration. The development was completed in 2001. B-PON could not claim any merits and the cost of the B-PON was much more expensive than the ADSL. And B-PON missed the chance to be deployed.

4.7 WDM-PON

4.7.1 NEXT GENERATION NETWORK

Spurred by the enthusiasm of Internet savvies and youngsters, broadband Internet service in Korea has penetrated into the daily lives of common people [6]. A recent survey of Internet usage patterns shows that users are now spending around 2 h/day on Internet applications [1], which is comparable to the average daily TV viewing time. Although encouraging, the survey also shows that the main Internet applications in Korea are still Web surfing, e-mail, games, and chatting using PCs at home. This shows that even though broadband Internet connectivity is becoming nearly universal in every household in Korea, applications are still limited to sending and receiving traditional text, image, and control information through the Internet, only at faster speeds. This is in contrast to the predictions of wide penetration of multimedia entertainment and communication applications, such as Internet broadcasting, video on demand (VoD), voice over IP (VoIP), and multimedia conferencing.

4.7.2 BANDWIDTH

Bandwidth demand is expected to grow rapidly as high-quality multimedia applications begin to be used in daily life. A typical service configuration for the home includes several high-quality video streams for broadcast and VoD sessions, a number of multimedia communication sessions, high-bandwidth peer-to-peer (p2p) sessions, and normal Web surfing. The bandwidth required for this service set can range from tens of megabits per second to 100 Mbps depending on the quality requirements for multimedia streams. To provide users with a multimedia experience comparable or superior to the current TV experience, it is envisioned that Internet-based multimedia streaming service will support high-definition TV (HDTV) quality for media broadcasting and VoD, and standard TV quality for multimedia communication sessions. Considering that the typical high definition media stream requires up to 20 Mbps, access network bandwidth of 50–100 Mbps is considered adequate to accommodate future broadband Internet traffic. To provide the required bandwidth as a universal service, service providers need to upgrade their access network. KT is taking a phased approach to provid bandwidth for next-generation applications – FTTC-based VDSL and fiber to the home (FTTH). VDSL has already achieved 50 Mbps within 300 m, and since fiber is available for apartment complexes, VDSL service is currently applicable to apartment areas. Apartment complexes currently under construction are also recommended to be equipped with UTP CAT 5 that enables 100 Mbps Fast Ethernet speed. *Moreover, upcoming apartments are expected to have optical fiber as basic cables, and every home in those optical apartments will have optical connection.* This means that FTTH can be provided to apartment complexes in the near future. For nonapartment areas (i.e., housing areas), VDSL deployment makes use of broadband access nodes that will be placed on the street, on a wall, on a pole, or somewhere near the houses. The broadband access node is connected by fiber from the CO and will be of varying size in accordance with location. Each house is connected to the broadband access node through VDSL technology. In the near term, it accommodates VDSL DSLAMs or access switches. VDSL or Ethernet on CAT 5 will serve as a preFTTH solution in short range for the time being. In the long term, the broadband access node will ultimately be responsible for FTTH. Regarding FTTH access technologies, passive optical network (PON) is considered at present the best candidate for an FTTH solution enabling point-to-multipoint fiber connection. While point-to-point fiber connection for FTTH requires as many optical ports as the number of subscribers, PON has an advantage in reducing this cost as well as cutting down fiber installation cost. Eventually, wavelength division multiplexing PON (WDM-PON) is expected to provide a dedicated 100 Mbps fiber connection to houses (Figure 4.16).

4.7.3 QoS

Compared to the current practice of providing multimedia services over a best-effort network, providing high-quality media services requires the network to meet the specific QoS requirements of applications. For example, high-quality interactive communication requires one-way delay of less than 150 ms and less than 1 % packet loss. On the other hand, a distributive streaming session (e.g., VoD) has less stringent delay requirements. From the service provider's standpoint, QoS provisioning implies building a managed IP network that can transfer the media information in a managed way. Also, the end-to-end nature of QoS

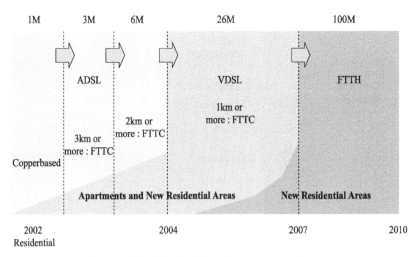

Figure 4.16 Bandwidth and access technologies.

requires quality handling not only at the access network but also at other parts of the end-to-end path, including the core IP network, the customer premise network, and even at the terminal itself. The control mechanism should be able to track the location/status of customer terminals and manage network resources for guaranteed performance. Considering that broadband Internet business until now has mainly focused on the access network part of the end-to-end path, this approach means a significant change of direction for traditional broadband Internet service providers.

Providing QoS at the network and premises sites has been a major technical issue for network service providers seeking revenue sources other than network connectivity services [7,8]. KT's effort to meet future service requirements involves the buildup of a premium network, as shown in Figure 4.17. The network consists of three parts: access, edge, and core networks. The access network is responsible for delivering information between customer equipment and edge nodes. For economic reasons, the network is shared between managed service traffic and best-effort traffic, requiring differentiated traffic handling at the access network. It has some differentiation mechanisms at the access network ranging from simple priority handling to more sophisticated ones such as virtual tunnels supporting QoS and security. The managed core network will consist of a set of MPLS tunnels that provides simple but high-throughput transport of packets between edge nodes. The edge network contains service intelligence in that it performs the mapping of QoS requirements between the access network and the managed core network, and different route control for best effort and managed service traffic. This requires the most sophisticated processing in edge nodes in that each of the incoming packets from the access network should be mapped to the proper tunnel in the core and vice versa, based on the QoS marking and destination address. Added to the complexity is the requirement to process each packet to identify source address, application type, destination address, and other information elements to perform proper routing and traffic management functions. The edge network also contains value-added service features that are best provided at the edge locations (e.g., managed security, content filtering).

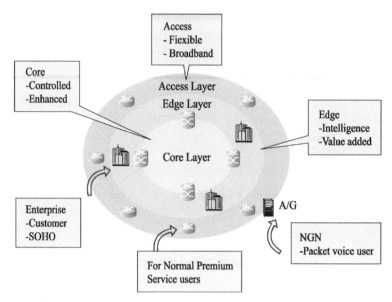

Figure 4.17 A conceptual view of the premium network for next-generation services.

Equally important is the QoS at the terminal where media encoding/decoding is performed. Terminals with less processing capability may not deliver satisfactory quality to end-users even though the network quality is above the required level. A typical problem arises when performing multimedia conferencing using a PC where the processing delay at the terminal is often more than 200 ms (compared to a 150 ms one-way delay requirement for a high-quality interactive session). From the service provider's point of view, providing a high-quality media application can include controlling not only the network but also equipment at customer premises.

4.7.4   DESIGN CONCEPT

Since the successful deployment of ADSL access network, the penetration of high-speed access network is increasing worldwide. Also the demand for higher speed is ever increasing. Initial application of the high-speed access network has been the internet access. As the number of internet user increases and the bandwidth improves, contents of internet web sites tend to shift from text-based page to picture and video-based pages. Also the usage pattern of internet shifts from web surfing to download of large files and streaming of multimedia contents.

A motivation for the implementation of higher speed network is the competition among the network operators. While KT is competing with Hanaro and Thrunet, higher speed was always the advertisement point. After introducing copper-based ADSL, in order to increase the speed, KT deployed DSLAM deeper in the vicinity of the customers. Eventually, ADSL modems were replaced by 13 Mbps VDSL, 26 Mbps VDSL, and now 52 Mbps VDSL modems. These ever-increasing speed upgrades were not initiated by customers rather by operators. This network upgrade was not followed by revenue increase.

The next motivation for the higher speed is the convergence of broadcasting and telecommunication. Currently, voice, TV broadcasting, and data are delivered to subscribers in separate networks, that is, public switched telephone network (PSTN), CATV network, and DSL. However, it is believed that the three separate services can be offered in the integrated form of Triple-Play service (TPS) which can be delivered by a single network. Furthermore, the services may be displayed by a single equipment, for example, a TV set. CATV operators are offering video and data service over their HFC network with almost half the cost of the telephone operators. They offer as low as 18 000 Won per month for the bundling service of CATV and internet access. POTS is not bundled with that package because in Korea telephone service is already mature, and VoIP has very little room. The same service will cost about 43 000 Won if offered by KT. So, CATV operators are very competitive and they are taking significant shares in the market. KT is thus facing a serious competition from the CATV operators, and needs a new strategy to compete effectively.

HFC is quite adequate for broadcasting service, and cable modem is also quite effective in delivering internet data. KT have owned HFC network till 1999; however, KT sold its HFC network partly because of the difficulty of maintenance and partly because of the regulations. Korea has very unique regulations about CATV. CATV industry is divided into three groups of companies: the first group is Program Providers (PP), second one is Service operators (SO), and the last one is network operators (NO). KT was NO, so KT has the responsibility to install and maintain the network to transport program to customers. According to the CATV regulations, NO cannot make use of the excess channels which are left after channels are assigned to specific programs. The right to use the remaining network resource was given to SO. HFC network was therefore not attractive to KT, so finally KT sold its HFC network. However, service operators eventually bought NO's HFC network and they are allowed to offer video and data bundled service. As mentioned above, HFC was not attractive to KT, but HFC network becomes a spearhead against KT. CATV operators now have more than 7.4 % of the broadband market share by the end of 2004, and furthermore, their average growth ratio is much larger than any other broadband service providers. To cope with this competition, KT needs new enhanced video services. Competing with CATV operators with the same services, KT will have no real advantage. However, if KT can provide new, better, enhanced video services, KT could compete effectively against CATV operators.

KT performed many IP multi-casting trials before. IP multi-casting was believed to be the solution for broadcasting over IP network for the telephone operators. While there have been several proposals for video broadcasting overlay over PON [11,12], IP multi-cast technology is bandwidth efficient and proved to be reliable and scalable. However, IP multi-casting is only the duplicate of the CATV broadcasting service, and no more. No additional value is given to the customers. That means KT can offer same service as the CATV operators with a more expensive infrastructure. Therefore, IP multi-casting alone is not enough to compete with the CATV operators.

However, IP-based video delivery allows new video services. IP-based video delivery can offer target advertisement (TA) by which advertisement is customized to each subscriber and the advertisement effect is enhanced. Also, IP-based delivery of video content allows the time-shifted viewing services. In the established broadcasting industry, the value of program has been determined not only by the content but also by the scheduling of time. The time-shifted viewing environment of TV programs eliminates the imposition of schedule and

Services	Bandwidth(Mb/s)
Three HDTV Channels	60
Internet	10
VIdeo Conference (Phone)	2
Telemetric / Remote Control	1
Total Bandwidth	> 75

Figure 4.18 Guaranteed bandwidth per user required for triple-play service.

creates additional value in the broadcasting. Therefore, IP-based video delivery is in more favorable position for the convergence service of telecommunication and broadcasting than the data overlay over HFC. These time-shifted viewing services and target advertisement service are basically unicast type services and are very difficult to implement over HFC. So, if and only if these applications (TA and time-shifted viewing) can be the killer applications of the TPS (triple-play service), telecommunication operators would have the chance to win the game.

In the multimedia communication environment such as TPS, the statistical multiplexing effect is not significant and the guaranteed bandwidth becomes more important rather than the maximum bandwidth. The guaranteed bandwidth requirement per user is estimated to be more than 75 Mbps in the near future as shown in Figure 4.18.

For the provisioning of the bandwidth required for TPS, the existing copper-based infrastructure such as twisted pair and hybrid fiber coax (HFC) network is not appropriate because of the limited bandwidth. Some very complicated VDSL technology can transmit data at 100 Mbps speed but within a very short range. That means we need to install ONU deeper and deeper toward the customers. That also means the total number of subscribers served by each ONU is decreased, and we need much more ONUs in the outside plant. That means fiber carries the signal almost all the way to the customer and only over the fraction of the distance copper pair carries the data. KT wants to get rid of the active element from the outside plant because of the maintenance problem, and if it cannot be avoided, KT wants to keep the number of the active elements to minimum. This arrangement is a very expensive solution too.

KT does not own the HFC. Even if KT owns the HFC network, KT does not have the privilege to use the network. Therefore, KT prefers to ignore the cable modem solution. However, the potential of the cable modem was carefully studied and economic analysis was done. Cable modem uses single TV channel, 6 MHz bandwidth, for data transmission. Downlink speed can be as fast as 40 Mbps. One hundred megabit per second transmission can be achieved by using three channels and inverse multiplexing the output of three cable modems. However, if we are considering streaming data, the whole bandwidth is occupied by a single subscriber and bandwidth sharing cannot be expected. We need additional cable modem terminating systems (CMTS) to use additional channels. However, CMTS is a very expensive network element (more than $10 000 USD per channel) and should be shared between subscribers. Usually, one CMTS is shared by hundreds of subscribers. If a subscriber needs to occupy the whole CMTS bandwidth, the system cost will be too high, and cable modem is not applicable in that case. So, neither VDSL nor cable modem can provide guaranteed 100 Mbps transmission. Therefore, FTTH development was started.

FTTH network architectures can be divided into two main categories according to fiber distribution architecture [13]. One is point-to-point star (also known as home run) architecture, and the other is double-star architecture. In the home run architecture, the required number of fiber core is the same as that of the subscribers, and it is expensive to install and handle the numerous fibers. Obviously, it is not appropriate for the massive deployment. In the double-star architecture, however, many subscribers share one fiber line through a remote node (RN) that performs one of active switching, passive power splitting, or wavelength (de)multiplexing functions. The RN is located between subscribers and central office (CO), and can be active or passive depending on whether the remote node is electrically powered or not. The double-star architecture with an active RN is referred to as active optical network (AON), and that with passive RN is referred to as passive optical network (PON). PON has advantages over AON in terms of installation, operation, and maintenance of network. PON can be divided into several categories according to multiple access schemes such as sub-carrier multiple access (SCMA), time division multiple access (TDMA), and wavelength division multiple access (WDMA).

In SCMA, the base-band signal of a subscriber modulates a radio-frequency (RF) carrier with unique frequency to the subscriber, which subsequently modulates a light wave [14]. It needs high optical power as the number of subscribers increases because of the clipping induced noise.

In TDMA, the collision of signals is avoided by access control protocol including ranging and cell allocation to each subscriber. TDMA has drawn interests from industries. ATM-PON was standardized by ITU-T G 983.3 [9] and Ethernet-PON (E-PON) is in the process of standardization IEEE EFM [10]. TDMA has merits in that it can utilize the bandwidth of optical link effectively by the statistical multiplexing of traffic for several subscribers. In TDMA, the downstream signals are broadcasted and optical signal power is split at RN. The upstream signals of subscribers are combined at RN. Therefore, security algorithms for downstream signals and collision avoiding algorithms for upstream signals are required. Even though TDMA has several advantages as an optical access network, it has several problems. Since optical power splitter is used in RN, the optical power loss of both direction signals increases as the number of ONU increases. And the splitting ratio is limited by the optical power of transmitters. Performance and speed of TDMA is restricted by the inherent characteristics of burst data transmission. First, the transmitter should turn off the signal power during transmission of other channels because background light will degrade signal to noise ratio (SNR). Also, there exists turn-on-delay before the data changes from low to high level. Second, the level of optical signal varies depending on the distance from the ONU's. It increases the complexity of the optical transceiver. Therefore, the burst mode transmission of TDM-PON makes it difficult to increase the speed high enough for TPS.

In WDMA, signal collision is avoided by allocating a dedicated wavelength for each subscriber. As WDM technology has matured during successful applications in backbone networks, there have been several suggestions for application of WDM to the access network. The proposals can be classified as hybrid WDM-PON solution and WDM-PON. The former proposals attempt to take advantage of a combination of merits of WDM-PON and those of other technologies. TDMA-PON partially employs WDM to add additional services. WDM/SCM PON was proposed to increase the utilization of bandwidth [15]. WDM-FTTC was proposed to use the existing copper lines and simplify the ONU [16]. However, the combination of technologies results in the combined effects of demerits as well as the merits.

In WDM-PON, the OLT in CO and ONT of each subscriber are virtually point-to-point connected through the dedicated wavelength. WDM-PON has inherent advantages over TDMA-PON in terms of bandwidth, protocol transparency, security, simplicity in electronics, etc. Also, the splitting ratio is not limited by splitting loss at RN. While several schemes of WDM-PON have been proposed, cost of the system has been of main concern. Even though distributed-feedback laser diode (DFB-LD) has been dominantly used in the long-haul WDM transmission, it is not yet appropriate to be used in access network. One of the problems is that it is not adequate to manage the wavelength of DFB LD's in each subscriber's premise. Management of wavelength of the optical transmitters in the subscribers' side is a significant burden for network operators. If we use the DFB-LD, wavelength-specific device, we should have N kinds of optical network units according to specific wavelength. This can cause problems in installation, maintenance, and logistics. And also the cost reduction will be limited due to the low production volume. Therefore, wavelength-independent characteristics is required at ONU if WDM is to be used in the access network.

Several approaches have been proposed to solve the problems. Spectrum slicing using a broadband incoherent light source such as a light-emitting diode (LED) or an ASE of the erbium-doped fiber amplifier (EDFA) have gained great attention [17–19]. The broadband light sources are used as transmitter and the wavelength for each subscriber was allocated by spectrum slicing at the arrayed waveguide grating (AWG). The LED can be fabricated at a low cost and modulated directly. However, its output power is insufficient to accommodate many channels by spectrum slicing. The spectrally sliced ASE light source provides much higher output power compared with the LED. Unfortunately, it requires an expensive external modulator. The PON employing a spectrally sliced F-P super luminescent laser diode (SLD) was proposed [20]. However, its performance is inherently limited by the intensity noise induced by the mode partition and/or the mode fluctuation [21,22]. The F-P SLD can be converted to a single mode laser by injection locking with an extra stable single-mode coherent light source, but it is not cost-effective, since we need a stable coherent source [23,24]. Reflective semiconductor optical amplifier (R-SOA) was used to modulate and amplify the spectrum-sliced broadband noncoherent light source and coherent laser light source [25,26]. Strained quantum well structure is employed to obtain high quantum efficiency and to alleviate polarization dependency; however, the cost reduction of R-SOA is limited.

4.7.5   WDM-PON SYSTEM BASED ON ASE INJECTED FP-LD SCHEME

Recently, a novel scheme of ASE injected FP-LD was proposed for WDM-PON [27]. Since it is expected to be economical and practical, the scheme was employed to implement a WDM-PON system. One of key devices for WDM-PON is the (de)multiplexing device.

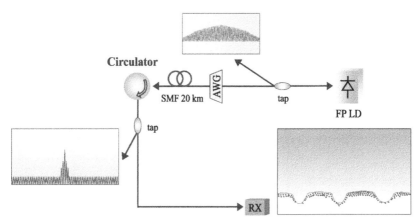

Figure 4.19 Operation principles of ASE injected FP-LD. Experimental setup for measurements of BER and spectrum of FP-LD without ASE injection.

There are many candidate devices, such as fiber type, thin film type, planar light wave circuit (PLC) type, etc. AWG is most favorable for (de)multiplexing a great number of wavelengths, since it can be manufactured by silica-based PLC technology. Also, temperature-insensitive AWG was developed and it seems quite reliable [28]. It eliminates the necessity for monitoring the wavelength shift of AWG which is to be placed outdoors.

It was shown that the wavelength of FP-LD can be locked to externally injected spectrally sliced ASE [27]. Figure 4.19 shows the experimental setup to demonstrate operating principle of the wavelength locked FP-LD. It consists of a FP-LD that was modulated by 155 Mbps pseudo random bit stream (PRBS) data. The broadband light source generated by pumped erbium-doped fiber (EDF) was transmitted through the transmission fiber before spectrum slicing. We used an AWG with wavelength spacing of 100 GHz for spectrum slicing. Then, the spectrum sliced light was injected to the FP-LD. The light from the laser was filtered by the AWG and transmitted through the transmission fiber. We can attach FP-LD at each output port of the AWG to have wavelength division multiplexed signal. The wavelength multiplexed signal can be de-multiplexed by using another AWG after the optical circulator that separates signals from the injected ASE. Then, bit error rate and eye pattern were measured.

When the ASE was not injected into the laser, the FP-LD laser shows multi-mode output as shown in Figure 4.19. After the modulated multi-mode light is filtered by AWG, we can observe single-mode peak in the spectrum analyzer. However, we cannot transmit data successfully, since the power of the particular mode fluctuates randomly with time. The origin of the fluctuation is randomness of the spontaneous emission coupled to each mode. The eye-diagram in Figure 4.19 is the measured result and it is completely closed. Thus, it is difficult to use the FP-SLD as a WDM source by spectrum slicing of a single mode among the multi-mode output. However, the laser output becomes almost a single mode when we injected spectrum sliced broadband light source (BLS) into the laser as shown in Figure 4.20. The typical side mode suppression is about 30 dB. We can see very clear opening of the eye. The measured bit error rate curves show error floor at 10^{-3} for the case of no injection as in Figure 4.19 and error-free in Figure 4.20 with the injection. By attaching a FP-LD at each

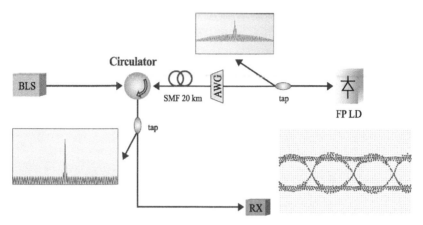

Figure 4.20 Operation principles of ASE injected FP-LD. Experimental setup for measurements of BER and spectrum of FP-LD with ASE injection.

output port of the AWG, we can generate the WDM signal. To increase the injection efficiency, we may need the anti-reflection coating on the front facet of the laser. In addition, the cavity length can be increased to have at least one lasing mode within the bandwidth of the injected ASE. We used FP-LD chip with 600 μm of cavity length and 1 % of front facet reflectivity. The length of chip was decided such that there is more than one mode within the transmission window of AWG channel. The reflectivity was reduced to increase the injection and output power of LD. We observed that if the injection wavelength is between the two lasing modes of the laser, the gain decreases about 5 dB. We have a 2 dB minimum gain when the detuning is ±20 nm and the injection wavelength is located between two modes. In other words, the minimum output power is −14 dBm. If the receiver sensitivity is −35 dBm, we have about 21 dB link budget. The measured bit error rate curve shows error-free characteristics whether the injection wavelength is matched with lasing wavelength or not. However, there exist maximum 1 dB penalty due to detuning of injection wavelength from the lasing wavelength. Thus, a single Fabry-Perot laser can be used for a WDM source within ±20 nm range. The operating range can be increased further by using a broadband FP-LD that has modified quantum well in the active region [29]. Since we can use the same kind of FP-LDs for all subscribers and communication wavelength is determined by the passive component (i.e., AWG), it is possible to have a single type of transmitter/receiver regardless of communication wavelength.

The WDM-PON system based on the proposed light source is shown in Figure 4.21. We have transmitter and receivers at CO. In addition, an AWG and two BL sources are located at the CO. The AWG is used for spectrum slicing and multiplexing for downstream signal. It de-multiplexes the upstream signals. Two different bands of BL sources are used for bidirectional transmission over a single fiber. C band and L band EDF amplifiers are used for upstream and downstream transmission, respectively. C band is used for upstream due to the fact that the optical loss has minimum value in the C band. We take advantage of periodic property of the AWG for the simultaneous usage of AWG for C and L bands. Therefore, the center wavelength in L band is off the ITU grid, while those of C band fits into the ITU-frequency grids. The different band signals are separated by thin film filter in the

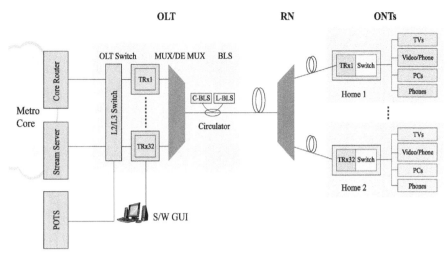

Figure 4.21 Architecture of WDM-PON network employing the ASE injected FP-LD.

transmitter/receiver module. It may be realized by a bidirectional module that consists of a laser, a thin film filter, a photodiode, and a trans-impedance amplifier.

The remote node consists of an AWG. We may need an athermal AWG for operation over a wide temperature range. At the subscriber side, each subscriber has a transmitter and receiver module to receive downstream data and to send upstream data. The transmitter and receiver module also contains the thin film filter to separate the two different bands. The broadband light from C-band BLS was coupled into transmission fiber via the optical circulator and sent to the AWG located at remote node. ASE is spectrally sliced by the AWG and injected to the FP-LD at ONT of each subscriber. The upstream transmission wavelength of the each subscriber is determined by the wavelength of the injected light. Hence, we can use a common type of FP-LD for upstream data transmission of all the subscribers, and the ONT is 'colorless.' The upstream data is multiplexed by the AWG located at the remote node. The multiplexed data is transmitted through the transmission fiber. At the CO, the received data is de-multiplexed by using the AWG located at the CO. Then, the receiver recovers the transmitted data.

For downstream data transmission, L band BLS is coupled into the AWG located in the CO. It is spectrally sliced and injected to the FP-LD. The lasers are modulated with the downstream data. The modulated signals are multiplexed by the AWG located at the CO. Then, it was transmitted through the transmission fiber, de-multiplexed by the AWG located at the RN, and sent to the receiver located at the customer premises. Since we allocate each wavelength for downstream and upstream, the customer can communicate with CO regardless of the status of the other customers. In other words, the system supports dedicated connectivity between the CO and the customers.

Developed WDM-PON system used C band for upstream signal and L band for downstream signal. The channel spacing is 100 GHz for C band. The maximum transmission length between ONT to CO is 20 km. The minimum injection power into the laser is -16 dBm/0.2 nm at peak. The measured BER curves show error-free for transmission over the 20 km transmission fiber. The measured sensitivity at BER of 10^{-9} was

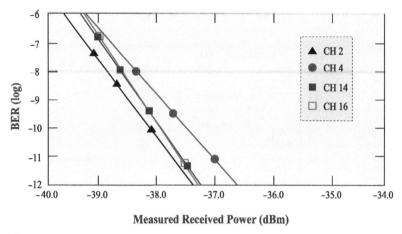

Figure 4.22   The measured BER of 4 channels among 32 channels for transmission of 155 Mbps over 20 km by the WDM-PON system.

about -38 dBm at 155 Mbps. Figure 4.22 shows the measured BER curves. No penalty was observed after transmission of 20 km.

A WDM-PON system was designed to support TPS including internet access, voice by both of POTS and VoIP, HDTV broadcasting, and video communication. The protocol stack used in the system is shown in Figure 4.23. OLT is placed in the CO and connected to ONT through optical fiber. In the optical layer, OLT and ONT are linked through WDM, opening point-to-point connections with speed of 125 Mbps. Since there is no signal collision in WDM-PON, WDM-PON MAC simply conducts the media converting functions, which lowers the overall system complexity.

Ethernet is adopted in layer 2, where Ethernet adaptation and switching is performed. Ethernet is the most widespread and low-cost technology. Also, it provides a practical way of combining data, voice, and video for TPS through class of service (CoS) and virtual local area network (VLAN) features offered in Ethernet protocol. Layer 3 of OLT performs routing, sub-netting, DHCP server function, and serves as V5.2 gateway for POTS.

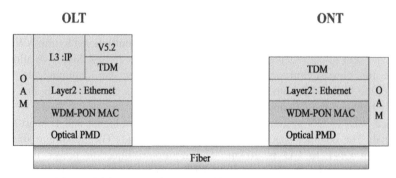

Figure 4.23   Protocol stack used in the WDM-PON system.

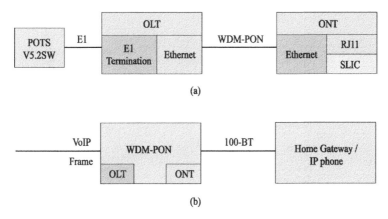

(a)

(b)

Figure 4.24   Architecture of network for voice service by (a) V5.2 protocol and (b) VoIP.

Figure 4.24(a) describes functions for supporting the POTS over Ethernet. OLT exchanges E1 voice data with V5.2 gateway by the V5.2 protocol. For the ingress voice data, OLT terminates V5.2 protocol and drops the E1 signal into the DS0E signals. Then they are cross-connected to an appropriate destination. The DS0E signals are encapsulated into Ethernet frames with the highest priority and transmitted to the ONT, where the Ethernet headers are stripped out and the recovered DS0E signals are converted into analog voice. Figure 4.24(b) shows the network architecture supporting VoIP service. In this case, ONT acts as a transparent pipe for VoIP packets which are passed to the IP phone through RJ45 port. VoIP packets coming through RJ45 are prioritized for scheduling with the highest priority. For voice service by either POTS or VoIP, QoS is of primary concern. In both cases, Ethernet layer should support priority queuing, fast scheduling, and packet classification.

IP video service can be provided in Ethernet frame format using IP multi-casting protocol and we need L2/L3 QoS functions, such as packet classification, 802.1p, Diffserv, VLAN tagging. Figure 4.25 shows how QoS information of L2 and those of L3 are interrelated.

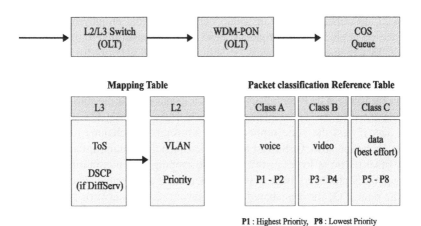

Figure 4.25   Packet classification and queuing in the WDM-PON system.

	OLT	ONT
L2	MAC Table Maintenance Capability Up to 4096 VLANs are supported STP(802.1d) Link aggregation(802.1ad)	MAC Table Maintenance Capability Up to 8 VLANs are supported
L3	Static Routing DHCP server : > 1000 IP Default Gateway : > 2	DHCP relay
QoS	Diffserv 802.1p Packet classification and marking Traffic management	802.1p Packet classification and marking Traffic management
Multicasting Functions	PIM-SM IGMP server IGMP snooping	IGMP snooping
Voice	V5.2	VoIP, POTS

Figure 4.26 Protocols adopted in the WDM-PON system.

Layer 3 supports multi-cast service protocol to increase the bandwidth efficiency. Protocol-independent multi-cast sparse mode (PIM-SM) and Internet group management protocol (IGMP) are used for the multi-cast protocol in OLT, and IGMP snooping in ONT. In case that the backbone network does not support any QoS scheme for multi-cast service, the rendezvous point should be placed as near as possible to the access network and a unified multi-cast protocol, like IGMP, could be supported. Figure 4.26 summarizes the protocols used in the WDM-PON system. The video services by the WDM-PON can be divided into a unicast and a multi-cast service. In the unicast service like VoD, users decide when and what contents are to be delivered, but in the multi-cast service users are served with scheduled contents of the broadcast service providers. The elements for video service include video headend, which broadcasts video contents and operates as a user/content administration and authentication, video data transmission networks, and set-top box (STB). Each connection for VOD and each channel for multi-casting forms a service-based VLAN.

Figure 4.27 describes the overall video service network employing the WDM-PON system. Video channels from the broadcast head end are transmitted to the rendezvous point placed in CO by either satellite network or dedicated optical network for broadcasting. Each channel are encapsulated into IP stream and sent to the rendezvous point, which is a high-speed L3 switch. If a subscriber of WDM-PON requests for a certain channel, IGMP join message is sent to the IGMP server. During the transmission of the join message, the node looks into database to decide whether the requested channel matches with that already in multi-cast, and, if yes, the node replicates and transmits the channel to the user who has requested it. In the scenario, QoS must be provided such that the multi-cast traffic flow is not affected by other traffic of lower priority, and the zapping time is minimized. Therefore, the rendezvous point is placed to the access network as near as possible. The back-end office is placed in the main CO, for it is shared among several access networks, and QoS is not an issue between back end and rendezvous point.

OLT of the proposed WDM-PON system consists of MCU, SWU, TIU, and PIU. Main control unit (MCU) controls all the operation and management (OAM) features of the OLT

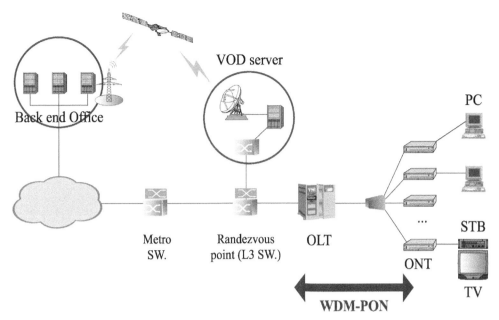

Figure 4.27   IP based delivery architecture of video service by WDM-PON.

and ONTs, and offers administration capabilities through the element management system (EMS). Switching unit (SWU) provides Gigabit Ethernet interface ports, switches receiving Ethernet frames to the destination port. TDM interface unit (TIU) processes V5.2 gateway functions and adapts receiving TDM traffic into the Ethernet frames (TDM over Ethernet) and vice versa. WDM-PON interface unit (PIU) transmits and receives Ethernet frames to/ from the ONT connected through WDM. The internal functions of the ONT consists of Optical transceiver, Ethernet switch, and user interface with RJ11 port for POTS and RJ45 port for data, video, and VoIP.

4.7.6   DEMONSTRATION OF TRIPLE PLAY SERVICE BY WDM-PON

FTTH can deliver TPS including several new services such as high-definition (HD) VoD, broadcasting of HDTV, private video recording (PVR), and time-shifted TV, etc. We have implemented the WDM-PON system and performed a field trial at Kwangju city for 100 subscribers since November 2004. The field trial offers the HD/SD TV broadcasting, HD/SD VoD, VoIP based on session initiation protocol (SIP), video telephony, and HD/SD time-shifted TV. KT implemented an IP-based HDTV broadcasting system. The HDTV content is provided from a satellite TV provider SkyLife' and encoded into 4.2.2 profile and level, 1080i resolution and CBR and VBR format by HD encoder. The streamer encapsulates streaming data into IP packet, which is multi-cast to subscribers through WDM-PON. For HDTV MPEG-2 decoding, we developed a HD IP set-top box (STB), which includes hardware HD decoder and runs Linux. For the implementation of videophony, we developed the video phone terminal that encodes and decodes video signal into MPEG-2 stream and

Home (Bandwidth)	STB	Equipments at Home	Services
Home#1 (34 Mbps)	IP HD STB 2 IP SD STB 1	ONT, L2 SW Notebook	1. HD VOD and SD VOD 2. Internet access by PC
Home#2 (47 Mbps)	IP HD STB 3	ONT, L2 SW Camera 1, VoIP(SIP) 1	1. Video Telephone 2. Internet access by TV 3. VoIP based on SIP
Home#3 (87 Mbps)	IP HD STB 4 IP HD STB 1	ONT, L2 SW Camera 2, VoIP(SIP) 2 Notebook	1. Video Telephone 2. Real time HD IP Multicasting 3. HD VOD 4. VoIP based on SIP 5. Internet access by PC

Figure 4.28   IP-based delivery of video service by WDM-PON.

vice versa. Real-time Transport Protocol (RTP) was implemented for the assurance of interoperability. The system successfully supported the services indicated in Figure 4.28. Even though the number of subscribers was limited, the trial showed the possibility of the HD-grade TPS by WDM-PON.

4.7.7   NOW WDM-PON

Although the development of WDM-PON is quite successful, the estimated system cost per subscriber reaches $350 US dollars. The cost of E-PON is expected as low as $200 US dollars. The fiber installation cost reaches $400 US dollars. FTTH is a promising future, but still very expensive. However, CATV operators are ripping KT's market and KT seems to have no chance to win the market except deploying FTTH and providing enhanced video services. E-PON seems to be cheaper than WDM-PON. However, WDM-PON is cheaper than E-PON if the bandwidth is taken into consideration. WDM-PON can provide 100 Mbps for each subscriber, while E-PON provides about 27–28 Mbps.

As there had been an argument in migrating from a PSTN modem to ISDN or ADSL regarding whether there were any applications or services that require high bandwidth as ADSL/ISDN in comparison to the PSTN modem, the same debate was recreated in validating the introduction of FTTH technology.

KT has recently demonstrated technology capability and feasibility to commercialize WDM-PON in Korea and decided to realize Fiber to the Home with WDM-PON in the second half of 2005 to provide new broadband converged services such as the Triple Play Service (TPS). However, KT finally chose E-PON instead of WDM-PON not for preparing its future but for wining the hyper competition.

Nevertheless, KT has made some tangible efforts to let WDM-PON technology be adopted as global FTTH standard technology. As part of its strategy that KT will foster WDM-PON as a national business growth engine, KT is now actively participating in the construction of 'FTTH model city' in Kwang-ju with a goal to commercialize WDM-PON on a large scale for the first time in the world.

REFERENCES

1. Korea Network Information Center, http.//www.krnic.or.kr
2. ITU-T Rec. H.610. System architecture and customer premises equipment, 2003.
3. ITU-T Rec. G.993.1. Very-High-Speed Digital Subscriber Line Foundation, 2001.
4. Choi Y, et al. Enhancement of a WLAN-based internet service in Korea. ACM Int'l. Workshop Wireless Mobile Apps. and Services on WLAN Hotspots.
5. Henry PS, et al. WiFi. what's next?, IEEE Commun. Mag., Dec. 2002; **40**(12): 66–72.
6. America's Broadband Dream is Alive in Korea, NY Times, May 5, 2003.
7. El-Sayed M, et al. A view of telecommunications network evolution. IEEE Commun. Mag., Dec. 2002; **40**(12): 74–81.
8. Fineberg V. A practical architecture for implementing end-to-end QoS in an IP network. IEEE Commun. Mag., Jan. 2002; **40**(1): 122–130.
9. ITU-T recommendation G.983.3, A broadband optical access system with increased service capability by wavelength allocation, 0/2001.
10. IEEE 802.3ah, Ethernet in the First Mile.
11. Frigo NJ, Reichmann KC, Iannone PP, Zyskind JL, Sulhoff JW, Wolf C. A WDM-PON architecture delivering point to point and multicast services using periodic properties of a WDM router. In Proc. *Conf. Optic. Fiber Commun. OFC'97*, Dallas, TX, Feburary 16–21, post-deadline paper PD-24, 1997.
12. Przyrembel G, Kuhlow B. AWG based device for a WDM/PON overlay in the 1.5 μm fiber transmission window, in *Techn. Digest of Conf. Optic. Fiber Commun./Int. Conf. Optics Optical Fiber Communications OFC/IOOC'99*, San Diego, CA, Feburary 21–26, Vol.1, 1999; pp. 207–209.
13. Kaminow IP. Advanced multiaccess lightwave networks. In *Optical Fiber Telecommunications IIIA*, Kaminow IP, Koch TL (Eds). Academic: San Diego, CA, 1997; pp. 560–593.
14. Sierens C, Mestdagh D, Van Der Plas G, Vandewege J, Depovere G, Debie P. Subcarrier multiple access for passive optical networks and comparison to other multiple access techniques, *Global Telecommunications Conference, GLOBECOM 1991*, Vol. 1; pp. 619–623.
15. Jintae Y, Lee M, Kim Y, Park J. WDM/SCM MAC protocol suitable for passive double star optical networks, *The 4th Pacific Rim Conference on Lasers and Electro-Optics, CLEO/Pacific Rim 2001*, Vol. 2, July 15–19, pp. 582–583.
16. Park SJ, Kim S, Song KH, Lee JR. DWDM-based FTTC Access Network. *J Lightwave Technol* 2001; **9**: 1851–1855.
17. Zirngibl M, Doerr CR, Stulz LW. Study of spectral slicing for local access applications. *IEEE Photon Technol Lett* 1996; **8**(5): 721–723.
18. Jung DK, Shin SK, Lee C–H, Chung YC. Wavelength-division-multiplexed passive optical network based on spectrum-slicing techniques. *IEEE Photon Technol Lett* 1998; **10**(9): 1334–1336.
19. Feldman RD, Harstead EE, Jiang S, Wood TH, Zirngibl M. An evaluation of architectures incorporating wavelength division multiplexing for broad-band fiber access. *IEEE J Lightwave Technol* 1998; **16**(9): 1546–1559.
20. Wroodward SL, Iannone PP, Reichmann KC, Frigo NJ. A spectrally sliced PON employing Fabry-Perot Lasers. *IEEE Photon Technol Lett* 1998; **10**(9): 1337–1339.
21. Petermann K. *Laser Diode Modulation and Noise*. Kluwer Academic Publishers: London, 1988.
22. Kobayashi S, Yamada J, Machida S, Kimura T. Single-mode operation of 500 Mbit/s modulated AlGaAs semiconductor laser by injection locking. *IEE Electron Lett* 1980; **14**(19): 746–748.
23. Iwashita K, Nakagawa K. Suppression of mode partition noise by laser diode light injection. *IEEE J Quantum Electron* 1982; **18**(10): 1669–1674.
24. Oksanen M, Hiironen OP, Tervonen A, Pietilainen A, Gotsonoga E, Javinen H, Kaaja H, Arnio J, Grohn A, Karhiniemi M, Moltchanov V, Oikkonen M, Tahkokorpi M. Spectral slicing passive optical access network trial. In *Proc. Conf. Optic. Fiber Commun. OFC'2002*, Anaheim, CA, March 19–21, 2002; pp. 439–440.
25. Healy P, Townsend P, Ford C, Johnston L, Townley P, Lealman I, Rivers L, Perrin S, Moore R. Spectral slicing WDM-PON using wavelength-seeded reflective SOAs. *IEE Electron Lett* 2001; **37**(19): 1181–1182.
26. Buldawoo N, Mottet S, Dupont H, Sigogne D, Meichenin D. Transmission experiment using a laser amplifier-reflector dor DWDM access network. *European Conference on Optical Communication ECOC 98*, Madrid, Spain, September 20–24, 1998; pp. 273–274.

27. Saito T, Nara K, Tanaka K, Nekado Y, Hasegawa J, Kashihara K. Temperature-insensitive (Athermal) AWG modules, *Furukawa review* 2003; **24**: 29–33.
28. Kim HD, Kang SG, Lee CH. A low cost WDM source with an ASE injected Fabry-Perot semiconductor laser. *IEEE Photon Technol Lett* 2000; **12**: 1067–1069.
29. Mikami O, Yasaka H, Noguchi Y. Broader spectral width InGaAsP stacked active layer superluminescent diodes. *Appl Phys Lett* 1990; **56**: 987–989.

5

Broadband Fiber-to-the-Home Technologies, Strategies, and Deployment Plan in Open Service Provider Networks: Project UTOPIA

Jeff Fishburn

DynamicCity Metronet Advisors, Project UTOPIA, Utah, USA

5.1 INTRODUCTION

The Open Service Provider Network™ operational model for municipal broadband networks is unique. Founded on the needs and requirements of the municipalities it serves, it separates ownership of the broadband network from delivery of services and presents a community-conscious solution to what is erroneously perceived to be exclusively a technical problem. Developed by DynamicCity, it was first deployed in the UTOPIA project and may well prove to be the defining model for future municipal broadband projects.

The first few sections of this chapter focus on the political, technological, and economic drivers behind the model. A study of these drivers and their critical interaction leads logically through their evolution into the Principles upon which all phases of the design, deployment, and operation of the UTOPIA infrastructure are founded. A brief description of the operational model gives additional understanding of how a municipality with little to no experience in communications networks can oversee the deployment and operation of an advanced fiber network infrastructure.

A thorough understanding of the core Principles ultimately dictates the network design. Presenting a survey of available technology options, the next sections of the chapter summarize the strengths and limitations of each, concluding with a set of technical positions that best conform to the Principles. With the technical positions defined, the templates for

Broadband Optical Access Networks and Fiber-to-the-Home: Systems Technologies and Deployment Strategies
Chinlon Lin © 2006 John Wiley & Sons, Ltd

the design of the outside plant are covered with an emphasis on the separation of the active electronics from the passive outside plant. Given the half-life of technology, the section on the outside plant describes the design of an active network that remains useful through multiple evolutions of technology.

Following the template for the outside plant is a discussion of the criteria for selecting the architecture and electronics that best fit the Principles. The architectural template describes the trade-offs of various technology options and drives towards the logical conclusions from that analysis. Focusing on UTOPIA, the discussion explains how UTOPIA analyzed solutions from more than 80 vendors to create the one that delivered the best fit to the underlying Principles of the network.

The section on Service Provider Interfaces explains how services and applications are delivered over these advanced fiber networks. It describes the service provider interfaces and the logical service mappings that minimize problems and that maximize utilization of network resources.

The final section of this chapter addresses asset management. Focusing on the utilization of the infrastructure, it looks at the effect advanced services may have on the spare capacity of the network. Finally, with an understanding of how advanced services can affect network capacity, the analysis concludes with suggestions about what may be required to evolve the infrastructure beyond initial deployments.

5.2 MUNICIPAL PERSPECTIVE

Much like the rail systems of the late 1800s, today's advanced communications infrastructures represent a means by which communities may participate in, or find themselves left out of, the global economy. Many communities are discovering that critical telecom needs in their business and residential markets are going unmet, and since these advanced communications infrastructures are essential for the current and future economic vitality of their communities, they have begun to act on these needs. Just as city councils have traditionally grappled with municipal infrastructure issues including roads, electricity, and water, they now find themselves adding broadband availability to that list.

Why are these needs going unmet, and why is Government stepping in? Because the private sector typically under-invests in infrastructure – to the point of developing their economic models around the management of scarcity. Government, on the other hand, has historically provided the infrastructure to support business and residents: directly, as in the case of highways and airports, or indirectly through the support of monopolies such as the telephone companies. The reason for the different approaches to infrastructure is easy to explain: return on investment. Private industry typically expects quarterly increases from investments; those with slow returns, such as telecommunications infrastructure, are usually abandoned in preference for those with higher and more rapid returns. Governments, by contrast, measure returns differently, at times looking at community impact rather profit. And where money is the primary metric for an investment, governments can wait for decades to see a return. Inasmuch as the private sector, with its focus on near term gains, may continually fail to deliver critical communications infrastructure, many communities have begun to consider communications to be equivalent to other essential community utilities and services.

Recognizing that private industry would not adequately address the needs of the communities for robust telecommunications infrastructure in a timely manner, and recognizing that they each had similar needs and goals, 14 cities in Utah decided to work together to address the issues that the incumbent telecom providers would not. Their goal was to have an infrastructure capable of current and future advanced services, deployed ubiquitously to every home and business, whose operations and services depended upon a partnership between government and the private sector. They created an interlocal agency – the Utah Telecommunications Open Infrastructure Agency (UTOPIA) – to guarantee their communities a high level of economic vitality and their citizens an improved quality of life.

As a political subdivision of the state of Utah, UTOPIA has all the rights, provisions, and obligations of municipalities to form and enter into interlocal operating agreements under the Utah Interlocal Cooperation Act, which grants cities (and other public agencies) the authority to form legal subdivisions that can do anything a single city can do, including providing telecommunications services. The 14 member cities have agreed to the UTOPIA by-laws, paid membership fees, and signed UTOPIA's Interlocal agreement.

The 14 municipalities which UTOPIA currently comprises are shown in Figure 5.1 as a map, along with a table indicating the populations.

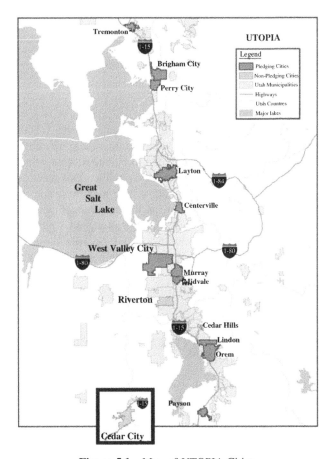

Figure 5.1   Map of UTOPIA Cities.

Member City	Residential population	Number of households	Business population
Tremonton	5592	1808	369
Brigham City	17 411	5840	494
Perry	2383	786	51
Layton	58 474	19 144	2435
Centerville	14 585	4238	269
West Valley	108 896	33 463	3906
Murray	34 024	13 303	2577
Midvale	27 029	10 729	1601
Riverton	25 011	6594	690
Cedar Hills	3094	1701	83
Lindon	8363	1977	527
Orem	84 324	24 156	2838
Payson	12 716	3869	371
Cedar City	20 527	7134	1223
Totals	422 429	134 742	17 434

Figure 5.2   UTOPIA Cities including populations and households.

Having decided to work together to address their unmet needs, UTOPIA communities considered the financing mechanisms that would allow them to move forward. After considering a variety of options, UTOPIA decided to pursue revenue bonds as its financing mechanism. Along with the ability to provide telecommunications services, Utah interlocal agencies have the ability to issue bonds on behalf of their members. Consequently, UTOPIA financed its project through revenue bonds or, as they are sometimes referred to, limited obligation bonds that obligate UTOPIA as an organization, but which do not impact the bonding capacity of the individual communities. And because these bonds are legally secured by the revenues projected to be derived over the network rather than by the city's power to tax, they do not increase taxes.

Cost for the UTOPIA project as defined in Figure 5.2 is projected to be approximately $340 million. Bonds were secured at a 6 % interest rate with a 20-year maturity.

5.3   OPERATIONAL MODEL: OPEN SERVICE PROVIDER NETWORK^{TM}

While the goals of UTOPIA are fairly simple and straightforward, they are largely incompatible with the business and operations models in use by the telecom industry today. UTOPIA's vision required a unique operational model, and they found it in the Open Service Provider Network^{TM} (OSPN^{TM}), a model created by DynamicCity to address both the immediate as well as the long-term needs of communities looking to build and operate advanced broadband telecommunications networks. At the heart of the OSPN model is the

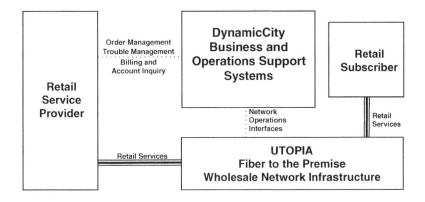

The diagrammed depicts these UTOPIA specifics of the aforementioned tenets of the Open Service Provider Network model:

- UTOPIA, as a representative for the End-Users, finances and owns the physical Network Infrastructure

- UTOPIA does not provide retail services; rather it 'sells' wholesale services to service providers

- Service Providers bring content and applications to, and develop the relationship with, the retail subscribers

- DynamicCity manages the design, deployments, and the ongoing configuration and asset management of the wholesale infrastructure

Figure 5.3   Open service provider model.

separation of retail services from infrastructure ownership. Additionally, to minimize any potential conflicts of interest, the model requires management from a neutral operator who neither owns the infrastructure nor provides retail services over the infrastructure. This tripartite structure allows for each entity to work towards the success of the others. An added benefit of the model occurs when the infrastructure is financed by a municipality: this makes the end-users the owners of the physical network. As owners of the network, end-users work for the most cost-efficient and diverse system possible.

The IEEE-USA Position Statement [1] is an excellent reference in support of and emphasizing the importance of the Open Service Provider Model. Figure 5.3 shows the relationship of the parties involved in making the OSPN so successful.

An analogy may help illustrates the concept of the OSPN: When cities realize the need to build a municipal airport, they come together to form an airport authority. That organization exists for the sole purpose of building and operating the municipal airport. It builds runways and structures, but it does not fly the airplanes. Instead, private airlines come and use the infrastructure. Because the high cost of the airport is spread over multiple airlines using the facility, the cost to use the airport becomes much lower than if each airline had to build its own airport.

When an airline sells tickets to passengers, the cost of the ticket covers runway fees, gate fees, and other costs which the airport authority assess them to use the airport and pay off the bonds it was constructed with. The airport authority does not interact directly with passengers – it does not charge the passengers fees directly, nor does it consider them customers. Instead, the airlines are the customers, and the arrangement allows the airlines to compete against each other, not against the airport authority. This competition helps airlines focus on things like value and services rather than on maintenance of the airport. It also benefits customers because airlines become innovative in their approaches to win and keep customers.

Similarly, in the OSPN, municipalities build and maintain the optical fiber infrastructure of their advanced telecom network, but they should not engage in selling services to the end-user. Rather, they open it up to private service providers. Ideally, multiple service providers compete against each other for market share. It is they who become the customers of the municipal network owners. In this model, the private sector still owns the relationship with the end-users and, being freed from concerns about maintaining the infrastructure, they are free to focus on their service offerings. This stimulates innovation as providers seek to differentiate themselves from one another, and it helps ensure that, through competition, prices remain at an appropriate market level. Additionally, since government financing for the network can secure lower interest rates than private industry can, the cost of construction is less than what it would be for private industry, and the cost savings benefit the service providers who end up paying lower access fees. Because their overhead is lower, service providers can price their services lower, thus benefiting the end-user.

To recap, then, this is how the OSPN operates:

1. The high cost of deploying the network is borne by the municipalities.
   - Because municipalities need not realize a rapid or high return on an investment, they are able to choose a financing mechanism that ultimately minimizes the cost of capital.
   - Since municipalities, through municipal bonds, have the ability to borrow at the lowest rates available, one of the prime costs associated with this capital intensive business is minimized.
   - With interest rates around 3 %–6 %, and with the ability to borrow on 20+ year repayment cycles, the cost of the infrastructure is reduced to more cost-efficient levels.

2. These savings are ultimately passed on to the retail service providers who 'buy' bandwidth over the network.
   - Because service providers have lower overhead, and because they need only invest on a per-subscriber basis (and not in infrastructure), they can offer better pricing for their retail services to end-users.
   - Separated from worrying about maintaining infrastructure, these retail providers are also more likely to focus their investments on the development of advanced services that differentiate themselves from their competitors' retail product offerings.

3. Finally, the municipalities outsource operations of the network to a neutral third party.
   - The management firm brings to the table the costly business and operational systems required for the infrastructure to operate smoothly. As the management firm is likely to leverage those costs across multiple projects, the operational costs for any one network are decreased.
   - By outsourcing the expertise required to operate the network, municipalities avoid the need to hire a large staff of internal technical professionals. The management

company is hired based on the expertise and professionalism of its staff, and it maintains its position as network manager only to the degree that it performs in a satisfactory manner.

With this system of checks and balances, distributed cost and work load, and diverse perspectives, the OSPN is an ideal model for effective public/private partnerships.

5.4   GUIDING PRINCIPLES

The logical directness of the OSPN sits in contrast to the complexity of the issues it is designed to address. In fact, the OSPN only becomes effective after clearly defined objectives and operational principles for the network are established. Like a Rubik's cube, the multitude of concerns inherent in municipal telecommunications projects are interrelated in a complex fashion: business modeling, financing, cash flow forecasting, legal issues, public relations, technology, maintenance, and operations all have solutions – both short-term and long-term – that are interdependent on one another.

To simultaneously solve each of these areas requires that they be addressed under a common set of goals or guiding Principles. Trying to solve one issue without consideration for all the others, that is, without using some type of guiding principles – may leave you with a superficially pleasing one-sided solution while the other issues still remain jumbled.

In the case of UTOPIA, careful consideration yielded four key Principles as the philosophical drivers for the project:

1. *It must be open.* By definition, 'open' implies that the system facilitate a true Public/ private partnership. Philosophically, UTOPIA was opposed to the idea of delivering services itself. Rather, it perceived for itself a more traditional role – providing infrastructure. The actual delivery of services was to be left to private service providers – as many as are qualified to service the market. The OSPN model ensures that a publicly owned infrastructure is made available to a wide variety of competing private firms for the delivery of goods and services.

2. *It must be carrier class.* This requirement, though obvious, was sustained as a guiding principle during initial market research. A scientifically administered survey was conducted in each of the individual UTOPIA communities to determine, by munici-pality, the characteristics of each of the markets. In every case, the number one concern for businesses, and the number two concern for residents, was reliability, along with security these are the fundamental characteristics of a carrier class system.

   Service providers also required the network to perform with carrier class reliability. From the smallest start-up to the global giants with international reputations, each is willing to entrust those reputations to the network only if they are guaranteed carrier class reliability. From the physical design to the overarching operational model, the infrastructure must deliver exceptional performance and offer absolute security.

3. *It must offer high, scalable bandwidth.* In addressing the first principle – opening the system to multiple service providers – the network had to meet the needs of multiple service providers simultaneously. In other words, it had to be capable of delivering all the current services available as well as higher-bandwidth consuming future services from all service providers on the network. Thus, the system had to start out with

tremendous bandwidth capacity and be able to grow larger still. And in growing, it had to evolve as well. In a way, this is a requirement to make the system 'future proof,' meaning that it is capable of adapting to new and emerging technologies that otherwise might obsolete the investment.

The value of incorporating this principle is obvious. Just as 'whistle stop' communities had an advantage over those bypassed by the railroad in the old west, cities with the ability to support multiple current and future services will have economic as well as quality-of-life advantages over other communities. And it ensures that the investment made today won't become outdated because the system is designed to scale to meet future demands.

4. *It must have an open and independent architecture.* While many proprietary solutions could be employed to deliver the first three principles, this principle aims at ensuring that the efficiencies of the system are always maximized. By requiring solutions to be standard-based and founded on open technologies, UTOPIA can 'shop around' for the best deals and is not beholden to any one particular company or proprietary invention. While there is often a benefit to a proprietary solution that can outweigh the negatives of diminished choices, UTOPIA determined that the ultimate benefits derived from vendors who are actively competing for its business would drive even greater benefits and more innovative solutions. Later we will demonstrate how selections based on global standards can lead to the lowest cost solution and deliver incredible flexibility in vendor choice.

These Principles were the guiding force in addressing a technical solution for UTOPIA. The solution to other areas, such as financing and the legal environment, are equally vital; however, they merit consideration on their own and are outside the scope of this discussion.

5.5 TECHNOLOGY POSITION: PHYSICAL MEDIA

In addressing the needs for advanced telecommunications throughout the UTOPIA footprint, the initial technical question requiring resolution was that of distribution medium. Which of the various technologies – copper, wireless, fiber, or others – will drive the rest of the technological solution. Any one of the options provides some type of resolution. But by applying the guiding principles, one choice became obvious – fiber. With one of the Principles being High Scalable Bandwidth, no other physical media comes even close to the capabilities of fiber.

Broken into Dedicated and Shared access solutions, the strengths and weaknesses of each solution (from a bandwidth perspective) are obvious. A quick look at the symmetrical bandwidth available in a neighborhood of a thousand homes reveals a profound difference in the available bandwidth.

| Dedicated Access for 1000 homes | Fiber, 150 000 Mbps | Copper DSL, 250 Mbps |
| Shared Access for 1000 homes | Coax, 40 Mbps | Wireless/Powerline, 90 Mbps |

Clearly, fiber has enormously greater capacity and, thus, greater ability to offer innovative services. Here is the reasoning behind these numbers:

Dedicated Access: Two technologies deliver dedicated connections: fiber and DSL.

- *Fiber's* 1000 home throughput of a symmetrical 150 000 Mbps is based on fiber deployments such as UTOPIA's, where the majority of the network connections are 100 Mbps and slightly more than 5 % of the connections are a full Gbps.
- *DSL's* numbers are premised on the copper infrastructure. A ubiquitous deployment of anything beyond the most basic of DSL services is beyond the 'reach' of current incumbent copper-based infrastructures. A basic DSL solution typically delivers 256 Kbps in the upstream; therefore a thousand home footprints would have a maximum of 250 Mbps of symmetrical bandwidth available.

Shared Access: Both cable modem and wireless/powerline systems are founded on shared access rather than on dedicated access. Shared network solutions sound promising when the throughput is stated without taking into account the shared nature of the network. The reality is that the access is not only *intended* to be shared, it always *is* shared, yielding less than the maximum potential.

- *Coax* networks have difficulty supporting any sort of bi-directional data transmissions without additional infrastructure investment. Traditional cable deployments were purely copper based and disallowed bi-directional communications. Recently, an infusion of fiber to support these coax networks yielded operational and services benefits, including the support of two-way data services. Unfortunately, the system upgrades were designed to squeeze in additional services on an already-crowded delivery medium. Reluctant to sacrifice video bandwidth for Internet use, most cable companies have allocated about 40 Mbps for data services for every node, that is, 40 Mbps to be shared among an average of 500 homes.
  The primary problem with today's shared networks is that, for optimal speed and performance, they depend on unchanging end-user traffic patterns. The networks are designed to deliver optimal speeds based on pre-defined usage expectations, and when usage increases, the performance degrades. When a hundred users initiate 'always on' traffic on these shared access networks, they fail to deliver much more than dial up speeds. As the availability of streaming video content increases, typical usage patterns will include more always on' flows.
  This will eventually create network engineering problems similar to those found in the historical changes experienced by the public switched telephone network. Class 5 switch engineers realized the inadequacies of their networks when dial-up modems exploded into people's homes. The shared trunk networks in the PSTN were vastly under engineered for the uptime characteristics of that new 'voice grade' application, leading to people's inability to even make a simple phone call.
- *Wireless* solutions are marketed the same way as cable networks and have the same inherent drawbacks. WiMax is an excellent solution for a small user base with little to no other access options available. A throughput of 90+ Mbps sounds comparable to a fiber connection of 100 Mbps until the end-user realizes they have to share that with the other 1000 subscribers in the neighborhood.

There are other options still, but each with its own set of limitations. Copper solutions could be re-deployed to deliver comparable speeds to fiber – but only if the single twisted pair were replaced with significantly shorter CAT5 cable. Coax networks could deliver the listed fiber speeds if the fiber nodes were pushed as deep as 25–50 homes and the whole RF spectrum on the coax was modulated with 1024 QAM. And wireless could approach the listed fiber speed only if every house had its own antenna. This actually makes sense – but it would only work if every house ALSO had its own fiber!

Ultimately, fiber is the only truly high-bandwidth and scalable solution. Consider the following analogy to illustrate the long-term scalability of fiber: if a standard drinking straw represent dial up speeds (56 K), then a pipe about a foot in diameter equals a 100 Mbps connection. Using the same scale, a Gigabit connection would roughly be a pipe 1 m in diameter. The fastest commercial connections for a single fiber would equal a pipe about 35 m in diameter and the theoretical capacity of a fiber would be represented by a structure over a half a kilometer in diameter. Clearly, if we are using a one-foot diameter straw today, we have room to scale a network given the theoretical capacity of fiber.

Appealing to the other guiding principles also leads to fiber as the obvious solution. For example, the Carrier Class characteristics of a nonmetallic conductor such as fiber are far superior to copper-based solutions, and even more so when compared with wireless:

- It is imperviousness to electromagnetic interference.
- It is impervious to corrosion.
- It does not 'leak' radiation, or signals, so it is more secure.
- It cannot be tapped without losing the signal, making it more secure.
- It is not subject to weather disturbances.
- It has no latency issues as some satellite-based systems do.

And the list goes on.

Finally, with respect to Open Service Provider Network, the capacity of a fiber network supports not only multiple service providers but also current as well as future services without the physical network owner needing to mediate disputes due to an inherently scarce physical network resource. While copper and wireless may provide sufficient bandwidth for a single carrier delivering a single service to a small number of dedicated users, neither can accommodate the vast amount of bandwidth required by multiple competing service providers offering a variety of services. Clearly, the guiding principles for the UTOPIA network suggest that only fiber provides the desired solution.

5.6 ARCHITECTURE TEMPLATE: OUTSIDE PLANT

Normally, in designing a fiber infrastructure, technologists and engineers are quick to engage each other in joyful discussions of photons, micro-optical switches, and so on. Such discussions are almost always premature. Before becoming enamored with a specific set of technologies it is important to step back and realize that the life of the fiber infrastructure will far outlive any of the technologies existing today – perhaps even the career of the technologists themselves. If the life expectancy of a fiber network is 20 years or more, then the strategy of deploying that fiber should focus first on the fiber itself: will the design of the outside plant support future technologies? When that is considered, only then can the focus shift to the technology choices available today.

5.6.1   TYPE OF FIBER

Of critical import is the type of fiber to use: single-mode or multi-mode. In the case of UTOPIA, an appeal to the Guiding Principles (particularly the principle of scalability and future-proofing the network) dictated using single mode.

- Single mode fiber leverages industry support for the currently broad deployment of single mode fiber. In other words – it is compatible with other existing fiber deployments and hardware solutions. This calls on the guiding principle of Open and Independent solutions in that competing vendors and solutions will help keep pricing low and options high.
- The carrying capacity of single mode fiber is much greater in practice than multimode fiber. A choice towards multimode fiber might make more sense if researchers had demonstrated the ability to successfully couple all of the modes of a multimode fiber across a broad spectrum. Again, this addresses the principle of scalability and high bandwidth.
- The deployed cost of single mode fiber tends to be lower than multimode. It is true – the higher cost of the lasers and connectors required in a single mode deployment versus LEDs, with their much simplified splicing costs, used in a multimode deployment, allows for relatively short runs to be more cost effective with multimode fiber. However, a single mode fiber-based optical solution can be shown to deliver a lower cost of deployment when the fiber runs are greater than 500 m and the bit rate is 100 Mbps or greater. The relatively short runs where multimode makes more economic sense are only relevant for intra-building connectivity. At greater link lengths and higher bit rates the single mode fiber is profoundly more cost effective.
- There are a number of types of single mode fiber currently in the market. Manufacturing techniques have been developed to deliver many different fiber types but for general transport purposes lower water peak loss fiber, modified or nonzero dispersion fiber, and large effective aperture fiber are most relevant. The latter two fiber technologies were developed for the long haul market where the long distances (100–3000 km) created a requirement for larger optical energies and the ability to tailor the dispersion. Due to the relatively short distances involved in FTTH, the fiber technology loss tends to be more relevant than dispersion. Therefore, fibers which deliver a lower loss around the 1400 nm window, by reducing the amount of water in the fiber, make the most sense.

5.6.2   ACTIVE VERSUS PASSIVE

Having selected the type of fiber, the next consideration should be the design of the network, specifically the placement of electronics and the fiber runs themselves. A fundamental tenet of any conductor is that as long as the media is dedicated to a specific end-user, then the full capability inherent in that media can be accessed by that end-user. This leads to the understanding that a fiber 'from' the home should be carried into the network as far as financially reasonable before that fiber is then terminated at some transition point, whether it is a optical combiner/splitter or an active electrical/optical switching device.

Many passive optical networks PON have been engineered with PON splitters deployed as close to the home as possible to minimize fiber costs. While this logic does indeed save a few dollars per home passed during the initial deployment, it is short sighted and will lead to scalability issues for the network in that it limits the deployment of electronics for future generations. And while the per-home passed cost is marginally reduced, a passive deployment actually increases the overall per subscriber electronics costs: unless every home on that edge split subscribes to services, the cost allocation of the OLT gear is over a smaller number of subscribers. In a typical municipal overbuild, take rates are less than the 100 % assumption upon which the capacity of the PON is designed.

5.6.3   REDUNDANCY

The next consideration in architecting the network examines the value/requirements of single versus dual physically redundant fiber paths to each end point. If there are two paths the fiber can take to get to the end-user, then the reliability of the solution increases – but so do the costs. An appeal to the guiding principles led UTOPIA to a determination to provide redundancy in all layers of the network down to, but not including, the home run.

During its analysis of potential revenue opportunity, construction costs, and the frequency of outside plant cuts, UTOPIA determined that the extra cost associated with providing a redundant path to every end-point exceeded available revenue streams. More specifically, physically diverse paths to most businesses could be offset by projected business revenue, but redundancy to support residential services could not.

The distinction between business and residential customers is not determined by zoning: businesses will coexist with private residences in many areas. Certainly the advent of the Internet has allowed just about anyone with the desire to develop a website to also develop a work-from-home economic model. This means that plans for what are traditionally viewed as 'residential neighborhoods' must allow for the inclusion of redundant fiber runs to home-based businesses – both now AND in the future. To support the ability to deliver redundant connections as well as to support general changes in needs, UTOPIA allocated roughly 10 % extra fiber strand beyond what was required to service residential customers in each neighborhood.

5.6.4   FIBER STRAND COUNTS

An additional question that needs to be answered is how many fibers should be taken from the different physical locations back into the network. For a typical residence, a single strand of fiber is sufficient. This conclusion is supported by the understanding that a single strand of fiber has over 150 Tbps theoretical carrying capacity, and that photons are unlikely to interfere with each other through nonlinearities in the fiber over the relatively short FTTH distances. On the other hand, though, businesses, MDUs, and MTUs will require, in most cases, not only redundant connections, but multiple fibers as well.

5.6.5   OPTIMAL FIBER AGGREGATION

After evaluating some basic decisions as to how the outside plant should be structured, the cost associated with deployment needs to be factored into the analysis. The largest component of capital outlay for an FTTH deployment is construction. Since construction costs are the single largest component of the business case evaluating the financial sustainability of a municipal fiber network, a careful analysis of the impact that design has on construction costs is critical.

The goal for a proper outside plant analysis is to gather and analyze information on all of the pertinent material, components, and labor costs related to both aerial as well as underground deployments. Such an analysis should identify the most cost efficient location to terminate the dedicated fiber strands and to implement some fiber sharing technology. In other words, how many homes will be serviced by a single aggregation point, or community cabinet? Such an approach has been applied to various network deployments for quite some time, including design work to support Digital Loop Carrier deployments, fiber node placement for HFC (hybrid-fiber-coax) networks, and more recently at the turn of the century in the WINfirst FTTH deployment in Sacramento and Dallas where this technique was used to layout the most efficient fiber aggregation cabinet locations.

In the analysis, it is important to note that costs for construction vary not only city by city, but even neighborhood by neighborhood. The prime influence on construction cost is related to the density of homes/structures per constructed mile and the type of construction required – either buried plant or aerial plant. Because of that constant variation in costs, the optimal engineering design needs to be flexible enough to be able to adjust the fiber routing and aggregation strategy during the initial engineering stages.

For the UTOPIA project, an optimal location for the fiber termination point was found to be between 700 and 900 potential end-users. In a high density area such as a tier 2 city this number would typically reach over 1500 and in a large metro area the number of fibers at the aggregation point is likely to be well over 5000 (Figure 5.4).

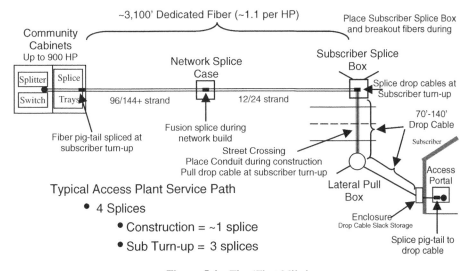

Figure 5.4 The 'First Mile.'

5.6.6   OUTSIDE PLANT DESIGN

For simplicity as well as for cost reasons, UTOPIA determined that network construction would end at the property line of every potential subscriber. In other words, fiber would be placed up to, but not beyond, the property line for every residence and business. In this scenario, sufficient quantities of fiber are deployed to each parcel based on the anticipated potential need of the zoning of that property (where there were no existing structures) or on the need for the actual structures on that parcel.

To reduce the cost of construction, the access and distribution network is run predominantly up one side of the street only, with a lateral crossing the street at every other property line. During the construction phases, these laterals are left as empty conduits, connecting the subscriber splice box to a lateral pull box. During engineering, engineers track the number of network splices per subscriber splice box: if it turns out that the design requires substantially more than one splice per home passed, the design is likely more costly than necessary and it is revised.

When an end-user requests services, a conduit is placed between the lateral pull box and the end-user's home. The drop fiber is then spliced at the subscriber splice box and pulled (through the lateral pull box if across the street) to the home for connection to the interior electronic termination point, the Access Portal.

5.6.7   THE DISTRIBUTION NETWORK

Identifying the Community Cabinet location where each of the dedicated access fibers terminates leads to the design of the distribution network. Since the fibers themselves will be functional long after the debates over active and passive have been settled (or at least until they have cycled back and forth a few times), the distribution fiber should be deployed with thought given to the Carrier Class Principle. In other words, even if a PON design is chosen for today, the design should support the possibility of changing over to an active design in the future when the need arises.

The most obvious approach should be to deploy physically diverse redundant paths from these access nodes back through to the core nodes. This Distribution network, which connects the dedicated access plant from the community cabinets through to the Hub locations, should have a fiber count based on the type of access technology chosen today: 1 fiber for every 8 homes passed for a PON and 1 fiber for every 30 to 50 homes passed for an active solution. However, in the case of a PON, by splitting the fibers (normally deployed single threaded) over two physically diverse paths, the outside plant can support both a PON deployment today and be prepared to implement the higher reliability inherent in the physical diversity of an active deployment in the future (Figure 5.5).

5.7   ARCHITECTURE TEMPLATE: STANDARDS

Implicit in the ability for the fiber to support the principle of servicing multiple service providers is the conclusion that the physical network owner needs to, in fact, light the fiber and provide the transport. If the physical network provider does not light the fibers of its network, then their infrastructure can only offer dark fibers. A dark fiber solution means that cities are involved only to the extent that fiber is deployed throughout the neighborhoods.

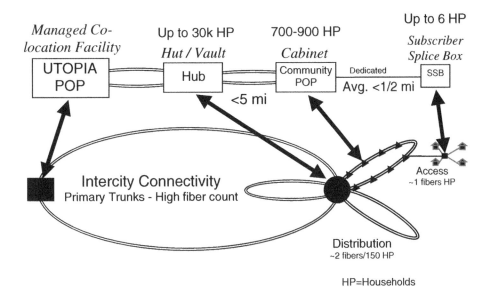

Figure 5.5 Fiber topology template.

This, after all, is not really community involvement and fails to provide the tripartite separation of ownership and responsibility critical to the success of a municipal network. In such a case, where the city fails to light the fiber and provide the transport, the supposed assumption is that by simply making a robust and ubiquitous infrastructure available, the benefits of the technology will also be automatically accessible.

This logic fails to consider that whenever a city leases or sells fiber to a private entity, it essentially delivers a monopoly to that profit driven entity. Owning and managing the transport of goods and services across a fiber provides that owner with unmitigated control and allows it to exclude access to that fiber by any and all competition. The case of multiple dark fibers is not much better: the limit of the number of physical strands of fiber in the network means only a limited number of service providers can purchase access the individual end-users.

There is nothing inherently monopolistic in private ownership of a fiber, but history and logic both concur to suggest that the drive for competitive advantages persuades private owners to move in that direction. While multiple providers could conceivably use a privately owned fiber in a multi-dark-fiber environment, the market would be extremely consolidated and little true competition would exist.

Since one of the Principles guiding UTOPIA is to create an Open Service Provider Network where abundant competition delivers greater choice and better pricing, then it is obvious that the municipalities, as the physical network owner, should light the fiber. This gives the municipalities monopolistic control over the infrastructure, but their motivation is to add as many competing service providers as the system can sustain, not to restrict access. Thus, by controlling and providing transport, cities can maintain the benefits of an open system.

Since it makes sense for the municipalities to provide the transport as well as the medium, the next question raised is whether the infrastructure should be involved in any of the higher layers.

A valid starting point to help sort through the options is the question of whether the network owner should manage only at the photonic level or at a more granular cell/packets level. Photonic networking is considered to be the ultimate optical solution. The logic of pushing an all fiber network to an all optical solution makes a sort of romantic sense but it is difficult logically to operationalize when the vast majority of the edge equipment and devices expect some sort of electrical cell/packet interface.

An 'all optical' network makes more sense in the backbone where the electrical interfaces have long evolved into straight optical interfaces. But even there the pure optical switch fabrics have not yet taken over the electrical cell/packet switch/router control planes and switching fabrics. Since the business model for these access networks should be defined first and foremost from the perspective of the most common denominator, the residential consumer, then designing a solution based on pure optical networking would only generate higher costs to access the majority of the end-users. Perhaps a few service and technology evolutions down the road a pure optical networking solution will be more obvious.

Since an all optical network is impractical today, we are faced with the question of what cell/packet transport protocol is best. By applying the guiding principles, we are inclined to make that analysis based on the Principle of Standardization.

By referring to 'Industry Standards,' we can mean many things. There are different levels of standardizations, which include, in descending order of influence:

- *Global.* An example of a global standard is Ethernet. Available globally in a wide range of applications, an Ethernet interface is commonly found on most servers and PCs produced and are even appearing on TVs and VoIP handsets.
- *Industry.* Industry standards such as ATM are certainly very common and accepted to the industries they are within, and yet finding direct ATM interfaces on computers and televisions and VoIP telephones is very difficult.
- *Consortium.* Sometimes consortiums are formed to help create standards. Although consortium solutions typically do not lead towards a global standard, this is a path individual companies will take as they attempt to join forces with others.
- *Proprietary.* Proprietary standards are typically pursued when a specific company has discovered a novel solution. The value of the unique advantages such a solution may offer need to be weighed against the sole source environment.

The distinction between these standards is important because the farther down the chain one progresses from an open global standard towards a proprietary solution, the greater the impact on:

- *Innovation.* It is difficult to foster innovation in services and applications if the technology required for that innovation is locked up in a single vendor's proprietary relationship. Unless a researcher buys in to the royalties of the solution, they will not have access to the technology, thus stymieing product and service innovation.
- *Cost.* Cost is a well-understood benefit of standardization. During an initial period of time when technologies are being developed, the cost of a solution is typically high as every effort, from design to manufacturing, is a customized one. Once a solution becomes a global standard, the cost of implementation drops as components become commoditized.

- *Competitive options.* When consumers have a choice between competitive offerings, prices stabilize at the appropriate market level. The more vendors supporting a standard, the more the competitive offerings available to the consumer.
- *Interoperability.* The more universal the standard, the greater the likelihood of interoperability between diverse solutions. Unfortunately, many vendors who claim to follow an industry standard find they are unable to interoperate with other vendors also claiming to follow the same standard.
- *'Leading-edge' versus existing technologies.* Leading edge technologies are sometimes developed for the sole purpose of creating a proprietary business model for a specific company attempting to develop a new standard. Often times these new technologies do not add a substantive improvement over the existing or established technologies – they are merely an attempt by a group of investors to capitalize on a growing market segment.

By examining the effect of each of these levels of standardizations it is easy to see that the more accessible a technological solution is to the global community, the more flexibility a systems operator will have. The technical position for municipal projects should be focused on global standards with a careful eye on new technologies which add a substantive evolution/revolution to the tried and true global solutions.

5.8 ARCHITECTURE TEMPLATE: TRANSPORT LAYER TOPOLOGY

As with all other key decisions, UTOPIA selected its transport layer topology by appealing to the guiding Principles. First, however, it needed an awareness of the options available, followed by a careful analysis of those options. A good reference for this type of analysis can be found in the Bell Labs Technical Journal [2]. To discover what options existed, UTOPIA solicited proposals from vendors through a public procurement process that culminated in a Request for Proposals. Over 80 vendors responded, affording UTOPIA an opportunity to scrutinize a variety of solutions. After careful analysis, and by appealing to the guiding principles, UTOPIA finally selected a vendor and a solution. Though exhausting, the analytical process was instructive.

The responses to the RFP fell into the following rough categories:

- Passive Optical Network
  - ATM PON OC3 RF video overlay required
  - ATM PON OC12
  - GPON
- Active/Dedicated Star Ethernet and SONET
- Active PON – 96HHP switch nodes with PON access to the client

Recognizing that all solutions could, in some way, address the stated needs of UTOPIA, reviewers of the proposals focused on the guiding Principles – particularly the principle of Open and Independent standards – as the defining criteria for evaluating the options. Not surprisingly, most of the responses supported the global Ethernet standard. The next smaller set of solutions supported industry standards rather than global standards, many focusing on the G.983 standard. Finally, the smallest grouping of similar solutions included vendors supporting FSAN's G.984 GPON standard. There were a few other responses that did not fit into any group – they included vendors representing unique and proprietary solutions.

As might be expected, vendors proposing anything other than the Ethernet or SONET compliant solutions were unable to commit to interoperability – even those compliant with the industry standards. Even though the FSAN standard has been under development for quite some time, none of the vendors proposing a solution using that standard could provide a list of other vendors with whose solutions theirs were compliant. The other industry standard represented, SONET, had vendors who were, in fact, interoperable. Of course the vendors representing unique solutions or solutions based on the developing GPON standard were unable to commit to any kind of interoperability whatsoever. Clearly, in appealing to the guiding Principles and by committing to adopt solutions that supported Open and Independent Network, only the global Ethernet standard, with dozens of interoperable respondents, made sense.

5.8.1   RELIABILITY

Another criterion used to sift through the responses was the guiding Principle of building nothing less than a Carrier Class network. Since reliability is at the heart of a Carrier Class network, the review committee filtered out those responses incapable of providing the required level of reliability – at least four 9's of guaranteed up time.

There are aspects to reliability over which the network owner has no control, and then there are those over which it clearly does. For example: while the occurrence of 'backhoe fade' (the cutting of a fiber line by a backhoe) is out of control of the physical network provider, the effect of backhoe fade on the delivered services is dependent on the topology of the solution. Figure 5.6 shows the effect in different network topologies (the unprotected

PON: Unprotected optical paths – Larger exposure to fiber cut failures

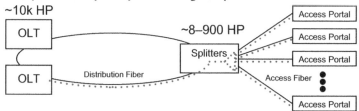

Physical diverse and switched Distribution Fibers reduces exposure to downtime due to outside plant failures by 10 Times

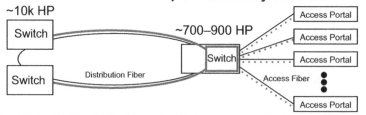

Active: Protected optical paths – Minimize exposure to fiber cut failures

HP=Households Passed

Figure 5.6   Physical diversity and switching in distribution networks.

portion of a signal's path is indicated as a dotted line and the protected path as a solid line).

As part of the consideration, the reliability of the network was weighed against ease of maintenance. While a passive topology requires fewer maintenance trips, there is no mechanism available to maintain connectivity when there is a cable cut in the distribution network. Furthermore, inasmuch as fiber cuts are an inevitable fact of life, the analysis of system reliability ultimately comes down to an assessment of the impact of the cut.

Assessing the potential impact of downtime in a PON solution first requires an analysis of the failure group size per cable and the probability for the cable cut. Estimating both the frequency of cable cuts and the average number of subscribers affected per unprotected optical path, we can determine the impact of an inevitable fiber cut. The larger number of unprotected subscribers carried on a PON solution near a hub makes the PON insufficiently reliable to be called Carrier Class. Applying the same rates for cable cut occurrences across all layers of the network; UTOPIA derived the following conclusions concerning network reliability in a PON (Figure 5.7).

PONs have up to 10 times more unprotected optical path than a comparable active network. Therefore, it is expected that the cut rate/failure rate of a PON would be as much as 10 times higher than its active counterpart, thus further reducing its reliability even below the three 9's indicated above.

5.8.2   TRAFFIC MANAGEMENT

The guarantee that services can be delivered in a carrier class manner is affected by more than fiber cable cuts; it also requires sufficient bandwidth and traffic management capabilities. UTOPIA scrutinized proposed solutions for traffic management support, including the ability to manage traffic flows in a sufficiently granular fashion to preclude one vendor's service from over-riding another service provider's services. ATM and SONET have been

Single Homed No Diversity	Topology	Failure Group Size	Annual cuts per cables	MTTRA (hrs)	Downtime per subscriber each year (minutes)
Access Cut	PON	96	0.0012	6	41.5
	Active	960	0012	6	41.5
Distribution Cut	PON	800	0.0012	6	345.6
	Active	N/A – Physically Diverse/Redundant Transport			
Maximum Availability	**Availability**				
	PON	99.93 %	THREE 9s		
	Active	99.99 %	FOUR 9s		

Active:
   • Order of magnitude less likely to fail due to fiber cuts.

PON:
   • Large exposure to Fiber cuts in   unprotected distribution plant.

   • Doubling cost is required to achieve minimum of Four 9's.

Figure 5.7   Typical cable cut impact on maximum reliability in active and PON architectures.

designed with a great deal of traffic management capabilities. With their deterministic cell and pipe characteristics, these protocols excel at traffic management and isolation. SONET is a bit less flexible; the lowest level of granularity for traffic management is by definition a virtual tributary, but SONET vendors offer a mix of Ethernet and ATM interfaces complete with switching fabrics with their products.

Native Ethernet has had the ability to identify or tag different types of traffic, but it was not expected to scale to city wide proportions and certainly not to provide bandwidth management. However, supplementing Ethernet with the standardized implementation of MPLS and label switch paths allows for the utilization of predominantly switching protocols to provide connectivity in a reliable and managed fashion without intruding on the service provider's quality of service management techniques. Previously to employing MPLS, Ethernet access infrastructures utilized ATM core switches or even SONET to ensure traffic handling characteristics suitable to maintain availability of services. Today, Hierarchical Virtual Private LAN Services support scalability, and Ethernet access infrastructures can be aggregated through MPLS core networks to allow for the entire pertinent network characteristics required:

- Availability.
- Bandwidth.
- Packet loss.
- Latency.
- Jitter.

From the perspective of traffic management, then, there was no clear differentiator: ATM, SONET, and Ethernet/MPLS all addressed the need for reliability as it pertains to building a Carrier Class network.

5.8.3   SCALABLE BANDWIDTH

Inasmuch as there is no practical difference in the reliability of managing bandwidth among ATM, SONET, Ethernet/MPLS solutions, the availability of high, scalable bandwidth itself remained as an area of prime differentiation.

Included in the Principle of High Scalable bandwidth is the requirement for symmetrical transmissions. Examples of networks that follow an alternative logic are the hybrid fiber coax (HFC) networks deployed by the cable TV companies. The HFC networks which were initially engineered with the assumption that the download usage would be roughly 17–19 times that of the upload, do not provide symmetrical transmissions. This assumption may have been the result of the cable TV industry's focus on their previous pure broadcast model, or it could have come about due to the belief that most people would be satisfied with browsing and downloading from the internet while only a select few would actually generate content to push. Shortly after these networks were deployed, capacity planners began to see ratios much closer to 3 to 1 rather than 17 to 1.

Today, the Internet has become a network for interactive communications. A predominant value of the Internet is its ability to bring people with similar ideas together, and they frequently form groups. These groups interact through their webpages, blogs, instant messaging, and email. Their interaction, however, is limited to what their bandwidth supports, and this is proving to be a growing source of frustration for many. In recognition of the growing need to push information out to the Internet in an interactive fashion, any practical solution to the demands of building a network with high, scalable bandwidth should include symmetrical high-speed transmission.

5.8.4 PLENTIFUL BANDWIDTH

Though fiber itself is capable of supporting incredible speeds, not all fiber solutions deliver on that promise – depending on the network architecture, the resulting available bandwidth can vary greatly. Some solutions start out with high bit rate optical speeds, but after those speeds have been shared across multiple end-users, the bandwidth that is ultimately available to the end-user is often no better than what is available to them today over traditional metallic systems. To benefit from the potential of fiber, UTOPIA sought to identify which solution has the most bandwidth and the most flexibility in applying that bandwidth.

A simple analysis of some typical solutions for a 1000 home neighborhood reveals some startling differences:

	Active/dedicated		Passive/shared		
			OC3	OC12	
	100 Mbps	1 Gbps	APON	APON	1 Gbps GPON
1000 Home footprint per user	100 Gbps	1000 Gbps	5 Gbps	19 Gbps	31 Gbps
	100 Mbps	1000 Mbps	5 Mbps	19 Mbps	31 Mbps

It is obvious that an Active solution is as much of an improvement over Passive solutions as Passive solutions are over existing DSL.

Even with the most rudimentary understanding of what a fiber network can offer, consumers understand that incremental improvements of bandwidth over what current solutions offer are not compelling – they expect a quantum leap in the speeds offered over a fiber network. We need look no further than the ill fated deployment of ISDN to see that consumers expect a clear differentiation in the offerings. Inasmuch as ISDN speeds were not substantially different from dialup, and given the greater speed offered by DSL, the incumbent phone companies quickly stopped deployment of ISDN and moved into DSL.

Furthermore, the network owner should expect the network to have capacity sufficient for significant growth. Having invested so much in its system, UTOPIA would not want to make additional, expensive, and major changes to the network architecture or to the corresponding technologies to accommodate the inevitable growth its network will sustain. The Active solution clearly provided the superior solution demanded of the High, scalable bandwidth principle.

5.8.5 COST TO SCALE

The next pertinent bandwidth concern UTOPIA considered was the cost to scale the selected solution. The table presented above demonstrates bandwidth in the access network. Since 90 % of the money spent on transport technology is spent in the access infrastructure, and since it is so difficult to change out the edge of the network, it makes sense to place a great deal of emphasis on bandwidth throughput at the access layer.

It is generally acknowledged that APON solutions do not scale well; therefore, they do not satisfy the requirement. Given their limitations, it is likely that the only evolutionary path for them is a complete replacement when increasing bandwidth needs exceed their capacities. GPONs, on the other hand, are more likely to evolve in the fashion that PONs were designed to be scaled. Over the next 5 years, the FSAN 984.x Gbps could see changes in the reinforcement of the uplinks and in the reduction of the splitter ratio. Still, a deployment of even a 2.5 Gbps GPON solution split 32:1, delivering 80 Mbps of dedicated capacity per user falls short of the bandwidth of even a 100 Mbps Active solution.

However, having tremendous amounts of bandwidth in the access layer is of little use if that bandwidth cannot be passed on through the next layer – the core network and its connection to the rest of the world. Whereas small networks may have only a single switch bringing together the community and connecting it to the rest of the world, a larger network, such as the one in the UTOPIA network, require a relatively large core network to aggregate the traffic from the edge and send it on.

In the initial deployment of the network, the access layers will not have reached capacity. The cost conscious tendency is to implement solutions that handle current traffic needs, ready to scale as the need arises. As the use of the network increases, and as applications begin to require greater amounts of bandwidth, the need for cost-effective scalability becomes obvious. Therefore, an important consideration is the ability for the network to scale beyond its initial deployment to accommodate the robust core bandwidth of a more mature network.

To evaluate the different scaling costs of GPONs and Active solutions we started with an initial bandwidth of 1 Gbps within the distribution network for every 100 subscribers. Each solutions cost was tracked as the amount of bandwidth in the distribution network was doubled again and again. A comparison of cost and an evaluation of the convenience of the scalability between the two potential solutions present a compelling reason for choosing an Active solution over a Passive one. The graph below shows how costs increase as the capacity of the network scales (Figure 5.8).

Besides being less costly to scale, the Active solution has the benefit of being very convenient, requiring nothing more than the addition of single mode GBICs. The GPON solution, however, can be problematic if, at the hub, the real estate is not sized to handle four times as much equipment. Finally, because the GPON solution is still a developing standard in the industry and is not yet a global standard, it is not inconceivable that the standard could evolve, thus making future equipment incompatible with the equipment deployed today requiring a trade out of end-user equipment.

5.8.6 ARCHITECTURE TEMPLATE – TRANSPORT LAYER TOPOLOGY

Discovering that an Active solution best addresses the principles of high, scalable bandwidth; carrier class functionality; and open and independent architecture, UTOPIA was able to bring to bear the last of the four Principles guiding its selection of a solution: support for an Open Service Provider Network.

It is worth re-emphasizing that this Principle may likely be the most important consideration for any municipal deployment. Publicly owned advanced communications infrastructures only make political sense if they help the private sector to thrive. A properly formed public/private partnership frees the fiber infrastructure from monopolistic business practices and facilitates the growth and development of private, competitive solutions, allowing private industry to flourish. Before selecting a technical solution that supports this

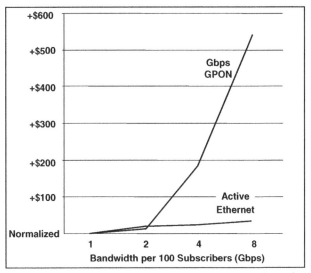

Assumes future standards allow for use of initially deployed proprietary technology – if not add an additional $500–$700 per subscriber

Figure 5.8   Comparison of scalability costs.

fundamental principle, it is important to understand how services should be delivered over the network.

The primary function of the infrastructure is the delivery of data. To many, that means Internet traffic only. Though not always considered such, voice and video services are also, ultimately, data. In fact, anything that can be digitized and delivered over the fiber optic network is, in essence, data. The current vertically integrated model for voice and entertainment video is evolving and converging with other data services. Today, VoIP services are sold independently from the PSTN network – and independent from the data provider itself. As we recall the way in which data service providers competed one with another through dial-up modems across the PSTN, we can envision similar competition and end-user flexibility as services delivered over the fiber infrastructure are separated from the infrastructure ownership.

Technically, services can be separated from the infrastructure quite readily. In fact the OSI model can be used to describe the demarcation point for the delivery of services. One of the best references for this can be found in an MCI Public Policy Paper [3]. In the OSI model, layer 1 represents the physical media fiber, layers 2–4 embrace transport and networking, and layers 5–7 deal with applications. Today service providers typically provide network services at the IP layer (layer 3) and depend heavily on TCP/IP for the connections. It is an effective model, but it has the disadvantage of being at a minimum management intensive and perhaps even intrusive to the services delivered.

If the municipally run network is deployed utilizing routing protocols, it is likely that the services delivered through that infrastructure will not be transparent to the infrastructure. While it is certainly possible to design, deploy, and operate a network that is based on packet handling through to layer 4, it is likely that services delivered in that way would be more limited than if those packets were transported only at layer 2. Ideally, a layer 2 network presents the greatest options for the service providers and the least amount of management for the owner.

Wholesale network services at layer 2 are very attractive to a municipal owned network from a variety of perspectives. Transparency to the retail service provider is more complete – but a transparency with respect to the popular hacks, viruses, and DOS (denial of service) attacks means the municipal owner can minimize its involvement not only in the retail products sold across the infrastructure, but it avoids the majority of the problems as well.

As anyone would, Municipalities place more emphasis on solutions that will be functional for longer periods of time. Because layer 2 is inherently simpler, the likelihood of technological obsolescence is reduced. If the network is deployed emphasizing layer 2, changes that occur at layers 3 and 4 are more likely to be transparent allowing the investment in network technology to be relevant for a longer period of time.

An emphasis on layer 2 does not mean that the network operates exclusively at layer 2. In fact, an Ethernet network supplemented with MPLS introduces higher complexity, through layer interaction, than a strict layer 3 or 4 network would. Despite that complexity, the access network needs some ability to influence traffic handling at some of the higher layers than just layer 2. To protect the network and minimize problems, the network needs to be able to support some layer 3/4 filters along with ACLs.

The reality is that a complete layer 2 network would be characterized as one large broadcast domain. Without some simple filters in place, all broadcasts from all users would reach all of the other users in the network. Simple DHCP requests would end up looking for the nearest server to grant an IP address. In addition to DHCP filtering, NetBIOS filters would be essential, especially when such common functions, such as the sharing of a hard drive between Windows machines, would also have the unexpected side-effect of letting everyone on the network see those drives, whether that was intended or not.

Transparency to services does not mean traffic cannot be directed according to at least a high level set of needs. Traffic that needs to be isolated while traversing the network may be switched in a point-to-point service similar to the virtual tributaries found in a SONET solution. Point-to-point connections allow for services such as transparent LAN service to be delivered through the infrastructure. For applications such as voice over IP point to multipoint connectivity gives the isolation between end-users. For general connectivity, a multipoint-to-multipoint network provides the greatest amount of flexibility for connectivity through that VLAN.

Despite the limitations described, a network emphasizing Layer 2 still presents the best option for longevity, ease of management, and transparency to the service provider. For these reasons, UTOPIA looked for solutions that emphasized Layer 2 (Figure 5.9).

5.9 NETWORK TECHNOLOGY: TECHNOLOGY AND VENDOR SELECTION

Focusing on an active Ethernet solution operating at layer 2 (with some filter/ACL capability at layers 3 and 4) allowed UTOPIA to select from a narrower field of solutions. Since most of those remaining solutions could satisfy, technologically, the Principles, UTOPIA next focused on cost – both initial capital cost as well as long-term operating cost. Additionally, the characteristics of the vendor itself were evaluated – number and size of comparable deployments, financial stability, technical support structure, and more.

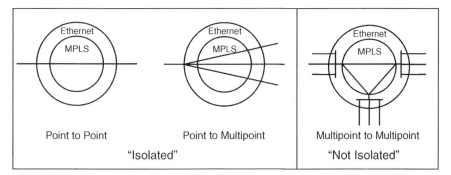

Figure 5.9　Connectivity options in a layer 2 network.

To review quickly, UTOPIA's selection process, to this point, included evaluations of:

- Physical/Environment.
- Protocols and Standards Compliance.
- Reliability characteristics.
- Security features.
- Regulatory certifications.
- Scalability model.

All these elements were addressed from the perspective of the Principles guiding the design and operation of the network.

Despite commonalities in these areas among the remaining solutions, UTOPIA evaluated the interaction of different vendor solutions across each layer. During this evaluation, an interesting debate presented itself: where to place the device (the Access Portal) that resides at the end-point of the fiber connection. Should it go inside the premises or outside?

Different deployments across the country show that both are reasonable options – it's really a question of balancing trade-offs. Placing it indoors leads to questions about ownership and liability, maintenance access to equipment, and so on. Placing it outside leads to questions about security from tampering, protection from the elements, and so on. Ultimately, the additional cost of hardening the enclosures for outdoor placement seemed a greater concern than an indoor placement with less access. UTOPIA decided on an indoor solution.

It should not be surprising that cost of implementation takes a defining role in the selection of a specific technology. While a cost analysis complicates an already tedious analysis of technology, it is important to remember that the ultimate solution to the Last Mile problem – to a Municipal telecom project – is more of a business issue (including politics as well as finances) than it is a technology issue. A thorough understanding of any technical solution requires an analysis of the near-term capital cost as well as the long-term operating costs. The technology review committee considered both the Principles established by UTOPIA as well cost data about specific configurations and solutions. In their analysis, UTOPIA's review committee used the specific data presented in the proposals to configure and reconfigure different alternatives to the proposed solutions in an effort to discover the best – and most cost effective – configuration for the project. Additionally, in order to evaluate long-term operating expenses, UTOPIA's RFP stipulated specific data be provided regarding maintenance agreements, professional services, liabilities, and more. Those responses were also compared and projected over the life of the project to determine which technical proposal offered the most cost-effective solution.

There was an additional element of complexity in analyzing the costs to implement the various solutions. Since the different solutions require different outside plant deployments, UTOPIA further 'normalized' the various solutions to make a fair comparison showing how each would compare if it were matched one against the other using comparable topologies and architectures. Using information gathered from both from existing deployments as well through requests for information and proposals, the review committee assimilated all direct costs into a single set of analyses for comparison:

- Fiber/conduit infrastructure.
  - Fiber/conduit/splitter.
- Labor
  - Splice cost.
  - CEV (controlled environmental vaults)/cabinet placement.
- Real estate.
  - Buildings/co-locations.
  - CEV/cabinet/enclosure.
- Technology.
  - CPE: voice and video converters/gateway (only as required).
  - Edge: access portal (AP).
  - Access: access distribution switch (ADS).
  - Distribution: distribution core switch (DCS).

With 9 different topologies to analyze and over 80 specific vendors participating in the procurement process, the UTOPIA review committee was severely challenged to complete the process within the allotted time. Fortunately, 6 months of previous analysis confirmed something not commonly understood in the industry: that the cost associated with the outside plant design need not be considered a first order influence on the cost of any topology. Here's why.

It is easy to determine the cost per home passed for Active and Passive designs. In defining the terms, we can summarize by saying that if the switch which terminates the subscriber's link is within 2 km of the end-user, the solution is active. If the terminating port is shared among multiple users over 10+ km of outside plant, the solution is a PON. To minimize outside plant costs a PON solution would have a 4 port splitter/combiner deployed near the end user and an 8 port splitter near an aggregation point relatively remote from the hub location which would house the terminating equipment (not more than 20 km from the most distant end-user.) A relatively simple cost comparison between that solution and a solution designed to support a dedicated active solution demonstrates that the cost of the outside plant infrastructure to support a PON solution is less costly than an active solution by roughly $9 per home passed (Figure 5.10).

That gives the reviewers one of their key variables in comparing costs and technologies. Now, since most municipal business cases are designed for less than 100 % penetration, it is important to also understand the potential burden of costs the electronics will have to an overall solution. Assuming a 50 % penetration rate, a high-level view of other relevant costs

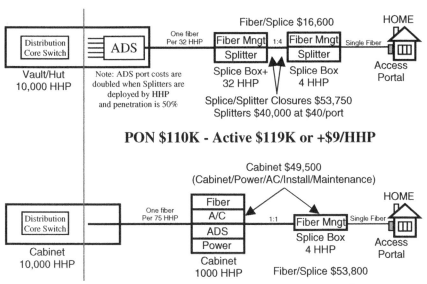

Figure 5.10   Outside plant cost comparison.

uncovers a huge difference between the two competing solutions. Using only the difference between the average costs of the five most cost-efficient technically compliant solutions, the analysis showed the following difference in cost:

	Cost Advantage: Active over PON
Outside Plant	(–$18)
ADS (OLT)	$189
Access Portal (ONU)	$408
ACTIVE Savings	**$579**

The $9 savings per home equals $18 per sub savings at 50% penetration

The cost difference inherent in the outside plant solution for the PON is completely overshadowed by the difference of the cost in the electronics. To extend those saving out over the entire project, a $580 savings per subscriber for the dedicated active solution is likely to reduce the cost per residential subscriber by nearly $15/month.

Since a large component of the cost difference is the direct result of the inefficiency of OLT port allocation, a different outside plant design could be implemented for the PON solution to reduce the overall cost of that solution. Figure 5.11 below illustrates an alternative.

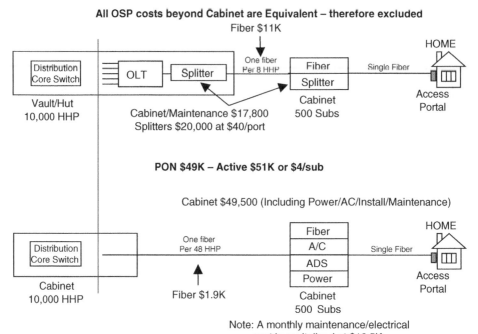

Figure 5.11 OSP cost comparison with optimized PON architecture.

Even with a relatively large change in the splitter deployment for the Passive solution, there is little change in the difference between the costs of the outside plants themselves. The biggest cost reduction in the redesigned Passive solution comes from much better OLT port utilization.

	Cost Advantage: Active over PON
Outside Plant	$4
ADS (OLT)	(−$3)
Access Portal (ONU)	$408
TOTAL Savings	**$409**

The result of the new approach is a cost reduction of nearly $170 per subscriber. This has a significant impact on the overall business model, but not enough to offset the $409 cost advantage that an Active solution has over a PON. Technologists nearly always take delight in discovering that the best technical solution is also the most cost-efficient solution. Certainly the UTOPIA review committee was pleased.

It is likely that as both the proprietary and the more standard PON solutions become more prevalent as network solutions, the cost of those solutions will go down. As for Ethernet – since the global standard for 100 Mbps solutions has already experienced a steep cost reduction, it is more likely that the next level of significant savings will focus on next generation dedicated active solutions – dedicated standard Gbps Ethernet link to each end-user.

5.10 NETWORK INTERFACES

A layer 2 Ethernet infrastructure is very well suited for native data applications. Since most applications are already designed to accommodate Ethernet's dominant global standard for data interfaces, the network needs to offer just a couple of options to support the bulk of these applications:

- The Access Portal (CPE) represents the media conversion point, and traffic mapping interface, for data applications.
- Optical Interfaces may be required for more sophisticated business, both via multimode as well as single mode fiber. Those interfaces will be handled directly off of the OSPN access distribution switches.

Other standard applications, such as telephony and voice, should not require end-users to change their telephony device. While a terminal adapter could certainly be deployed by the service provider, currently the most dependable solution is to integrate the voice gateway function into the Access Portal, which is part and property of the OSPN. This allows for such features as an uninterruptible power supply as well as such services as E911 to be implemented with a higher degree of certainty. Figure 5.12 illustrates the applications interface.

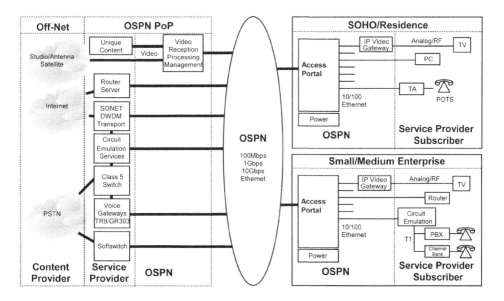

- Data Services interfaced via standard CAT5 and optical Ethernet
- Traditional POTS/T1 services through terminal adapters
- Video received into IP Head-end and through video gateways

Figure 5.12 Network demarcation.

As with voice and telephony, video-based services should not require end-users to upgrade all of their equipment to support a digital data only delivery mechanism. Using video gateways in the premises to convert and deliver standard video signals to existing TVs and receivers, the network can control ingress and egress of traffic and still remain open.

Service providers benefit from these common interfaces as well as end-users. If a retail service provider was required to build a head-end, integrate the middleware and digital rights management and purchase and deploy their own residential video gateway, then the barrier to enter the video market on the OSPN would be too high for all but the most resource-rich providers. By integrating these functions, based on global standards, into the design and scope of the OSPN, content providers who do not have any interest in those devices would still be able to deliver their services to the end-user.

5.11 NETWORK OPERATIONS: CAPACITY MANAGEMENT

As stated before, the correct operations model is more important to the success of a Municipal network than is any particular technology decision. Once an operational model is selected and the technology has been chosen and deployed, the business of asset management, and in particular capacity management, becomes the architect's primary focus. With the utter failure of the mystical, metaphysical, and alchemical industries to deliver a working crystal ball, most engineers find they are only able to guess at the actual demands and performance of the network or determine how the infrastructure is going to behave with the deployment of advanced broadband services. Using some imagination and a good spreadsheet, Dynamic City made a few basic assumptions to illuminate the potential capacity issues UTOPIA's infrastructure might experience in the future.

An overview of a typical municipal deployment covering roughly 200 000 homes shows the four main paths along which traffic will flow (Figure 5.13).

One of our methods for analyzing capacity was to break potential services into three groups: Existing, Imminent, and Possible. Each of the services was given defined characteristics that allowed a general impact assessment on the network. The typical usage cycle, bandwidth required, and traffic connectivity (point-to-point and level of locality) were sufficient to discover some relatively obvious conclusions.

Advanced services were defined only for the purposes of stressing the network resources. An analysis was performed by network layer to discover which layers would need

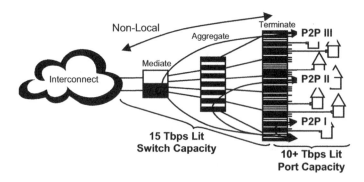

Figure 5.13 Basic traffic flow.

reinforcing based upon various services mixes and associated traffic patterns. By introducing different scenarios of service penetration we got a sense of how the links between the layers in the network are stressed beyond their initially deployed levels.

As we interpreted the results of our analysis, we noted that the bandwidth available on the dedicated link to the home, with only one subscriber sharing that link, is the least likely portion of the network to fail first. The various service penetration scenarios show adequate capacity. UTOPIA concluded that the most difficult part of the network to change – the portion deployed to every end point – will be the last part of the network to require an upgrade. As would be expected, the more subscribers that share the network resources, the more likely it is that that portion of the network will require the deployment of additional, more advanced technologies. Evolving the connections between switches in the core of the network from a 20 Gbps link to an equivalent of a 200 Gbps link is likely to be the most important evolution after the initial deployment of the network resources has been accomplished.

5.12 CONCLUSIONS

Municipalities have the opportunity to address various community broadband services needs by providing a new access network infrastructure. As an essential service sponsored by the community itself, an advanced fiber network infrastructure can enhance economic opportunities for its business sector and improve the quality of life for its residents for many years.

The solution to meeting these needs comes in the form of a public/private partnership where the public sponsors and owns the infrastructure, competing private service providers sell services to the customers, and an independent overseer manages the operations of the network. This tripartite system provides the checks and balances that ensure innovation, economy, and quality of service. And by defining a set of guiding principles that form the template for selecting the right technical solution, decision makers can reduce the complexity of selecting an appropriate technology.

The Open Service Provider Network model sets the standard for municipal involvement in the deployment of advanced networking services.

REFERENCES

[1] IEEE-USA's Committee on Communications and Information Policy, "Accelerating Advanced Broadband Deployment in the U.S.", http://www.ieeeusa.org/forum/positions/broadband.html, February 2003; 1–3.
[2] Weldon M, Zane F. The Economics of Fiber to the Home Revisited. *Bell Labs Technical Journal* 2003; **8**(1): 181–206.
[3] Whitt R S. The Network Layers Model. *Federal Communications Law Journal* 2004; **56**.

6

High-Speed FTTH Technologies in an Open Access Platform – the European MUSE Project

Jeroen Wellen

Lucent, Bell Labs Advanced Technologies, The Netherlands

6.1 INTRODUCTION

To obtain a picture of next-generation Fiber-to-the-Anything (FTTX) networks, it is necessary to analyze the direction that the current generation is going to take on a wider scale than today. Currently, mostly Passive Optical Networks (PONs) and Point-to-Point (P2P) networks are being deployed only on a limited scale. In order to connect the majority of the subscribers in the future, FTTX networks must meet some specific requirements. These will differ with geographic circumstances, the installed base and the service portfolio being offered to the end-user. Future services are a rewarding subject for speculations, and it is likely that in due time network requirements will change.

As an objective, the Multi-Service Access Everywhere (MUSE) project [1] targets first-mile solutions that are capable of providing 80 % of the European end-users with 100 Mb/s by 2010 [2]. Obviously, the challenges here are found in minimizing the deployment cost in this timeframe rather than the bandwidth itself.

6.1.1 A DIFFERENT VIEW OF NETWORKS

The design of traditional communication networks was optimized for one main purpose: to efficiently connect every telephone on this planet. A somewhat arbitrary three-segment hierarchy can be identified, which is illustrated in Figure 6.1: a Transport or Core segment responsible for the global interconnection of communities and cities, a Distribution or Metropolitan section for the aggregation and distribution of traffic in these areas, and an Access or Last-Mile segment that provides the connectivity to homes and offices. The blueprint of this network view originates from already a century ago, but remains valid, since

Broadband Optical Access Networks and Fiber-to-the-Home: Systems Technologies and Deployment Strategies
Chinlon Lin © 2006 John Wiley & Sons, Ltd

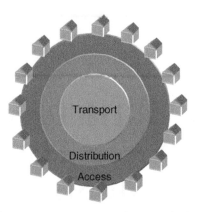

Figure 6.1   Transport centric network view.

its geographic foundation of continents, cities, and customer locations is more or less ageless. At the same time, however, about everything else is changing. Apart from the drastic changes that network operators have faced during recent privatizations, many of the traditional conditions and requirements are shifting into a new era.

The first important change includes the deployment of data services in the 80s, generating traffic beyond voice modem rates. Public data services, preceded by progresses in consumer electronics and data technologies triggered an irreversible transition to digital media. This has accumulated into the current upgrading of legacy voice networks, and to some extent also TV cable networks, into broadband triple-play service networks. With improved Cable and Digital Subscriber Line (DSL) transmission techniques, this could until now be accomplished with the available installed base. It may, however, be more difficult to address the changes that were initiated by the second important event: the introduction of mobile telephony. Cellular phones have irrevocably changed the communications habits of users and their expectations of future services. Again, this was made possible by an accompanying development in terminal equipment – miniaturization of radio technology and batteries.

Where traditional networks were designed to provide narrowband single-service connections between buildings, commercial operators now have to provide multiple broadband services to a moving target. This has turned the network view inside out as shown in Figure 6.2, where the end-user is not only the focus of interest, but actually becoming a part of the network itself.

Personal Area Networks (PANs) or, more intimately, Body Area Networks (BANs) have until now been the subject of mainly physicians and academics. With the ongoing miniaturization of appliances and the advances in wireless technology however, these type of individual networks may become real. Interfaces and storage components will increasingly occupy our personal surroundings, cloths and may eventually even end up inside our bodies. Recent announcements even claim harmless duplex transmission up to 10 Mb/s through Human Area Networks (HANs): the body as an integral part of the network [3]. Getting in touch with someone or a handshake will never be the same again.

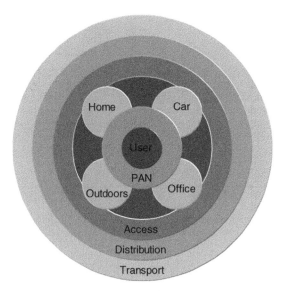

Figure 6.2 User centric network view.

6.1.2 CHANGING ARCHITECTURES

Figure 6.2 also indicates an elementary change in the network hierarchy. By resolving the bandwidth need in access networks, not only triple-play services became possible. The Internet is gradually expanding by another, local Access segment with some unprecedented characteristics. LANs come in many different flavors and are in most cases owned and managed privately. Home, Car, Train, and Office Networks will nonetheless provide ambient services also to visiting users, at the same level as the public Mobile and Wireless Access facilities that Base Stations and WiFi Hotspots provide today [4]. The appearance of many different commercial and noncommercial network providers will affect the service supply chain and the business models. Moreover, the Broadband architecture will need to incorporate this anarchistic edge of the network, including the potpourri of LAN technologies and topologies that come with it.

As shown in Figure 6.3, Access Networks will need to connect all kinds of Local Networks through Local Nodes of various sorts, and as such form the focus of the physical network convergence. Although these Access Points, Base Stations and Residential Gateways, may well be equipped with a uniform Network Termination device, that is an Optical Network Unit (ONU) or the like, the Control and Management of this converged Access Network is becoming very complex. Issues such as quality of service (QoS) provisioning, Authentication Administration and Accounting (AAA or Triple-A) to name but a few require a uniform approach throughout the network. Especially Residential Gateways are currently the main topic in Broadband Architecture studies [5], since they occupy a key position between the Communications and Consumer Electronics arenas.

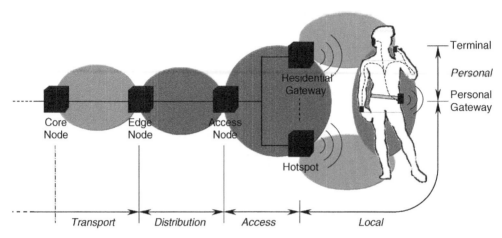

Figure 6.3   Broadband access architecture extended with local and personal networks.

6.1.3   FUTURE APPLICATIONS AND SERVICE REQUIREMENTS

In the recent past, advocates of fiber deployment in Access Networks have argued that the demand for bandwidth will soon outdate any copper-based access technology. This demand is attributed to, for instance, heavy data traffic from peer-to-peer file sharing and video. Especially video services, such as broadcast television and Video on Demand (VoD) are regarded as the 'killer application' satisfying the business case for fiber access. Opponents to this view contend, however, that advances in compression techniques, supported by the ever-increasing processing power of media equipment, will allow moderate line rates of DSL and Cable networks to last for decades to come.

There is no reason to assume that video media will lose its popularity in the future. The associated data traffic, by either file sharing or streaming, will more and more become a dominating load to multi-service access networks as a popular medium. As illustrated in Table 6.1, it is possible to stream DVD video with less than 5 Mb/s. Current compression techniques and codecs already include most of the geometric and perceptual redundancy of frame and color/luminance information. It is therefore questionable if DSL will ever be able to support 3D and High-Definition Television (HDTV) with a quality that appeals to the end-user (Figure 6.4).

New access infrastructures and upgrades are meant to provide future-proof solutions and for a safe prediction of the coming access loads; also here a more user centric view is preferred. A simple multiplication of 15 Mb/s HDTV streams times an average of 2.5 TV sets per household may not hold for very long. On one hand, the size of modern Home Cinema systems, LCD television sets, and surround sound speaker sets prevent duplicates in most homes. Individual virtual communications will more and more become the source of traffic, in both residential and working situations, and also through mobile connections. Although picturephones and teleconferencing have had little interest outside the office, videoconferencing is gaining public interest due to the possibilities of broadband access and popular communication Web applications and Internet sites. Miniturization will again be the enabling technology for improved quality and mobility. Especially the availability of

Table 6.1 Video formats and their bandwidth requirements for moderate picture quality.

Video	Format codec	VCD MPEG1	DVD MPEG1/2	HDTV MPEG2/4
Frame	Width [pixels]	352	720	1440
	Height [pixels]	288	567	1134
	Rate [f/s]	25	25	25
	Compression (lossy)	3 %	3 %	3 %
Pixel	Depth [bits]	24	24	24
	Compression (YV12)	50 %	50 %	50 %
Stream	Raw rate [Mb/s]	60.8	244.9	979.8
	Compression rate [mb/s]	0.9	3.7	14.7
2 h movie	Size [Gb]	0.8	3.3	13.2
Transmission time [hh:mm:ss]	Voice modem (53 kb/s)	34:25:47	138:38:51	554:35:25
	ADSL (1 Mb/s)	1:49:29	7:20:54	29:23:36
	xDSL (10 Mb/s)	0:10:57	0:44:05	2:56:22
	Fibre (100 (Mb/s)	0:01:06	0:04:25	0:17:38
	Fibre (1 Gb/s)	0:00:07	0:00:26	0:01:46

portable and even wearable displays [6] will cause an explosion of vision data from a new breed of virtual services: professional, educational, and entertainment (multimedia, gaming). Even conventional broadcasting characteristics may change. From a network point of view the issue may shift from 'how many households are watching what' to 'who is watching where.' Although individual visual communications allows for further compression of the data flow (by tracking the focal point interactively), this will require a very low network latency. All things considered one can expect that, apart from a load that is roughly 500 times as high, the requirements for future fixed and mobile access networks will resemble those of today's most important communication service: voice. This would involve familiar latency requirements and QoS in general. Also network loads would inherit much of the stochastic

Figure 6.4 Devices for multimedia and virtual communications.

nature of voice traffic rather than the deterministic load of peer-to-peer networking that is observed in broadband networks today.

6.1.4 NETWORK CONVERGENCE AND DISTRIBUTION OF INTELLIGENCE

From the previous discussion it is clear that future Access Networks will have to provide two main service properties: bandwidth and mobility. Since the end-user will turn into a Cybernaut some day, he or she will need the space to communicate through a virtual layer on top of the real world. At home, at work, and on the road people will require ambient services transparently and without any knowledge of the local network conditions. Office Routers, Residential Gateways, Access Points, and Base Stations have to become the portals of a uniform Access Platform. Although it would make sense to collapse aggregation networks for mobile and fixed access, most of the existing infrastructures have evolved independently since the introduction of cellular phone services. Nonetheless, a steady migration towards Ethernet and IP-based platforms can be observed today in both fixed and wireless networks.

Access Network topologies, as they have emerged in the past, are optimized for aggregation and distribution of data traffic between the central office and the edges of the network, either Customer Premises or Base Stations. One of the challenging questions is whether this model is sustainable for the high-speed scenarios drawn before. A massive increase of bandwidth requires upgrading of the access transmission technologies to fiber, much like metropolitan networks today. Also the network equipment, switches, and multiplexers may have to be upgraded when, for instance, IP and higher layer functionality is required locally in Base Stations and Department Buildings. Operators prefer to put everything that is powered and requires maintenance in a single location, resulting in Central Offices for hundreds of thousands of connections. But with increasing bandwidths the range of wireless links drops and active radio equipment will need to be scattered wherever people go. With changing applications and habits, and appliances that become more interactive, the traditional dominance of long-distant calls and FTP sessions may be taken over by local traffic caused by online gaming, remote monitoring, security sessions, and the like. A Mesh type of topology as shown in Figure 6.5

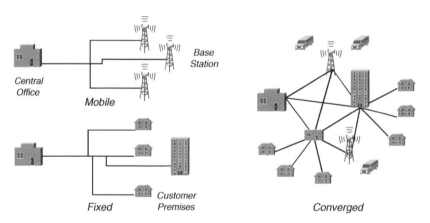

Figure 6.5 Convergence of fixed and mobile access networks.

could off-load a central office from local traffic and, by distribution of Network Intelligence such as AAA and wireless handovers, also match the performance requirements for the mobile users.

The convergence of Broadband Access networks will involve more than just a blurry collection of very different edge devices. The mobility of the end-user will also introduce an unprecedented volatility to the Network architecture. Nomadicity causes end-users to pop up and disappear at different locations in the network. Users, Personal Networks, and cars will cause dynamic changes in the architecture itself, including swapping hierarchical relations between gateways. All this will require fundamental changes to the operations of Access Networks, the functionality of Network Nodes, and the architecture itself.

6.1.5   MIGRATION OF ACCESS NETWORKS

The introduction of optical fiber access to the majority of the end-users will unavoidably require a transition from existing copper infrastructures. This so-called 'Brown Field' scenario will dominate for the simple reason that coaxial cable, twisted pair, or both already connect the bulk of the Customer Premises (CP) locations. The upgrade to high-speed access according to predicted service acceptance rates will outnumber any new CP builds in a 'Green Field' scenario which are estimated to account for only a few percent of a total of about 200 million European households in 2010, according to Figure 6.6, but will also include rebuilds in eastern Europe. Another driver for a gradual introduction is the investment involved: For roughly 1 k€ per CP, this would require an amount in the order of $200 billion euros, which operators simply cannot bear overnight.

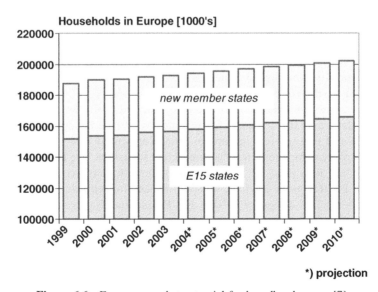

Figure 6.6   European market potential for broadband access [7].

To defray the installation of fiber and to stimulate the acceptance of new broadband services at the same time, it is important that existing access networks are utilized to their physical limits. This can be seen in the current interest in digital video and broadcast TV services over enhanced Digital Subscriber Line (DSL) formats. In order to push the line rate beyond 10 Mb/s, the current loop reach is insufficient. DSL Access Multiplexers (DSLAMs) will therefore move out of central offices, into the outside plant. To ensure efficient utilization of central office equipment in time, the required feeder network would ideally coincide with FTTH networks. Remote DSLAMs need, however, to be small, and consume a minimum of power when they are also remotely powered through available copper lines. Therefore, Very High speed DSL (VDSL) modulation over optics is also studied in MUSE [8].

From the previous findings it can be concluded that, with a gradual replacement of copper access by optical fiber, the access network will unlikely be a specific homogeneous FTTH system. Apart from DSLAMS located remotely either in dedicated cabinets in the street, or in the basements of apartment buildings, other types of access may be integrated as well to provide a converged Wireline/Wireless Access Network. Public wireless services such as WiFi [9] and 3rd and 4th generation mobile services suffer from range limitations similar to DSL and will require feeder networks as well. This illustrates that the convergence of these different access networks into a single fiber infrastructure would be beneficial, especially when utilization of individual services is still low. This has been the subject of several studies in the past [10–12], but the key issue remains however at which layer this convergence should take place. Small, cheap, and low-power modems, and access points in the field could require dedicated channels at the physical layer, by means of Wavelength Division (WDM) and Frequency Division Multiplexing (FDM), or even a dedicated fiber (Space Division Multiplexing or SDM). In addition to DSL over optics techniques, MUSE also studies microwave signal delivery for wireless access [13].

But when DSL, wireless, or any other access terminals would be connected through identical Optical Network Units (ONUs), the convergence would manifest itself upward, allowing for more efficient use of Optical Line Terminations (OLTs), switches, and possibly routers in the Central Office (CO). This is, of course, under the assumption that all first-mile technologies share the same link and network protocols. A uniform Access Platform as studied by MUSE is, therefore, an important requisite for FTTX.

Similar to the access network itself, the convergence and evolution of services will also predominantly determine the design of future Gateways. Since the end-user will make or break the success of any service, in particular the Residential Gateway at the customer premise (CP) is a key component in access migration scenarios. The Residential Gateway should connect not only new IP services but preferably also existing telephony and broadcast radio and television services, without the need for replacing present phone sets and television sets. The design of ONUs and Residential Gateways should therefore ideally allow for a convergence layer that will move from the physical layer upward in time.

This is schematically drawn in Figure 6.7, where voice and TV are initially offered on separate connections, then multiplexed by TDM and (Coarse) WDM, while eventually new IP services can be introduced nondisruptively.

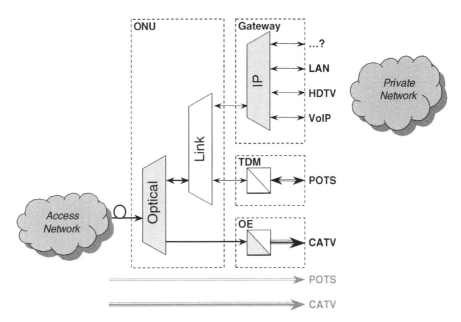

Figure 6.7 Evolution and convergence of services from separate modems and terminations at the physical layer towards Residential Gateways at the network layer.

6.2 FIBER ACCESS NETWORKS

The key function of Fiber Access networks is to provide broadband network access to the end-user. In practice this means that the network connects CP and apartment buildings, Office Buildings, Base Stations, and Access Points to a CO. In the CO the different traffic flows are multiplexed/demultiplexed for further uplink connection to Metropolitan and Transport Networks or, when it concerns local traffic, Switched or Routed back to the Access Network.

Although optical fiber access technologies can provide transport solutions for bandwidths well beyond the current needs of Access Networks, it is mainly the cost aspect per connected subscriber that is the key issue. Techno-economics is therefore one of the main topics of Access Network development [14]. Apart from the running costs or Operational Expenses (OPEX), including management and maintenance of the network, mainly the installation costs or Capital Expenses (CAPEX) are considered. These include not only the cost of equipment (OLTs and ONUs) but also the outside plant: the optical fiber cables, ducts and their installation costs. Ducts are generally dug from subscriber to subscriber with a number of optical fiber cable distribution points or aggregation nodes (AGNs) to allow bundling of optical fiber cables in a shared duct of the Feeder Network between AGNs and the CO (Figure 6.8).

6.2.1 ACCESS NETWORK DESIGN

Designing Access Networks involves a very complex tradeoff of pros and cons regarding the Operational and Capital expenses mentioned before, which can be translated into requirements for the choice of network topology and technology. These requirements apply to the two parts of the network:

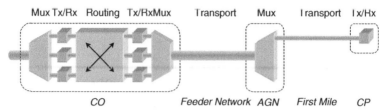

Figure 6.8 Location of the basic access network functions: multiplexing and transport in Feeder sections and First Mile of the outside plant.

- The Outside plant: fibers, ducts, and remote cabinets
- Network element: Access Nodes, remote Multiplexers, and CPE

For both, the major issues involve:

- Initial cost:
  - material
  - installation
  - housing
- Running costs:
  - management
  - maintenance and repairs
  - upgrades and reconfigurations
  - powering
- Performance
  - Throughput and QoS support
  - Security
  - Robustness and redundancy
  - Safety

In the past decades, FTTH networks have been studied intensively, initially mainly focusing on the optical transmission and distribution techniques, but gradually more on the economic aspects involved of the complete system. This has resulted in several promising topologies and technologies, which unfortunately always appear to provide a compromise. The key issue is that infrastructure resources can be minimized by using shared media, but the multiplexing involved requires sophisticated equipment or optics. Moreover, robustness and security issues are inherently poor in shared media. Single medium technologies suffer less from this, and save on equipment and component cost, but obviously at the expense of higher infrastructure costs.

6.2.2 TECHNO-ECONOMIC MODELING

The economics of access networks have been the subject of many studies in the past years [14–16]. As commonly acknowledged, the costs involved in the outside plant, both material and installation, weigh heavily on the costs per subscriber and in fact dominantly determine the feasibility of any access technology. Most studies however simplify the costs of the outside plant by assuming average lengths. Although sufficient for global estimates, and

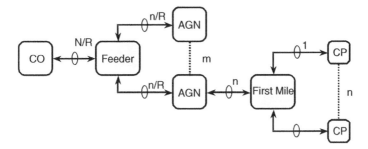

Figure 6.9 Basic outline of a Techno-Economic Access Network model.

rough comparisons of different solutions, it is not accurate enough for determining optimal design rules and dimensions. In reference [17], a model was presented that does take the two-dimensional geographic properties of first mile access into account. The part of the outside plant connecting any remote Aggregation Node (AGN), the Feeder network, is neglected since it would represent only minor contributions to the cost per user.

When relatively small numbers of users are connected to the concentrator, however, and many concentrators are connected to the CO, the Feeder contributions should be included. As illustrated in Figure 6.9 the most important parts of Techno-Economic (TE) modeling consist of the Network Node at the CO, a feeder network connecting the remote Aggregation Nodes, which in turn connects the Network Termination at the CP. The model should be generic enough to support comparison of different technologies (also copper and wireless) and topologies (rings, trees, and meshed networks). Another point of interest is the distribution of functions across the network: for the economic evaluation of technologies, different functions such as switching may be positioned at either AGNs or CO with associated consequences for transmission requirements. Moreover, powering is, apart of being a point of concern in general, a significant expense factor in especially small concentrator designs and will depend on the complexity of nodes. The model is therefore modular on a functional rather than on a nodal basis. The cost model is based on a single CO, connecting a total of N subscribers through m AGNs. Both m and the number of subscribers per concentrator, $n = N/m$, are considered design parameters (see Table 6.2). AGNs are connected to the CO by n/R fibers each, with R being the multiplexing ratio of the concentrators. Here it is assumed that customers are connected to an AGN by a single wire (fiber or twisted pair) or none (wireless).

6.2.2.1 Outside Plant

The outside plant is modeled as shown in Figure 6.10, where n subscribers (or m AGNs) are distributed in a two-dimensional area according to a rectangular grid with spacing l. Also for nonregular topographic street plans, this model is accurate when l is treated as the characteristic average distance between homes or AGNs of a specific representative

Table 6.2 Network model dimensions.

N	Number of subscribers per CO
M	Number of concentrators per CO
n	Number of subscribers per AGN

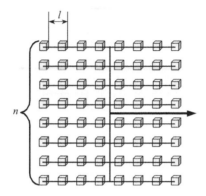

Figure 6.10   Outside plant topography model.

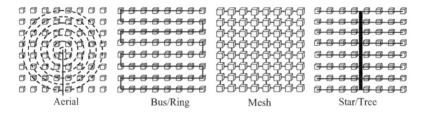

Figure 6.11   Outside plant topology models.

topography. Each subscriber is connected through a dedicated medium (fiber, twisted pair, or cable) to the concentrator in the center of the rectangular area. Media of different subscribers are collocated in ducts where appropriate and exit the duct at taps.

Specific network topologies (Figure 6.11) need to be considered for accurate calculation of the duct media lengths. Also different topologies may be used for Feeder and First Mile sections of the access network.

Since ducts often follow the street or pavement plan, instead of running through the basements of homes, a small correction l' to the characteristic length can be added to account for the last feed to the home (Figure 6.12). To keep the model simple however, and to allow for a realistic allocation of deployment costs, this part may is included in the fixed CP cost. The metrics resulting from this model are then as shown in Table 6.3:

The associated costs involved are split up into four components: the digging or trenching costs of a main duct, the installation of a dedicated tube per endpoint, the costs per wire that is installed in this tube, and costs associated with tapping the duct and making splices.

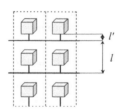

Figure 6.12   Driveway correction to characteristic length.

Table 6.3 Outside plant metrics.

L_t	m	Total tube length
L_d	m	Total duct length
n_t	—	Average tubes per duct
n_s	—	Number of splices/taps

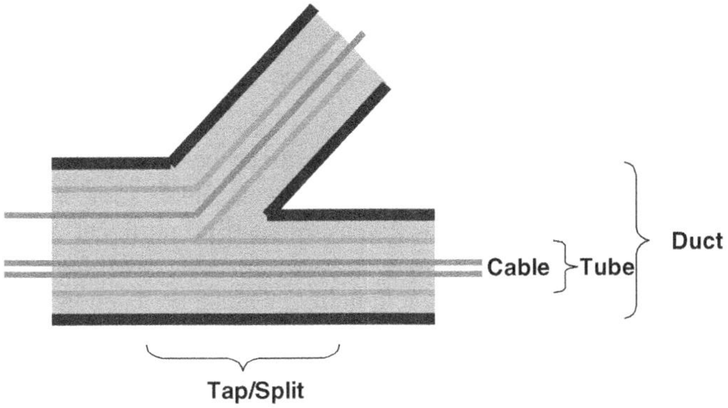

Figure 6.13 Duct/tube/wire model.

This model (Figure 6.13) is capable of representing most wireline situations by selecting some or all of the different parameters. Since ducts, tubes, and wires are cheaper when they are installed in bundles, two parameters are used: a basic installation cost per unit length and a price per duct/tube/wire (Table 6.4):

The total plant costs are calculated according to:

$$C = (c_{di} + n_d c_d)L_d + (c_{ti} + n_t c_t)L_t + (c_{wi} + n_w c_w)L_w + n_s C_s \tag{1}$$

6.2.2.2 Network Nodes

Network Nodes involve any Node in the Access Network where aggregation and distribution of traffic takes place, regardless of the multiplexing technique. Nodes therefore include both active equipment (Routers, DSLAMS, Gateways) as well as passive Cable Distribution

Table 6.4 Cost parameters of the outside plant model.

c_{di}	€/m	Duct installation costs
c_d	€/m	Costs per duct length
c_{ti}	€/m	Tube installation costs
c_t	€/m	Costs per tube length
c_{wi}	€/m	Wire installation costs
c_w	€/m	Costs per wire length
C_s	€	Cost per split/tap

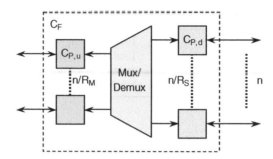

Figure 6.14 Network node model.

Points. Network Node costs are broken down into fixed costs, and costs associated with uplink and downlink ports. Fixed costs include material and installation of housing, closures, and chassis costs. Port costs include transceiver costs and proportional costs for powering. Either upstream or downstream ports may account for the costs for multiplexing and demultiplexing, or in fact any other functionality, whichever is more appropriate. The number of downlink ports is calculated by the number of downlinks divided by the number of shares per downlink port. This allows for shared media (PON, wireless). (See Figure 6.14 and Table 6.5.) The number of uplink ports is calculated from the number of downlinks divided by the multiplexing ratio. By default this value is calculated from the uplink bitrate divided by the downlink rate, but this value may be reduced to account for applied multiplexing gain. Administration and software costs per user can be included, most importantly for the CO.

The total costs are calculated according to:

$$C = C_F + nC_u + \frac{n}{R_S} C_{P,d} + \frac{n}{R_M R_S} C_{P,u} \tag{2}$$

6.2.2.3 Deployment Scenarios

For the calculation of deployment scenarios, the costs are calculated for a given dimensioning: $n, m, N = nm$, and the actual numbers $\bar{n}, \bar{m}, \bar{N} = \bar{n}\bar{m}$ in a given year according to a given take ratio or service adoption. Although $\bar{N}(i)$ may be retrieved from historical numbers for previous broadband introductions (ADSL), there is some degree of uncertainty about the actual number of AGNs.

Table 6.5 Cost parameters of Network Nodes.

$C_{P,u}$	Uplink port costs	€
$B_{P,u}$	Uplink bitrate	Mb/s
C_F	Fixed Costs	€
C_u	Costs per subscriber	€
R_M	Multiplexing factor	—
R_S	Downlink maximum shares	—
$C_{P,d}$	Downlink port costs	€

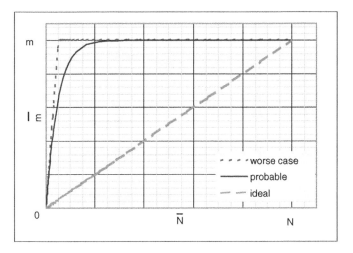

Figure 6.15   Number of required AGNs as a function of the take ratio.

Ideally, users fill up AGN by AGN, and in the worst case every new user requires installation of a new AGN (Figure 6.15). Actual distribution can be modeled by using:

$$\bar{m}(i) = \frac{\bar{N}(i)}{k} < m \quad 1 \le k \le n$$

$$\bar{n}(i) = \frac{\bar{N}(i)}{\bar{m}(i)}$$

(3)

Most often however, a worst-case scenario ($k = 1$) is assumed.

To prevent excessive interest costs, deployment of any network should be postponed as long as possible. Although any change in network configuration often will require additional initial costs, especially when outside plant and remote AGN are concerned (installers will have to travel and sometimes excavate remote locations), these cost can be outnumbered by long-term interest payments.

For upgrading of active equipment, especially at the CO, deployment of the network is straightforward by simply adding the appropriate number of uplink and downlink ports (with the associated switch capacity). For the outside plant this is less obvious, since ducts will be shared with future connections. To minimize initial costs, one would prefer to install the complete duct infrastructure at once, and install tubes and/or wiring later on demand. But the cost of the duct is mostly dominant, and especially when low take ratios are expected in the initial phases of deployment (when only one or two users are connected), it would be better to also perform the rollout of the ducts in stages.

A realistic scenario would be to start deployment with a main duct running through the area of the CO or AGN, and by adding street ducts one by one, connecting individual AGNs or CPs on demand (Figure 6.16):

$$L_{\mathrm{d}} = \sqrt{\bar{n}} L_{\mathrm{d,street}}$$

(4)

Again, for a worst-case scenario, one can assume that the first users will all be located in different streets. Note, however, that practice with DSL and Cable Modem services has

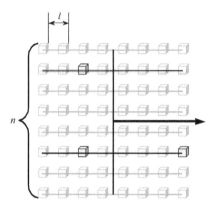

Figure 6.16   Phased rollout of the outside plant.

shown that new broadband services are often adopted block by block, representing the community groups corresponding to early adopters and trend followers.

6.2.2.4 Examples

As an example, Figure 6.17 shows the cost breakdown for PON installation in a Central-European suburban scenario ($l = 49$ m) for a 100 % take ratio. The model calculates the cost components for different AGN sizes. The parameters of this and the following calculations are based on the parameters of Table 6.4, Figure 6.14 (Network node model).

Tables 6.5 and Table 6.6. represent momentary cost estimations which depend strongly on the situation. Even with a minimal number of parameters, as presented in this model, some

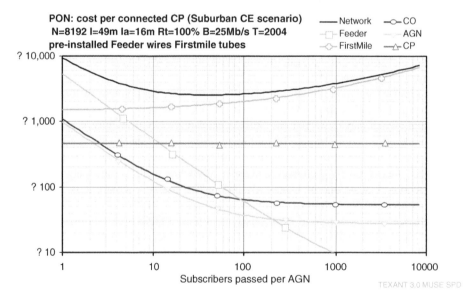

Figure 6.17   Example 1: calculation of the cost per connection for a PON access network, broken down into the various components, for different AGN sizes.

Table 6.6 Equipment costs used.

Node	description	symbol	unit	AON	P2P	PON	VDSL	FWA	Comment
CO	Uplink linecard costs	C_{Cu}	€	1000	1000	1000	1000	1000	⎫ 2.5G WDM Metro
	Uplink maximum ports	R_u	—	1	1	1	1	1	
	Uplink maximum portspeed	B_m	Mb/s	2500	2500	2500	2500	2500	⎭
	Fixed costs	C_F	€	10 000	10 000	10 000	10 000	10 000	Router
	Downlink linecard costs	C_{Cd}	€	400	1000	2000	400	400	
	Downlink maximum ports	R_d	—	1	20	1	1	1	
	Downlink reach	LB_{max}	Gb/s m	20 000	20 000	20 000	20 000	20 000	
AGN	Uplink linecard costs	C_{Cu}	€	400		50	400	400	
	Uplink maximum ports	R_u	—	1		1	1	1	
	Uplink maximum portspeed	B_m	Mb/s	1000	−1	1000	1000	1000	−1 : P2P
	Fixed costs	C_F	€	5000	500	500	5000	2000	
	Downlink linecard costs	C_{Cd}	€	100	25	800	50	400	
	Downlink maximum ports	R_d	—	1	1	16	1	10	
	Downlink reach	LB_{max}	Gb/s m	20 000	20 000	20 000	15.0	5.0	
CP	Uplink linecard costs	C_{Cu}	€	100	100	250	50	75	NT1 ONU
	Uplink maximum ports	R_u	—	1	1	1	1	1	
	Uplink maximum portspeed	B_m	Mb/s	100	100	1000	50	50	
	Fixed costs	C_F	€	200	200	200	200	200	NT2 Gateway

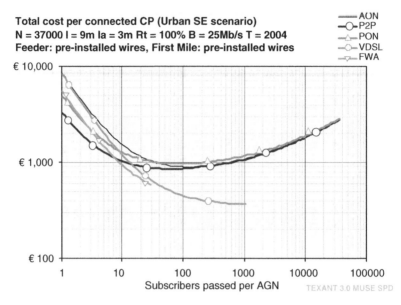

Figure 6.18 Example 2: Comparison of different technologies for different AGN sizes. AON: Active switched Optical Network; P2P: Point-to-Point; FWA: Fixed Wireless Access.

amount of speculation is unavoidable. Nonetheless the model is capable of providing an understanding of the impact of many variables, including cost components as well as dimensional properties. As indicated by the plot, the costs of the CO, Feeder Network, and Aggregation Node drop per connected CP when the AGN size increases. For a given population density, this means, however, that a larger area is covered per AGN, and the connection lengths and costs in the First-Mile section increase. This results in an optimum AGN size around 40 Subscribers per AGN.

This effect is also visible with other Access technologies, as shown in Figure 6.18 where the total costs are calculated in a Southern European Urban scenario ($l = 9$ m). As in the previous example the costs and optimum AGN size are dominated by the outside plant costs. Note that both very high-speed DSL (VDSL) and Wireless LAN-based Fixed Wireless Access (FWA) techniques, shown for comparison, are cheaper since they do not require new installation of a First Mile network. With their physical connection range, however, the maximum AGN size is limited also for urban densities.

The most interesting subject for the MUSE project is what improvements can be made to the available technologies. Figure 6.19 shows again a comparison of the installation cost of the different technologies, now broken down for nodes and network segments, and for the different plant components.

With the computed scenario and AGN size, FWA is again no viable solution even at high population densities. The figure also shows that although the Feeder and AGN costs of PON are minimal, the total costs are more expensive than AON and P2P solutions. This is mainly due to the ONU costs at the CP (including 200 € Gateways).

As shown in Figure 6.20, the comparison depends strongly on the scenario. In a suburban Central European scenario with lower population densities, VDSL is hardly an option when

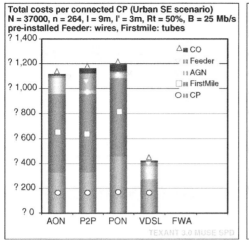
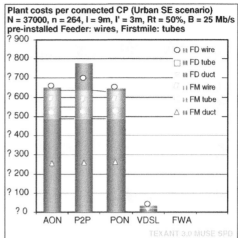

Figure 6.19 Breakdown comparison of total costs (left) and outside plant costs (right) of different access technologies.

available AGN sites are to be reused. Now PON solutions are cheaper than P2P where the amount of fiber in the Feeder outweighs the savings in CPE and CO.

The figures show that for long-term deployment, AONs do provide cost-efficient solutions, provided of course that no large additional investments need to be made in the remote powering and maintenance of Access Concentrators. In other words, the figure corresponds to a situation where the Concentrators are located in present cabinets that are powered and well accessible.

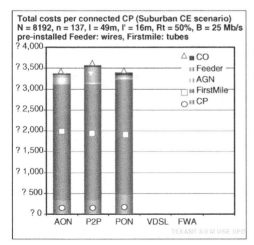
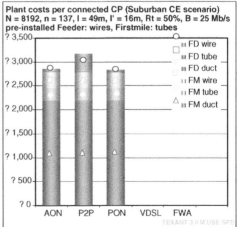

Figure 6.20 Breakdown comparison in a suburban Central European scenario.

Table 6.7 Installation profiles used for different topographic scenarios.

Installation profile	component	unit	Urban	Suburban
	Duct installation costs	€/m	25.00	20.00
	Costs per duct	€/m	1.00	1.00
Installation	Tap costs	€	25.00	25.00
	Tube installation costs	€/m	0.50	0.50
	Costs per tube	€/m	0.50	0.50
Fibre (standard Single	Cable installation costs	€/m	1.00	1.00
Mode)	Costs per cable	€/m	0.05	0.05

In Tables 6.6 and Table 6.7, the different costs components that were used in these calculations are shown.

6.2.3 FIBER ACCESS TOPOLOGIES

The optical fiber cable network between CO and CP allows for several topologies (Figure 6.21) ranging from Star topologies, where each subscriber has a dedicated connection to Ring (or Bus) networks, where all subscribers share a single optical cable. Tree topologies can be located in between these two, with a shared connection between the CO and a multiplexer/demultiplexer located at the Aggregation Node in the field from which individual optical cables connect the CPs. Tree and Ring networks can come in active and passive versions. Passive Optical Networks (PONs) deploy multiplexing at the physical layer with unpowered components such as passive optical combiners/splitters and optical filters. Active Optical Networks (AONs) multiplex at higher layers (Ethernet or even IP) and require powering. Active rings often deploy regeneration of a shared TDM link, with the associated risk of malfunctioning ONUs. Passive rings can avoid this risk by deploying passive Optical Add/Drop Multiplexers, similar to Metropolitan networks today, but require optics that are still expensive today.

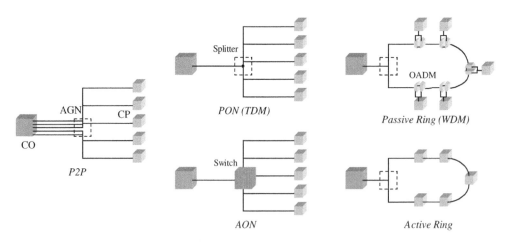

Figure 6.21 Most common Optical Access Topologies and implementations.

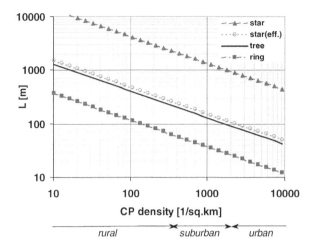

Figure 6.22 Comparison of the optical cable length per subscriber as a function of the CP density for different topologies.

In Figure 6.22, the optical cable length per subscriber for Star, Tree, and Ring topologies are shown as a function of the CP density. The plot shows the average cable length between CO and CP, according to the model described earlier. For these calculations, a CO size of 6400 subscribers and an aggregation node (AGN) size of $N = 32$ subscribers are assumed. Under these conditions, it is clear that P2P networks require more than ten times as much fiber as PON networks with a deep split (located at the AGN), and about 40 times as much as Ring networks.

The plant costs per subscriber C can be expressed as a function of the cable length L as:

$$C \equiv (c_i + c_m)L \tag{5}$$

It must be noted, however, that for fiber, the installation costs c_i exceed the material costs c_m (both per unit of length) also by factor in the order of about 50, a number that is increasing with increasing labor costs and dropping fiber prices. Therefore, by blowing fibers collectively from the AGN to the CO, the equivalent installation length L_i of P2P networks approaches L:

$$L_i \equiv C/c_i$$
$$L_i = \left(1 + \frac{c_m}{c_i}\right)L \approx L \tag{6}$$

This is indicated by the dotted line in Figure 6.22. For a complete comparison, the cost of Tree topologies must include an additional multiplexer/demultiplexer (WDM component) or power splitter, and their installation in the field.

Ring networks definitely minimize the fiber cable costs in the network. This lack of redundancy, however, also imposes risks of service interruptions due to fiber cable breaks and malfunctioning AGN and CP equipment. Some of these issues can be resolved by deploying protected bi-directional CWDM rings and passive Optical Add/Drop Multiplexers (OADMs) at least at the AGN location [18].

Apart from the installation costs, various other operational criteria need to be considered. The application of a deep PON split at a remote location for instance may save fiber, but this assumes a more or less simultaneous as well as static upgrade scenario for the subscribers that could obstruct upgrading of individual subscribers to higher bitrates in the future. The choice of topology will also be affected by the First Mile technologies sharing the Feeder Network, and especially the convergence layer that is considered. For multicast and broadcast services, Tree topologies provide the best support. Not only by means of analog WDM overlays, but also IP media applications may benefit from a shared downstream channel.

6.3 FTTX TECHNOLOGIES

Currently, mostly Point-to-Point Ethernet (P2P) and Passive Optical Networks (PON) are applied in still small-scale FTTH deployment projects throughout the world. The reason for this are obviously the attractive costs involved. For P2P mainly the use of technology has been deployed in Data Networks with high volumes which compensates for the slightly higher fiber costs for especially high-density, urban, scenarios. In less dense regions, where the fiber cost become more significant, PON appears more favored, but especially non-private initiatives may be less critical. Both technologies are by now mature and can benefit from years of research and promising upgrades towards very high-speed access services. PONs, in many variations, have been the subject of recent European research projects [10–12].

Figure 6.23 shows the Dynamic WDM/TDM PON studied in the Harmonics Project. In this approach, a passive WDM overlay was used to multiplex a number of PONs on a shared OLT pool to enhance resource sharing. This was performed dynamically with an Optical Cross Connect (OXC) consisting of Semiconductor Optical Amplifier (SOA) Gate Arrays. Although this system may provide efficient resource utilization in the future, the required optical components will only become available or otherwise cost-effective in the future.

The MUSE project focuses on enabling technologies for Broadband Access. Apart from a focus on the higher levels of the Access Platform, it studies technologies for cheaper deployment of FTTH in various scenarios and conditions. This includes studies on cheaper integrated transceivers for P2P Ethernet, but also innovative systems and architectures.

An example of this is a double CWDM Ring network [18] where a packet-oriented optical access network is considered which can deliver bandwidths of 25 Mb/s (mean) to 100 Mb/s (peak) to an average end-user. It can also accommodate high-end customers with higher bandwidth demands. It is assumed that the optical access network area consists of two major parts: *Feeder Area* (FA), covering the section between central office (CO) and splitting nodes (remote nodes, RN) placed in the field. The second section is the *Distribution Area* (DA) for linking customer's premises to the FA. Both parts of the network consist of ring structures (Figure 6.24). An example with eight RNs is shown, where the CWDM system is completely exploited by two wavelengths per RN.

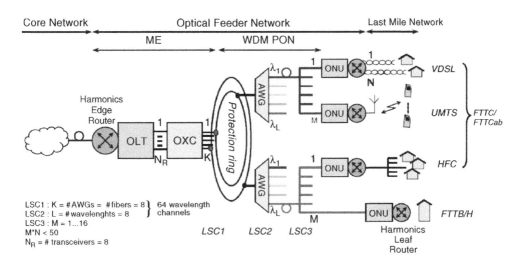

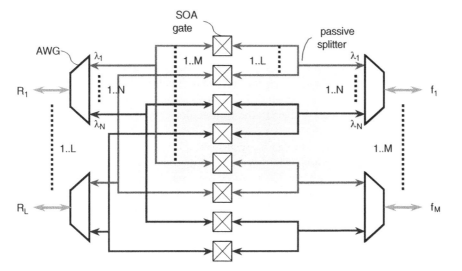

Figure 6.23 Dynamic switched WDM/TDM PON (top) and the applied optical cross connect (OXC, bottom) studied in the Harmonics project.

Other approaches investigate inexpensive multiplexing techniques, such as Sub-Carrier Multiplexing (SCM), and advanced technologies for transmission of Digital Subscriber Line data across optical feeder networks [19].

The MUSE project also addresses improvements to established technologies such as Optical Ethernet [20]. Research here focuses on lowering the running costs of housing and powering of network equipment. Major improvements can be obtained with integrated components to reduce the power and footprint requirements of Access Multiplexers as discussed next in more detail.

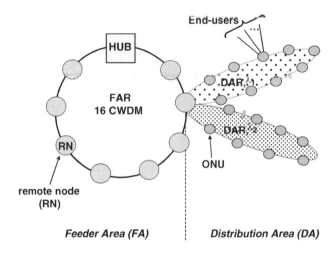

Figure 6.24 Studied CWDM double ring network concept [20].

6.3.1 *IMPROVEMENT OF OPTICAL ACCESS MULTIPLEXERS*

To address the issues discussed in the previous sections, the Asymmetric P2P/P2MP Optical Access Network (AsPON) depicted in Figure 6.25 is being developed for prototyping and end-to-end demonstration in MUSE. In the upstream direction, subscribers are connected through an optical link with a line rate of 100 Mb/s, while in the downstream direction all CPs share a PON and receive their traffic from a single transmitter operating at a Gb/s line rate.

The added value of this system can be seen from the features in Table 6.8, comparing PON, AsPON, and P2P solutions. The fiber cost in the outside plant per CP are listed in the second column, expressed in the CO to AGN segment, L, and the AGN to CP segment, l. Again, when regarding the effective installed fiber length (see also Figure 6.22) there is little difference between solutions. This indicates that also for PON there is less reason to locate power splitters in the field: for almost the same cost they can be positioned at

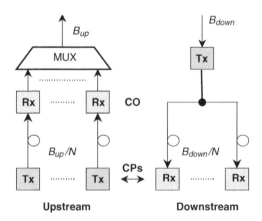

Figure 6.25 Asymmetrical P2P/P2MP operation of FTTX Access Multiplexers.

Table 6.8 Comparison of PON, P2P, and AsPON solutions: Plant and CO costs per subscriber, and service support.

System	Plant	CO	Service
PON	$l + \dfrac{L}{N}$	$\dfrac{T_x + R_x}{N}$	in/out band
AsPON	$l + \dfrac{L'}{N}$	$\dfrac{T_x}{N} + R_x$	in/out band
P2P	$l + \dfrac{L'}{N}$	$T_x + R_x$	Mux/Demux

the CO, requiring less field installation, and allowing for more flexibility. It also makes clear that the advantages of the AsPON are situated not in the field, but in the transmission technology.

The associated CO cost, footprint, and power consumption per CP are listed in the 3rd column, expressed in the amount of transmitters Tx and receivers Rx. Not explicitly shown in this table is the assumption that we are dealing with single fiber, bidirectional (Bidi) connections. For a fair comparison, the cost of Bidi splitters must be added to the Rx. It is in any case clear that PONs definitely have an advantage. The AsPON solution only requires receiver arrays and a single laser, and therefore has better properties than P2P solutions. Not shown in the table are the CPE costs. Since the AsPON ONUs use synchronous transmitters with moderate line rates, much like P2P networks, the associated costs are significantly lower than those of PONs, which utilizes a high (shared) line-rate upstream for Burst-Mode transmission. The right column finally shows the service support for multicast and analog service overlays. Here PON and AsPON solutions have the advantages of a shared medium in the downstream direction in common.

To quantify the savings possible with receiver arrays, Table 6.9 shows a comparison between the receiver components required for P2P (Small Form Factor, SFF, Fast Ethernet

Table 6.9 Comparison of dissipation and footprint of transceivers for P2P (left) and AsPON (right) solutions, for an Access Multiplexer connecting 48 subscribers at 100 Mb/s each.

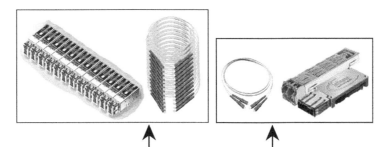

48 downlinks @ 100 Mb/s		P2P	AsPON	Saving
Tranceivers		48 SFF FE	5 SFP GbE + 4 PORx	81%
Dissipation	Watt	48	7	86%
Footprint (width)	cm	67	15	77%

Transceivers) and AsPON (Small Form factor Pluggable, SFP, Gigabit Ethernet transmitters, and 12-Channel Parallel Optical Receiver arrays) Access Multiplexers.

In this comparison, the dissipation and footprint (here the width of the connectors) is calculated for Multiplexers connecting 48 subscribers at 100 Mb/s, composed of standard Ethernet components commonly available today. Since the transceivers spend a significant part (up to 50 %) of the total power budget, it is clear that drastic savings can be obtained from deploying receiver arrays.

From this rough comparison, it can be concluded that the asymmetric P2P/P2MP approach can provide an attractive compromise between PONs and P2P solutions, by sharing downstream transmitters at the Access Multiplexer and maintaining synchronous transmission upstream. It is capable of supporting broadcast services and overlays with relatively cheap standard components for both Access Multiplexers and CP equipment with minimal power and footprint requirements.

6.3.2 FUTURE DEPLOYMENT

Today's research challenges lay primarily on enhancing the rollout of fiber in access networks, but the question is if by the time FTTH dominates the access networks, the conditions are still the same as they are now. Internet traffic is nearly doubling every year, and the current generation of Routers is expected to run out of steam within a surveyable timeframe. Research has started on next generation of Terabit Routers. These will need to employ optical rather than electrical switch fabrics and backplanes in order to manage the required switching speeds while limiting the dissipated heat [21]. For access networks, this may not as soon become too much of a problem, especially if the distributed Mesh approach is used (see Subsection 1.4). Still, when in the future the bitrate needed exceeds the 100 Mb/s range towards 1 Gb/s, Central Offices that are serving more than 10 000 users will require Terabit Edge Routers as well. A possible future FTTX network would look similar to Figure 6.26.

The big question is then how to link thousands of CPs at 1 Gb/s and by which technology. For gigabit speeds, there are already P2P technologies available, for example, 1000 Base LX Ethernet [22]. However, although migration of analog broadcast TV will probably be no issue by then, PONs will still be attractive for efficient broadcasting and multicasting of IP media. Research has started on 10 G PONs, but it remains to be seen if this will result in (economically) fit systems. Especially in the upstream direction, the power budget will impede burst-mode transmission and without optical regeneration in the field it will be hard to maintain the split ratios (1 : 128) and ranges (20 km) that are aimed at by the Gigabit PON standard [23]. For the described Asymmetric Optical Network, applying power splitting only

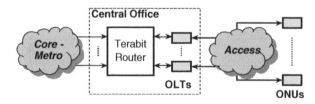

Figure 6.26 Gigabit access network deploying a Terabit Edge Router.

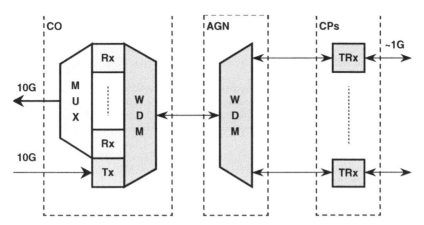

Figure 6.27 A 10 G Asymmetrical WDM/TDM passive optical Access Network. Single fiber configurations are possible by assigning a specific shared WDM channel for the downstream transmission.

in the downstream direction, the conditions are again less restrictive. A system should be feasible with 10 G components that are available today. In Figure 6.27 a possible AsPON configuration is shown with a TDM and WDM PON in, respectively, the downstream and upstream direction. The key components here will be an integrated Gigabit WDM receiver with low cross talk for the CO and cost effective, preferably wavelength agnostic, WDM transmitters for the ONUs.

6.4 CONCLUSIONS

In this chapter, the challenges of the current and next generations of FTTX networks were addressed. Key issues that need to be resolved first are the evolution of existing copper-based infrastructures toward FTTH and the inclusion of remote Access Multiplexers, wireless Access Points, and Base Stations in a common feeder. Additional requirements were identified for the smooth transition of existing voice and television services. Several technologies are studied in MUSE to address various deployment issues. Combining the overlay and broadcast properties of PONs with the low costs of P2P networks, an Asymmetric Passive Optical Network, AsPON, is developed with small low-power Access Multiplexers. It is expected that in the future, this solution will also need to be scaled up to 10 G operation, capable of supporting even high-speed broadband services well into the next decade.

REFERENCES

1. IST 2004-507295 MUSE project. Multi Service Access Everywhere, http://www.ist-muse.org.
2. Peter Vetter, Jeanne De Jaegher, Michael Beck, Kåre Gustafsson, Jeroen Wellen, Michel Borgne, Les Humphrey, Andreas Foglar and Rob van den Brink. Fiber access in MUSE: An End-to-end approach to achieve broadband for all, ECOC, 2004.
3. RedTacton: An innovative human area networking technology that uses the surface of the human body as a transmission path, NTT Press release, February 18, 2005.
4. IST 2004-001889 OBAN project. Open Broadband Access Networks, http://www.ist-oban.org
5. Home gateway: Key to convergence of BB and consumer industry, Alex De Smedt, BB Europe, Brugge, December 8–10, 2004.

6. MicroVision, http://www.mvis.com
7. Estimation for 27 European countries based on Eurostat data.
8. J. Lepley, C. Hum, M. Thakur, M. Parker, S. Walker. Super DMT: A concatenated discrete multitone modulation scheme for efficient multiplexing of xDSL signals in Access Networks, ECOC 2004, September 5–9, 2004, Stockholm, Sweden.
9. Wi-Fi or IEEE 802.11 Wireless LAN, http://grouper.ieee.org/groups/802/11/
10. AGN028 TOBASCO project, Towards broadband access systems for CATV optical networks.
11. AGN349 PRISMA project, Photonic routing of interactive services for mobile applications.
12. IST 1999-11719 HARMONICS project, Hybrid access reconfigurable multi-wavelength optical networks for IP-based communication services.
13. Ton Koonen, Anthony Ng'oma, Maria Garcia Larrode, Franz Huijskens, Idelfonso Tafur Monroy, Giok-Djan Khoe. Novel cost-efficient techniques for microwave signal delivery in fiber-wireless networks, ECOC 2004, September 5–9, 2004, Stockholm, Sweden.
14. Thomas Monath, Nils Kristian Elnegaard, Philippe Cadro, Dimitris Katsianis, Dimitris Varaoutas. Economics of fixed broadband access network strategies *IEEE Communications Magazine September* 2003; **41**(9): 132.
15. European Research project TERA.
16. European Research project TONIC.
17. Marcus K. Weldon, Francis Zane. The economics of fiber to the home revisited. *Bell Labs Technical Journal* 2003; **8**(1): 2003, 181–206.
18. J. Grubor, M. Schlosser, and K. D. Langer, Protected ring network, in optical access domain, in Proc. NOC 2004, Eindhoven, The Netherlands, 2004.
19. J. J. Lepley. High data rate DSL over optics, NOC 2005 conference, July 4–7, 2005, London, United Kingdom.
20. Mickelsson H. Ethernet Based Point-to-Point Fibre Access Systems, ECOC 2004, September 5–9, 2004, Stockholm, Sweden.
21. Jurgen Gripp, Marcus Duelk, John Simsarian, Ashish Bhardwaj, Pietro Bernasconi, Oldrich Laznicka, Martin Zirngibl, and Dimitrios Stiliadis. Optical switch fabrics for terabit-class routers and packet switches, *Journal of Optical Networking*, 2003; **2**(7): 243–254.
22. IEEE Standard 802.3-2002, Carrier sense multiple access with collision detection (CSMA/CD) access method and physical layer specifications.
23. ITU-T Standard G.984, Gigabit-capable Passive Optical Networks (GPON).

7

Residential Broadband PON (B-PON) systems

Frank Effenberger

Motorola, Andover, Massachusetts, USA

7.1 INTRODUCTION

Passive optical networks (PONs) have long been seen as an important part of many fiber-to-the-home (FTTH) strategies. Primarily, PONs are attractive because they economize on fibers leading from the central office out to the served communities, and reduce the number of optoelectronics at the central office, bringing direct and indirect savings. However, a long time has elapsed from the original development of PON until the large deployments happening today. There are both technical and economic reasons for this.

PON was invented at British Telecom in the late 1980s. The original concept was to use time division multiplexing to divide the available link bandwidth over many subscribers. The fiber network between the central office equipment and the customer's equipment would be entirely passive (nonelectronic). This was strongly motivated at the time by the relatively high cost of lasers (costing well over US$1000 at that time) and the low rate of users' bandwidth (telephony was the main application). For this reason, a great amount of research was initiated to study PONs.

From the initial concept, many variants were devised, including PONs based on WDM, FDM, CDMA, and hybrids of the above. In addition, many different component concepts, such as tunable lasers, modulators, broadband sources, EDFAs, and wavelength filters of all kinds, were applied in conceptual networks. Yet, despite all the myriad varieties of PONs disclosed in the literature, it is the TDM/TDMA architecture that is far and away the most popular. This is due to its low cost and technical simplicity. It is difficult to imagine a network with fewer components than a PON, as each node has one laser and one detector.

When PON systems were first introduced, the technology required to execute the designs were relatively primitive in terms of performance, power consumption, level of integration, and the manner in which they were manufactured. For these reasons, the systems tended to be impractically large and power hungry, often requiring large lead-acid battery packs at subscriber's homes. The systems were also costly, being well over US$5000 of total deployed cost per home, while the generally acceptable cost of an access network is ~$1000.

At the same time, the services commonly sold were telephony and video. Telephony is a narrow-band service, and fundamentally does not need fiber access bandwidth. Video, on the other hand, required excessive bandwidth in 1990. Video compression technology was in its infancy, and a video program might consume anywhere from 50 to 150 Mb/s. Considering that the highest practical data rate at the time was 600 Mb/s, video over digital medium was impractical.

So, for all these reasons, FTTH in 1990 failed. However, the situation in 2005 is markedly changed. Optics have matured to a point where telecommunications lasers can be had for US$50, and they are more reliable and efficient than ever. Electronics integration and low-cost manufacturing have become commonplace. These advances have pushed the costs of deployment below US$1000 per home. On the services side, data have opened new revenue opportunities for access systems that can be best served by fiber. Lastly, video compression has advanced to the point where a program consumes 1 ~ 6 Mb/s (using standard definition encoding rates) thereby fitting nicely within a digital transmission system. Last but not least, the regulatory and business conditions have become amenable for widespread deployment of FTTH and PON, which is what we see today in several markets in the world.

7.2   BRIEF HISTORY OF ATM-BASED PONS

There were many experimental PON systems in the early 1990s. Most of these were based on a TDM multiplexing concept, and delivered the digital equivalent of telephone lines (DS0's) to the subscribers. Such interfaces as integrated services digital network (ISDN) basic rate interface (BRI) and primary rate interface (PRI) were derived. However, these systems were quickly outmoded by the advent of data transport. Even before the advent of the Internet, engineers could see that connectionless packet transport was the future of telecommunications, and TDM would be forever fading in importance.

Hence, a new series of PONs were developed, based on asynchronous transfer mode (ATM). At the time, ATM was the focus of the world's attention, and was developed to be the network layer for broadband-ISDN. Hence, it became the natural choice for the network layer of both PONs (FTTH) and digital subscriber line (DSL) copper access systems. Both of these systems were standardized in the International Telecommunications Union (ITU), with the help of the Full Services Access Network (FSAN) Group.

7.2.1   ORIGINAL A-PON STANDARD AND THE EARLY JAPANESE DEPLOYMENTS

In 1995, a group of telecommunications providers (operators) joined together to develop a new PON system design. The leaders of this activity were BellSouth, British Telecom, Deutsche Telecom, France Telecom, and Nippon Telegraph and Telephone Company. The working plan was to pool their technical resources to help direct the vendor community. This arrangement was formalized in the creation of the FSAN group. This was an *ad hoc* association of operators and equipment vendors with the goal of developing draft standards for the ITU, publishing common technical specifications, and promoting the advancement of broadband telecommunications access.

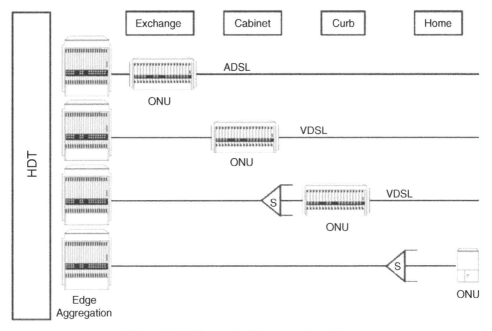

Figure 7.1 Fiber to the X conceptual architecture.

One of the initial products of FSAN was the Fiber-to-the-X concept. This is schematically shown in Figure 7.1. The brilliance of this architecture is that it combines in one system CO-based DSL, remote-based DSL, and FTTH. Politically, this merged the diverse network plans of all the operators inside of FSAN behind a single architecture. The major components are the central office-based optical line terminal (OLT), the field-based optical network unit (ONU), and the customer-based optical network termination (ONT). Joining these units is the optical distribution network (ODN), which is based on standard single mode fiber and splitters. DSL would be used to connect customers to the ONUs.

FSAN considered several proposed ATM-PON systems in late 1996, including contributions from the US, France, and Germany. In the end, the scheme proposed by Alcatel was adopted as the basic framework in early 1997. Through 1997, the technical details of the protocol were worked out through extensive negotiations and practical experimentation. In February of 1998, an official contribution was made to the ITU, and it was consented in November 1998, as G.983.1 [1].

This standard provides the basic specifications of the physical and transmission convergence layers. Armed with this standard, it was clear how to develop a basic ATM-PON system that could transport ATM cells to and from the customer to the service provider's core network.

From 1998 to 2000, NTT began to deploy ATM-PON across all of Japan. This used a symmetrical 155 Mb/s speed ATM-PON. Over 2M lines of capacity were installed at their central offices in preparation for a major roll-out. The service interface offered was ATM-25 or ATM-155 user network interface (UNI). This interface would then be connected using an ATM virtual path (VP) or virtual circuit (VC). Typically, the connection would run between customer locations, thereby forming a leased-line type of service.

Also peculiar to the NTT trial was the placement of splitters in the central office. Because the service was being sold to business customers, there was no way to predict where the customers would be located. Hence, it was more efficient to run direct fibers from the customers to the CO, where they were combined with a large single stage of splitting. While this eliminated the fiber savings of PON, this was not of primary importance in Japan, where the fiber plant is implicitly subsidized.

The shortcomings of this arrangement became evident from the start. First of all, most customers were interested in connecting their data networks together. This required them to install an ATM interworking device at their location to convert the ATM UNI to an Ethernet interface. This added cost and provisioning complexity to the service. Second, the ATM connection through the network was very expensively tariffed, partly because it locked up bandwidth unnecessarily, and partly due to the way these services were sold at the time in Japan. So, for these very real practical reasons, the deployment never really took off, and was eventually stopped.

Despite this failure, the early Japanese deployment proved that PON was a practically deployable system. The equipment performed as advertised, and many of the devices used in PON reached moderate volume production. This included planar waveguide optical sub-assemblies and application specific integrated circuits (ASICs).

7.2.2   EARLY INTEROPERABILITY AND THE BELLSOUTH DUNWOODY TRIAL

At around the same time as the Japanese deployment, BellSouth began a trial of ATM-PON for residential data service, in Dunwoody, Georgia. This used the same symmetric 155 Mb/s system as in Japan, but it was unique in several respects.

First, it was a residential PON system, and used a distributed splitter ODN. BellSouth's objective was to evaluate the system for FTTH applications. In FTTH, the network is deployed to cover an entire geographical area, and a good deal of planning can be done to optimize the layout of splitters and fiber cables. In the case of the Dunwoody deployment, 1 : 3 splitters were placed near the homes, while 1 : 10 splitters were placed at the entrance of the distribution area.

Second, it provided Ethernet services directly to the customer. Because of its residential focus, it was clear from early on that ATM UNI service was not going to be popular. The design of the ONT used in the trial had pluggable card interfaces, and this permitted the quick development of an Ethernet interface.

Third, it was a PON that demonstrated interoperability. Lucent Technologies developed the ONT, while Oki Corporation developed the OLT. While the two companies worked together extensively, and perhaps did not implement every detail of the standard, it was an important step towards interoperable PONs.

Importantly, the joint Lucent–Oki effort underscored the need for a management interface for ONTs. As a result, FSAN started work on the aptly named ONT management and control interface (OMCI) recommendation. This was consented in the ITU in 2000. The recommendation specifies the management information base (MIB) for the ONT core and for all the user interfaces and other features of ONTs. It also specifies the method for communicating between the ONT and the management function, using the ATM transport provided by the PON. G.983.1 and G.983.2 taken together provide a concrete basis for building interoperable PON equipment.

7.2.3   SYSTEM IMPROVEMENTS: WDM, DBA, SUR

After the Dunwoody trial was over, it was clear to the North American operators that FTTH would require all three services (voice, video, and data) to be supported. In this way, the fixed costs of the fiber infrastructure could be shared over more revenue-bearing services, and would be more economic.

However, video was problematic. If video was to be served over the digital PON transport system, the costs of the customer premises equipment (CPE) and home networking would be quite high. In addition, the network bandwidth requirements would be greatly increased. Lastly, digital delivery would require a wholesale redesign of all the video equipment, from the head-end to the set-top-boxes. So, a new means of transporting video in its 'native' radio frequency (RF) format was desired.

So, in 2000 FSAN began to consider improvements to the basic ATM-PON. First of these was a wavelength division multiplexing (WDM) scheme, primarily aimed at video delivery [2]. The primary design challenge was the re-allocation of the wavelength plan for the PON, as illustrated in Figure 7.2. The upstream wavelength band of 1260–1360 nm was untouched. The downstream basic band wavelength band was contracted to 1480–1500 nm. This was wide enough to allow the use of uncooled DFB lasers, and the center frequency of 1490 nm coincided with the coarse WDM grid of standard wavelengths. This contraction then opened a window from 1539 to 1565 nm for other signals, namely, video overlay. This addition of a third wavelength is commonly referred to as the element that changed APON into 'Broad-band PON', or B-PON.

Continuing on from this first improvement, the FSAN group worked to address other minor improvements to the protocols in B-PON. Dynamic bandwidth assignment (DBA) was described in G.983.4, so that the PON could shift upstream bandwidth from idle ONTs to busy ONTs in real time. The messages used to control optical survivability (protection

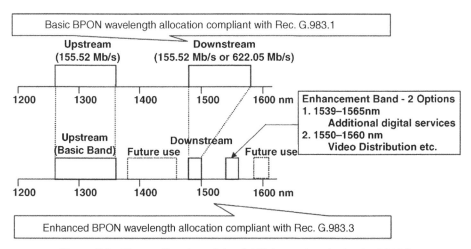

Figure 7.2   The re-allocation of the G.983.1 wavelength plan in G.983.3.

switching) were defined in G.983.5. Higher bit rates were defined in two amendments to G.983.1, and a new more secure mode of data encryption was added. Finally, continuous improvements to the OMCI recommendations (G.983.2, .6 to .10) were developed. The single fact that describes all of this work is that B-PON is the most widely followed and studied optical access system in the world today.

7.3   TRIPLE PLAY B-PON SYSTEM ARCHITECTURE

After the initial experiments and trials, the system operators and equipment vendors converged on a common 'recipe' for a successful B-PON system. This varied somewhat from country to country, but its main ingredients were conformance to G.983 standards, and the provision of marketable triple-play services. The data rate of the PON was 622 Mb/s downstream, and 155 Mb/s upstream. The PON either supported the three wavelength transmission directly, or at least kept the 1550 nm window open for future use. The 1550 wavelength is used to provide RF modulated video signals.

This section attempts to fully describe this common system. This is complicated by the fact that there are many implementation details that vary from vendor to vendor. In these cases, the variations will be indicated as aspects that should be studied when evaluating B-PON systems.

7.3.1   SYSTEM OVERVIEW OF TRIPLE-PLAY B-PON

A diagram of a common B-PON system is shown in Figure 7.3. This includes the OLT, PON, and ONTs. The OLT has the functions of (1) supporting the B-PON line interfaces, and (2) adapting the bearer traffic from the PON side to the service node interface (SNI) side. The ONT has similar functions of (1) supporting the B-PON line termination, and (2) adapting the user interface traffic for transport over the B-PON.

Note that the 'OLT' is typically composed of two separate pieces of equipment: the B-PON OLT proper, and the erbium-doped fiber amplifiers, EDFAs (sometimes such an OLT is called the video-OLT) that provide the video signals for the PON. The two outputs are

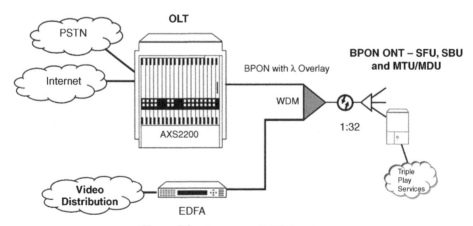

Figure 7.3   A common B-PON system.

combined using a WDM multiplexer component that is a separate from either of the OLT active components.

An ONT can have many variants to serve different customers. The variants include single family unit, multiple dwelling unit, small business, multiple business, and high-capacity ONTs. All of these ONTs can join a single PON simultaneously, and there should be no restrictions on the combinations possible except for the total bandwidth of the PON. Also note that in many cases, the ONT will operate in concert with various customer electronics (such as home networking routers, wireless adapters, etc.)

7.3.2   OPTICAL LINE TERMINAL

The OLT is the hub of the PON network, and for that reason is the most critical piece of the network. Most OLTs are based on a central ATM multiplexer or switch design, where the switch combines all the PONs supported on the chassis into a single interface for connection to the core, as shown in Figure 7.4. This design reduces the number of switch interfaces required and allows statistical multiplexing of user traffic flows. In the case shown, the switch is $1 + 1$ protected, which is why there are two switches and each switch has an independent connection to every line interface. However, some OLTs do not implement a central switch, and instead produce a SNI port for each and every PON interface. This reduces the cost of the OLT at the expense of requiring more switch interfaces in the core network.

Commercial OLTs have typically been developed as offshoots of existing telecommunications products. Because of the ATM basis of B-PON, the natural starting points of most OLTs have been either ATM switches or DSL access multiplexers (DSLAMs). In some recent cases, even digital loop carrier (DLC) platforms have been pressed into service as B-PON OLTs. The DSL and DLC platforms have the advantage that they are already deployed; however, they often have bandwidth limitations that can impact the speed and variety of services offered.

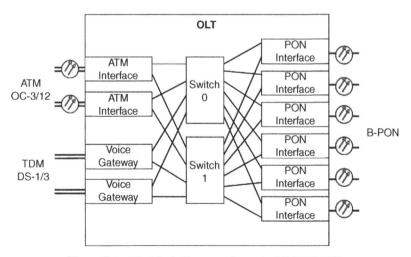

Figure 7.4   The block diagram of a typical B-PON OLT.

The simplest OLT will conduct all the service traffic from the PONs to an ATM interface towards the core network. This tends to require service adaptation boxes inside the network: for data, a Broadband remote access server (BRAS); for voice, an ATM-to-Voice gateway. More fully featured OLTs tend to integrate these functions into the OLT equipment, generally as special purpose interface cards.

7.3.3 PASSIVE OPTICAL NETWORK (PON)

The passive optical network is composed of standard single mode fiber (ITU-T G.652), wavelength-independent optical splitters, WDM filters, and the ancillary necessities of splices and connectors, of course. The most important design aspect of the PON is its loss budget. The B-PON standards specify three classes of loss (A, B, and C), but these classes are rather 'abstract' in that they specify budgets that have no practical relationship to either component availabilities or fiber plant requirements.

Therefore, the industry has informally converged on an optical budget that is somewhere between class B (25 dB of loss) and class C (30 dB of loss). The budget tends to have a little more margin in the upstream 1310 nm link than in the downstream link. This produces a better match in practical PONs because the fiber loss at 1310 nm is higher than at 1490 nm. So, most implementations provide a so-called 'Class B+' performance.

A typical practical network is shown in Figure 7.5. Fiber jumpers are used to connect the OLT and EDFA equipment to the WDM component. Another jumper is used to connect from the WDM component to the feeder plant fiber. All of these jumpers/connectors add significant loss which must be accounted for. The feeder fiber is generally fusion spliced along its route, and eventually appears in a service area interface cabinet. Splitters are usually spliced onto the feeder fibers, and then the distribution fibers are spliced onto the splitter. Both distributed and centralized splitter arrangements are possible. The distribution

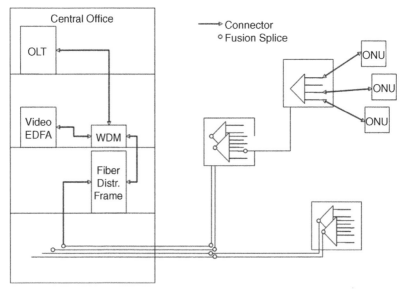

Figure 7.5 A practical PON network design.

fiber cables pass the subscriber locations, through which drop access points are installed. The drop is often connectorized at both ends, and plugs into the drop access on one end, and into the ONT on the other end.

Reflections are another important factor to consider, particularly with the analog video overlay service. The analog signal does not tolerate reflections well, and the fiber plant must be engineered to reduce these reflections. The two main approaches to this are to (1) reduce the number of connectors by using fusion splicing wherever possible, and (2) use angle polished physical contact (APC) connectors throughout the video path.

7.3.4   OPTICAL NETWORK TERMINATION

The ONT is the customer's gateway to the optical network, and is responsible for creating all the services for which the subscriber pays. Hence, it tends to be a very complex device that encompasses a large number of protocols and interfaces. On the other hand, the ONT is dedicated to a small number (many times, only one) of subscribers. This requires the ONT to be as inexpensive as possible to maintain the viability of system deployment economics. The conflict of ONT performance and cost is probably the most common theme of PON development, and is still the key aspect of most PON systems today.

There are many ONT variants available today, but we will consider a few key types that have become most popular in deployments. The major variability factors are the number and type of user service interfaces, and the modularity of the ONT (fixed or flexible slotted profile).

The most important is the single-family unit (SFU) ONT. This provides four POTS interfaces, one 10/100 Base-T Ethernet data interface, and one RF video interface. The ONT is an integrated fixed type, in that there are no field serviceable modules. The design approach is targeted to produce a unit at the lowest cost. An example of the physical design of a typical unit is shown in Figure 7.6. Note that mechanically, the unit is hardened for outside plant conditions, and has two access doors to partition the customer-side interfaces from the network provider-side interfaces. Such ONTs are rapidly becoming a commodity item.

Figure 7.6   A single-family unit ONT [3].

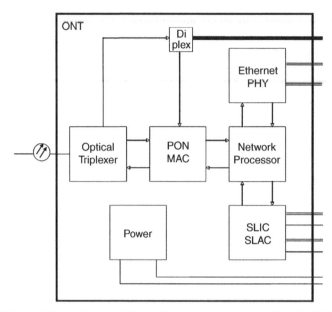

Figure 7.7   A high-level block diagram of a single-family unit ONT.

The block diagram of the SFU ONT is shown in Figure 7.7. The key elements of electronics are the optical triplexer (for multiplex/demultiplex of the three wavelength signals used, 1310, 1490, and 1550 nm), B-PON media access controller (MAC), network processor, and user interface physical layer devices. The triplexer provides the basic optical-electrical conversion, and often also provides the RF amplification needed for the video interface. The B-PON MAC drives the triplexer with the G.983.1 protocol, and produces an ATM Utopia interface for the processor. The network processor provides the ATM interworking functions for the interactive services. The voice services use ATM adaptation layer 2 (AAL2), and the data services used AAL5. The processor then drives the user physical layer devices (subscriber line interface circuit (SLIC) for voice, and Ethernet PHY for data). Beyond the electronics, the ONT needs the usual infrastructure components: weatherproof housing, power supply, connectors, indicator lights, and safety features. While these may seem mundane, they often drive both capital and operational costs, and should never be neglected.

Another ONT type is the multiple dwelling unit (MDU) ONT. This unit is designed to support 8 ~ 16 subscribers from a single ONT, such as in an apartment building. In contrast to the SFU ONT, the MDU ONT is typically a slotted chassis-based design, with flexible interface cards. This allows the service provider to install the mix of service cards required to serve the demand, and it also provides the opportunity to develop new service cards to support emerging services as they become available.

The MDU unit also supports the Fiber-to-the-Curb (FTTC) application, where the unit is installed at the curb. In a typical residential neighborhood, such a placement can reach 8 ~ 12 homes with good cabling efficiencies. The key difference is that the device becomes a piece of network equipment (and ONU), and the power is generally supplied from the network (usually using a current limited −130 V DC feed from a centrally located supply).

Last but not least, probably the most widely deployed ONT type is the Ethernet-only ONT. NTT has deployed over one million of these units. In this design, the ONT is stripped down to only the essential components for Ethernet service. In some cases, the B-PON MAC device produces the Ethernet interface directly, and there is little more than a optical diplexer and a single semiconductor chip in the box. Additionally, the unit is designed for inside the home environments, thereby reducing housing and power supply costs. Roughly speaking, this ONT can cost about half as much as the SFU ONT.

7.4   EVOLUTION OF BROADBAND SERVICES

The 'B' in B-PON stands for broadband, and that serves an excellent role as a name because it is vague enough to cover nearly any service imaginable. This permits the B-PON system to evolve and adapt itself to the changing commercial environment. This is important because the service set being offered is rapidly changing today. Even more importantly, the way that existing services are provided is changing. B-PON equipment is following these trends.

7.4.1   VOICE: FROM AAL2 TO VOIP

The traditional voice service from the customer's perspective is a voice grade transmission path, with associated signaling, on a pair of copper wires delivered to his home or office. To carry this over a broadband network, the analog channel must be adapted into a standard set of protocols. The two common methods for this are Voice-over-ATM, and Voice-over-IP, as shown in Figure 7.8.

The Voice-over-ATM service is more traditional, and attempts to fit into the existing voice network. In this regard, the ATM portion of the network extends from the ONT to the voice gateway in the OLT, and the class 5 switch in the central office does not know that anything special has happened to the customer's voice call. The protocol used between the ONT and voice gateway is AAL2/Broadband Loop Emulation Service (BLES). The ONT, OLT, and

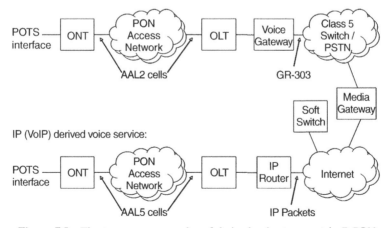

Figure 7.8   The two common modes of derived voice transport in B-PON.

ATM switches (if present) support ATM transport service. The connection between the ONT and the voice gateway is set up at provisioning time, and remains in place permanently. Note that although the ATM connection is always there, actual bearer traffic cells are only sent during a call, and so bandwidth is not consumed by telephones that are on-hook.

The Voice-over-IP (VoIP) service is newer, and attempts to replace the existing voice network. In VoIP, the processing of the user call data and the call signaling are separated, and are housed in server-like devices typically placed deeper in the network. The ONT becomes the customer endpoint of the VoIP session, and it must contain an IP host as well as a SIP agent application. When a call is initiated, the SIP agent communicates with the soft-switch in the network, which mediates the IP portion of the call. The target VoIP endpoint (either another IP telephone or a media gateway) is located and connected with the originating endpoint via the IP network. The soft switch is not in the media path of the call.

The earlier deployments of B-PON tend to use the AAL2 solution. This is motivated by the desire to leverage the installed base of class 5 switching equipment, and to simplify the provision of voice. Given the general contraction of the wireline voice market after 2001, there is a lot of spare class 5 line capacity. This would seem to argue for tapping into this otherwise stranded resource. However, often times the spare capacity is in the form of copper (DS0) interfaces. This is not useful capacity, in that voice gateways cannot produce a cost efficient DS0 handoff to class 5 switches, and a DS0 handoff would be bulky and deliver poor performance. So, AAL2 has limited usefulness.

The future solution that all operators are looking to is VoIP. There are many reasons for this including convergence on IP, added service flexibility, and switch cost reductions. Because VoIP uses a generic IP transport service, it can ride over any number of access and interoffice networks. This means that the dedicated voice network can be capped, and someday eliminated. Because VoIP is based on a 'smart end-point, dumb network' model, new VoIP services can be created quickly at a few endpoints, without any network upgrades. Finally, because soft switches are based on commodity server hardware and software, they are far cheaper than the class 5 switches that were based on specialized hardware and proprietary software. So, VoIP has become the 'objective architecture,' for incumbent local exchange carriers as well as new entrants.

To follow this trend, an increasing number of B-PON systems are implementing the VoIP feature on their ONTs. This is relatively easy to do, as the ONT's central processor can implement the VoIP agent software. The biggest hurdle to overcome is the interoperability of the ONT with the soft switch. Because VoIP is much newer than TDM voice interfaces, there is more variability in the interface, and more room for interoperability problems. Hence, it will take some time for the widespread rollout of VoIP over PON.

7.4.2   VIDEO: FROM ANALOG TO QAM TO IP

Standard video service has traditionally been defined as the delivery of about 100 RF carriers with signals (channels) to the customer. Each of these channels contains either an analog composite video signal (AM-VSB in North America), or a QAM-modulated digital signal. The analog channels can be used directly by television sets, and is identical to the service provided over the air by broadcasters. The QAM channels are received by set top boxes (STBs), which are specialized equipment that receives the digital data, decrypts, decompresses, and transforms it into an analog video signal. This signal is then delivered to the television set. Currently, video service is dominated by the operators of hybrid fiber coax (HFC) networks in cable TV companies.

Video has long been a goal for telephone operators. Their legacy copper network; however, is not very well suited to carrying it. It has a small bandwidth, and it cannot carry analog signals in their native form. This, in fact, is one of the big driving factors towards B-PON. Because B-PON has a third wavelength channel, the traditional RF video service can be delivered in a manner similar to the cable TV company's video delivery.

Providing the broadcast signals is the first step, but video services are now shifting towards more interactive offerings. To enable services like pay-per-view and video-on-demand (VOD), a return path is required. The traditional HFC method for return is a QPSK RF signal in the 5–40 MHz band, described in SCTE 55-1 or SCTE 55-2. Nearly all STBs today implement either one or the other of these QPSK protocols. Hence, ONTs often provide a feature to demodulate these protocols, and send the data up the digital path of the B-PON.

More recently, the trend has been to move towards IP. At first, the return path would be implemented over IP/Ethernet. The reason for this is to (1) avoid choosing either of the SCTE return path standards and thereby getting locked into a particular STB vendor, and (2) begin down the path of IP integration into video. Once the STB has Ethernet connectivity, it is a short leap to put VOD streaming video into that Ethernet. The key aspect of VOD is that it is a point-to-point stream, and appears to the network as ordinary data service.

To handle these trends, B-PON ONTs are appearing that include extra data interfaces to support the IP-video demands. Home networking is a key issue with video IP integration because STBs are not typically located next to Ethernet wiring. While Wireless LAN, HPNA, and power-line carrier have all been explored for this purpose, the market seems to be converging on a data-over-coax technology. This provides the best fit for video because the coax is by-definition always available at the STB/TV location. Also, coax provides more bandwidth with a higher reliability than the others.

7.4.3   DATA: ETHERNET TO EVERYTHING

The standard interface for data services is, of course, 10/100 Base-T Ethernet. This is by far the most ubiquitous data interface in the world, and its cost is very low (nearly zero) because it is built into nearly every processor manufactured today. So, every ONT has at least one Ethernet interface, and simple Internet access service is becoming a commodity service.

There are two directions in which data services are being extended. The first is to provide more Layer 2 and Layer 3 services. Vendors are working to differentiate their ONTs by adding Ethernet features, such as 802.1p QoS priority or 802.1q VLAN services. The ONT can also act as PPPoE proxy, to multiplex a single service provider PPPoE session across multiple customer computers. Vendors can also add layer 3 services, such as DHCP server, IP firewall, and Network address translation services.

The second extension of data service is to provide a wider array of interfaces and networks inside the home. This is normally done by placing a home gateway router device inside the home. This router can provide all the protocol services mentioned above for the ONT, and can also be a wireless LAN base station, a hub for video-over-IP distribution, and a general application server for intelligent home services.

At present, it is clear that value-added data services will be added to the B-PON system. However, it is not clear whether they will be implemented on the ONT, or on an adjunct home router. It is likely that some essential features (like data-over-coax) will be pushed into the ONT. In contrast, the home router will probably exhibit a large variability of features to support the wide variety of customer demand profiles. In this way, the separation of ONT and home router actually serves an important role in system flexibility.

7.5   FTTP ECONOMICS

The situation of B-PON in the overall optical communication market should not be overlooked. In the late 1990s to early 2001, the entire communication market received massive over-investment, particularly focusing on long-haul WDM transport systems. Optical access received a relatively small share of this largesse, as the 'smart money' thought it better to invest in the traditionally lucrative low-volume high-cost world of long-distance. Such investments were poorly placed. The long-distance market has become one of cut-throat commodity competition, with major bankruptcies threatening to destabilize all.

In the mean time, the access market has become more competitive also, but in a different way. In North America, there is a natural duopoly of the telephone company and the cable television (CATV) company. In the mid 90s, the CATV companies spent huge sums modernizing their access networks to the hybrid fiber coax (HFC) architecture, and by 2000 they were ready to provide high speed data services.

Only when the data service threat was on their doorstep did the telcos begin to deploy DSL (digital subscriber line) systems in a significant way. There were technical and operational problems stemming from the natural characteristics of DSL (its reach is limited, and it does not tolerate impaired loops very well.) As a result, the telcos have fallen behind in the race to provide broadband data service. And yet, even with this significant loss, the telcos were not primarily concerned because data was not part of their traditional revenue base, and was not a high penetration service, since perhaps at that time 20 % of homes take broadband data services, as compared to 98 % taking voice service.

More recently, the cable TV companies have begun to deploy voice services over their HFC networks. This was initially spearheaded by deployment of TDM-based voice over HFC equipment, but now has been superseded by voice-over-IP (VoIP)-based equipment. These deployments have been very successful, and in some areas have taken 20 % market share from the incumbent telco. This, finally, was a serious competition that could not be ignored.

At the same time, the collapse of the optical transport industry has resulted in a glut of optical devices and components, sub-assembly, and equipment manufacturing capacity. This has depressed prices for optics of all kinds to levels that were unheard of just a few years ago. This has made optical access less costly to the telcos, just at the time when they are feeling threatened by the cable TV companies. This, primarily, is the business driver for the current interest in B-PON systems and deployments.

7.5.1   COMPETING BROADBAND ACCESS TECHNOLOGIES

There are many studies that compare one optical access technology against another; however, these analyses miss the bigger picture of industry wide competition. It is important to consider that DSL and HFC as perfectly viable alternatives to broadband optical access networks in the residential market, and SONET/Ethernet in the business market. These wide comparisons reveal certain market segments where B-PON is relatively better than other systems.

For example, consider network costs as a function of data rate. DSL and HFC both start at a much lower cost than B-PON, and remain lower for bandwidths lower than about 5 Mb/s. At rates above that, however, DSL cost starts to increase rapidly, due to the fact that the DSLAM equipment will need to be located in carrier serving area cabinets. HFC also begins to suffer because of the need to subdivide the optical nodes. In contrast, B-PON remains a fixed moderate cost until bandwidths become very high. So, B-PON will do better in high bandwidth demand situations, and do poorly otherwise.

Another important point is customer density and 'take rate.' The DSL and HFC networks cost is inversely proportional to the homes per mile, while the B-PON costs are more constant. This is due to the fact that B-PON has intrinsically longer reach, and has a great deal of cost in terminal equipment, while both HFC and DSL require active amplifiers or DSLAM equipment to reach farther. Thus, B-PON will do better in suburban or rural areas, and not as well in the cities.

7.5.2 NEW BUILD APPLICATIONS

The first area where B-PON has been deployed has been new-build areas ('green fields'). This is only natural, since the costs of installing cabling are going to be spent regardless of what type of cable is deployed, and the network operator wants to deploy the newest technology and system. A typical cost analysis for a suburban buried deployment is shown in Figure 7.9. In this analysis, we assume the take rate is 100 %, and the total volume per year is 100K homes. (This is a rather small volume, and so these costs should be considered conservative.) This shows that the biggest cost is in the installation of the cable plant. The ONT and its installation (including the fiber drop) is the next largest contribution. Central office equipment is relatively small.

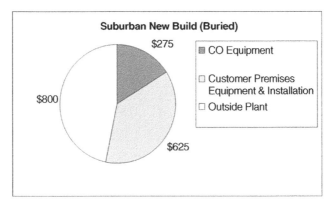

Figure 7.9 A typical cost profile in a new-build application.

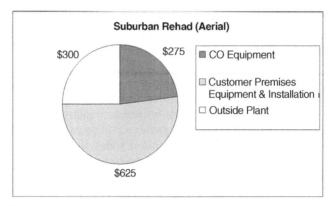

Figure 7.10   A typical cost profile for a rehabilitation application.

7.5.3   REHABILITATION APPLICATION

The next area where B-PON will be deployed is in neighborhoods where the existing copper plant is dilapidated and needs replacement (so-call 'rehab' or 'brownfield' deployment). Just as in new-builds, the cabling costs are unavoidable. However, these areas tend to use aerial plant instead of buried facilities. This changes the cost breakdown, as shown in Figure 7.10. The take rate here is also assumed to be 100 %, and the same equipment volumes. Here, the equipment costs have remained the same, but the fiber deployment costs are reduced by more than half. This makes B-PON more cost effective.

7.5.4   OVERBUILD APPLICATIONS

The last scenario to consider is overbuilding of existing, working networks. Here, the take rate is likely to be much lower, since customers may choose to remain on the older network, or with a competitor. The cost breakdown for an aerial overbuild is shown in Figure 7.11, for the same suburban area as was shown for the rehab case, but with a 33 % take rate. The cost for ONTs has been cut to 33 % of its previous value because they are only installed at 1/3 of

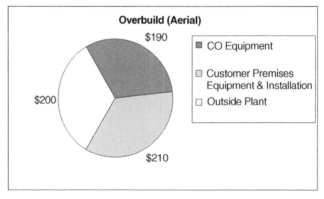

Figure 7.11   A typical cost profile for an overbuild application.

the homes. However, the fixed costs of outside plant and central office equipment do not fall nearly as much. This is due to the inevitable inefficiencies of less than 100 % take rate.

7.6   FTTP DEPLOYMENT PRACTICAL CONSIDERATIONS

A fiber-based optical access network is a new design different from traditional copper and coaxial outside plant networks. The fiber network is more demanding in terms of the amount of engineering required to produce a design that meets the requirements of PON equipment. It also takes a certain skill of craft to handle the fibers, since fiber connections are much more fragile than copper connections. However, these obstacles have been overcome with a combination of cable innovations and craft training. Currently, millions of access fibers are being constructed in the US, and millions are already in service in Japan.

7.6.1   OUTSIDE PLANT CONSTRUCTION

The basics of access network design are really not changed that much for PON deployments. The design must begin with a physical map of the areas to be served. The map must be detailed, showing all the existing rights-of-way, streets, houses, and other ancillary structures (ducts, cabinets, etc.) With this map, the engineer can establish the locations of the three major divisions of the network: the feeder, the distribution, and the drop.

Once such a map is obtained, the designer must locate the subscriber locations on the map. In many residential applications, this task can be as simple as claiming every house is a point of demand, which is the case if the penetration is largely 100 %. If penetration is not complete, one must consider the uncertainty of where the subscribers will be. In business applications, it is more difficult because there is not only the uncertainty of where the customers are, but also what services they will take.

Given the physical and service profile of an area, the designer must consider how much of the network will be built at first, and how much will be built as demand warrants. From a capital cost perspective, it is better to defer construction as long as possible; however, many small incremental builds greatly increase operating costs, and also degrade the efficiency and quality of the plant (more wear and tear on connectors, increased likelihood of craft errors). So, a balance must be struck. In many cases, the optimal situation is to fully deploy the feeder and distribution network, but defer the drop installation until the subscriber in question takes the service.

At the same time, the designer must consider the physical architecture of the PON, and the physical components and piece-parts that will go into its construction. Key decisions include: splitter sizes, single or multiple stages of splitting, and where to use connectors versus splices. Also, the designer must consider the complexity of whatever design he produces, and insure that it does not become too complicated for use in the field.

When considering splitter sizes, the obvious choices provided by the component manufacturers are 1 : 4, 1 : 8, etc. However, less common sizes can often lead to improved utilization efficiency in certain geographic areas. They can also reduce the loss of that stage, providing more loss budget for other portions of the network.

The choice of single or multiple stages of splitting usually comes down to the issue of take rate. If an area has a high take rate, then multiple stages splitting can save fibers. However, if the take rate is low, then multiple stages prevent any deferment of shared equipment because all the splitters must be activated.

The decision on connectorization is the most controversial. Connectors are useful because they provide an access to the fiber for troubleshooting, and they provide for the simple re-arrangement of the network. On the other hand, connectors add loss to the network, they cost more money than splices, and they can fail due to wear or environmental effects. In some early PON deployments, the PON network had up to 12 connectors in the path from OLT to ONT. Currently, most deployments consider having connectors at both ends of the drop cable, and potentially one at the distribution area cabinet. The drop connectors are mainly there for deployment operations reasons: they allow the drop to be installed by nonskilled craft at any time (i.e., upon a service activation). The cabinet connectors are there to concentrate active subscribers on a smaller number of splitters than would be possible with a spliced configuration.

An example of a deployment area is shown in Figure 7.12. This is a suburban area in the US, with a rather typical layout of homes on cul-de-sac streets [4]. The distribution plant is aerial, and each home is connected to the nearest pole via a drop cable. This results in groups of about three to four homes linked to a drop point. In this particular design, connectorized drops are used. The deployment anticipated less than 100 % service take rate, and so all the drop points are served with a 1 : 3 splitter. The chance of all four homes on a pole taking service is low, and if it did happen, a drop could be re-routed from a neighboring pole. The

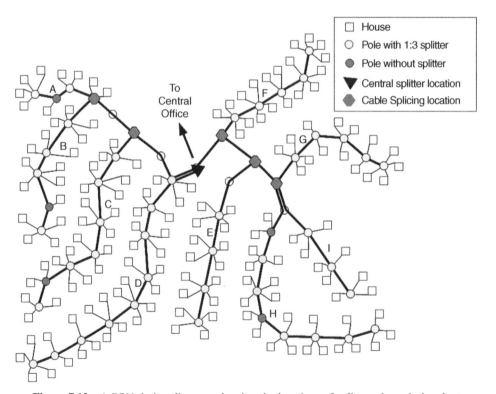

Figure 7.12   A PON design diagram, showing the locations of splitters through the plant.

1 : 3 splitters are enclosed in a weather-resistant housing that has the mechanical arrangements for accepting the drop cables, including connectors. The distribution side of the 1 : 3 splitters is fusion spliced onto the street cable fibers, which lead back towards the central splitter location. At certain points, two or three street cables are merged together to reduce the number of cables along each span of plant. The central splitter location is a street cabinet, and in this design, 1 : 10 splitters were used to multiplex the 1 : 3 splitter fibers into the final 1 : 30 split ratio PON.

7.6.2   ONU AND POWERING

The ONU and its associated power supply are an important part of practical PON engineering. It is important that the ONU has the correct balance of performance, efficiency, and cost to make it a compelling offer for the end customer. Some of the key requirements involved include:

- Size. The unit ideally would be the size of a conventional network interface device (NID), which are about $7 \times 15 \times 20$ cm. Current designs are approximately $10 \times 20 \times 30$ cm.
- Weight. Lighter is better, but a good ONU should not be so heavy that it puts a damaging load on the home's siding.
- Power consumption. The ONU gets its power from the customer's supply, so the power drain of the ONU must be modest, so that it adds a negligible amount of usage to the customer's electrical service.
- Ease of service. The ONU must be easily replaced in case of failure. The ONU can be mounted on a wall bracket with quick disconnects to facilitate this.
- Cabling access. Just like a NID, the ONU needs two separate enclosures for routing cables, one accessible to the customer, and the other to the craft person.
- Diagnostics. This generally involves LEDs that indicate the current status of the ONU.
- Safety features. The ONU is the termination of the network, and as such it has responsibility for protecting against electrical faults. The use of appropriate electrical fuse-like devices and a bonded ground conductor usually suffices.
- Environmental hardness. The ONU is generally mounted outside the home, and must pass a battery of environmental tests.

Figure 7.6 shows a picture of an ONU that meets all these requirements.

No matter how good the ONU is, it will be useless without a good power supply. Nearly every ONU uses a low DC voltage (48 V or less) primary supply voltage. This allows the use of low-voltage wiring, which is intrinsically safer and cheaper to use. A device is needed to covert the 120 VAC customer supply into the desired DC supply. This device is placed inside the home, to be close to the AC outlet and allow normal line-cord connections.

To provide protection against loss of AC power, some alternative power source must be placed at the home. The conventional answer to this problem is batteries for back-up. Most systems use lead-acid, but increasingly the trend is towards nickel metal-hydride and other newer types. Regardless of chemistry, battery life is greatly reduced by hot temperatures, and their effectiveness is reduced by cold, so they are placed inside the home if possible. Even so, they have a limited life (5 years is typical), so replacement is inevitable. The most cost effective way to accomplish this is to have the customer replace the battery, but this presents some human factors issues. The service provider must try to make the battery replacement as simple and convenient as possible. Means towards these ends include making the battery a 'slide in' module (rather than using screw-down terminals), and mailing the battery to the home (rather than using service craft to hand-carry them).

Alternatives to the powering/reliability problem are being considered. On the power generation side, alternative power sources such as fuel cells and flywheels are being studied. These sources are currently only suitable for larger emplacements (like a neighborhood-wide power supply), but they might be miniaturized. On the power consumption side, ONUs are being developed with special 'sleep modes' that reduce consumption during an outage. Finally, looking beyond the PON system itself, several providers have considered the idea of using their wireless telephone network as a kind of back-up in case of power failure. As cellphones increase in popularity, this may indeed become the dominant answer to 'catastrophic' power failures that last more than a few hours.

7.6.3   RF VIDEO ENGINEERING

From a technical perspective, the most difficult part of a B-PON deployment is engineering the RF video overlay. The analog video channels in the overlay are very sensitive to noise, and require a carrier to noise ratio (CNR) of 48 dBc. To produce this high CNR, the ONU must receive a high optical power (greater than -5 dBm, typically). In contrast, the digital side of the system can tolerate a signal to noise ratio of 16 dB, and requires only -28 dBm of optical power.

The high requirement on optical power causes two issues to become important. The first is receiver overload. Most receivers can tolerate no more than $2 \sim 3$ dBm of input power before they saturate. This puts the dynamic range of the video receiver at $7 \sim 8$ dB. The typical PON design considers the worst case losses of each element, and the sum of all these is devised so that the receiver will get its minimum required power at the worst case loss.

However, when a real PON is constructed, the optical power observed will be higher than the worst case design as a result of several factors, including component variability, design approximations, and design margin. An analysis of these factors was given in Reference [4], which is summarized in Table 7.1. This table considers the usage of four different design methods and two grades of components, and estimates the total power variation on the ONU for each of the eight combinations. The results show that there are some feasible design methods available, but that care must be taken to avoid receiver overload.

Table 7.1 Optical power variations in B-PON video overlays.

Design method	Optics grade	Component (dB)	Design (dB)	Margin (dB)	Total (dB)
Cookie Cutter	Normal	3.6	5	1.4	10
	Premium	2.5	5	1.5	9.0
Splitter-Distance	Normal	5.6	2.65–3.85	1.05–1.25	9.5–10.5
	Premium	4.5	2.65–3.85	1.15–1.35	8.5–9.5
Measure-Splitter	Normal	3.6	2.9–4.1	1.4–1.6	8.0–9.0
	Premium	2.5	2.9–4.1	1.5–1.7	7.0–8.0
Measure-Splitter-Pad	Normal	3.6	0.5	1.4	5.5
	Premium	2.5	0.5	1.6	4.5

The second issue that analog video raises is nonlinear effects in the PON fiber. To reach the ONU at −5 dBm power, the launch power must be around 20 dBm. This high power normally results in stimulated Brillouin scattering (SBS), which will prevent link operation. However, SBS can be suppressed by broadening the analog laser line-width. Conventional SBS suppression allows launch powers of up to 17 dBm, and extra suppression can allow launch powers of well over 20 dBm. So, the SBS problem is well controlled by the use of the appropriate transmitter.

More troublesome is the Raman gain related coupling of the data and video wavelengths. The 1490 nm data wave and the 1550 nm video wave are copropagating, and the 60 nm wavelength separation sees a significant Raman gain cross-section in the silica glass fiber. The data wave acts as a pump for the video wave, but because the data wave is modulated with a nonreturn to zero (NRZ) signal, the gain is a randomly varying white noise process. This impresses noise on the video carriers. Because of fiber's chromatic dispersion, the noise exhibits a low-pass characteristic that is dependent on fiber length and other properties. As studied in Reference [5], the worst case length is about 8 km of fiber, and the Raman effect impacts the low channels (2–6 in North America) the most.

There are a few methods to combat the Raman effect. The first is to reduce the digital power as much as possible. The G.983.3 recommended OLT transmitter power is too high, and in most cases is unnecessary for successful digital transmission. So, practical equipment tends to lower and more tightly control the OLT output. The second method is to pre-emphasize the lower channels' optical modulation index. This raises the impacted channel's power above the Raman noise floor, with only a minor impact on video quality.

7.7 SUMMARY

The concept and goal of fiber-to-the-home (FTTH) has had a long and varied history stretching back to the invention of practical optical fibers. While there have been many false starts and reversals, the industry has converged on PONs as the favored architecture for residential fiber services. B-PON is the first standardized PON system to reach mass deployment levels and a level of interoperability that enables plug-and-play flexibility. The B-PON system has evolved over time, and will continue to change to follow the market for advanced broadband residential services. The economics of fiber to the premises has also made progress, making it not only technically achievable but commercially viable. The mass deployment of B-PON has taught many valuable lessons on the design of the facilities and equipment to make them reliable, easy to install, and operate. Lastly, while B-PON is likely not the final step in residential fiber service, it sets the foundation for its successor systems, and will remain relevant for many years to come.

REFERENCES

1. Broadband optical access systems based on Passive Optical Networks (PON), ITU-T Rec. G.983.1, International Telecommunications Union, 1998.
2. Effenberger F J, Ichibangase H, Yamashita H. Advances in broadband passive optical networking technologies. *IEEE Communications Magazine* 2001; **39**(12): 118–124.
3. Concept drawing provided courtesy Motorola, Inc., 2005.
4. Effenberger F J, McCammon K, Cleary D. Analog Video and PON Optical Loss Variations, NFOEC'03.
5. Aviles M, Litvin K, Wang J, Colella B, Effenberger F J, Tian F. Raman crosstalk in video overlay passive optical networks, OFC'04, February 2004.

8

Optical Networks
for the Broadband Future

David Payne, Russell Davey, David Faulkner and Steve Hornung
BT, Adastral Park, Martlesham Heath, Ipswich, UK

8.1 INTRODUCTION

At the turn of the new millennium the broadband revolution was just getting underway. At that time mass deployment of broadband was in its early stages. In Korea and Japan there were high levels of take-up, while in the US and Europe penetration and growth was more modest. In the UK we were at the beginning of the broadband revolution with very modest take-up levels. Also at this time much of the bandwidth usage, certainly in the UK, was dominated by a small proportion of very heavy users using the network for mainly peer-to-peer applications (particularly music file transfers).

Since then there has been massive growth in take-up of broadband with little sign of any slow down in demand. There has also been a clear trend towards higher bandwidths, which is driving demand for higher speed access technologies. Both Japan and Korea are adopting fibre to the home (FTTH) strategies to deliver 100 Mb/s or more capability, while in the US some major FTTH deployments are planned over the next few years. Although Europe has not yet embraced fibre as the access technology of choice, serious consideration is now being given to fibre access solutions with small-scale deployments and trials taking place.

The relentless progress in consumer technology has also continued unabated with processor speeds continuing to increase roughly in line with Moore's law. At the same time very high capacity low cost storage has become readily available to match this increase in processing power. Video and imaging services via mobile phones has also seen remarkable growth, showing that there is real latent demand for new innovative services and some of these really will require higher bandwidth connectivity as expectations rise and frustrations with delay and low quality limit usage. Image and video file transfer, richer content on the web, increasing popularity of online gaming (not just in Korea and Japan) and the need to back up large amounts of personal data will all drive the demand for increasing access speeds and network capacity.

Broadband Optical Access Networks and Fiber-to-the-Home: Systems Technologies and Deployment Strategies
Chinlon Lin © 2006 John Wiley & Sons, Ltd

As operators and service providers move forward trying to meet these ever-increasing demands for high-capacity services, a major obstacle to sustainable profitability arises. The issue is that the cost of installing the additional network capacity to meet the predicted growth in demand for bandwidth can easily exceed the subsequent growth in revenues. Bandwidth growth exceeding revenue growth is not a new phenomenon but has previously been balanced by the normal price decline of equipment that occurs in any industry as product volumes increase. The problem for the telecommunications industry with the broadband future, particularly when moving beyond today's DSL speeds and contention ratios, is that many of the projected possible service and growth scenarios produce such large bandwidth growths that traditional price declines will not be sufficient to keep cost growth in line with growth in revenues.

This chapter will briefly review the history of optical access, in particular the Passive Optical Network (PON) architecture (still the most promising candidate for the mass-market fibre to the home (FTTH) solution). It will examine how it might evolve in the future [1] and maybe finally become the access network technology of choice for all service providers by tackling the problem of bandwidth growth costs outstripping growth in revenues.

8.2   BRIEF HISTORY OF FIBRE IN ACCESS

Fibre to the home and business was a consideration from the earliest days of optical fibre technology development. In the late 1970s point-to-point replacement of copper by fibre was being considered as a way of delivering broadband (mainly video) services to customers. These early systems were predicated on multimode fibre technology, the only viable solution at that time. More sophisticated versions of these early systems were also studied including remote electronic multiplexers sited at the street cabinet location with fibre feeders from the exchange and point to point optical links from the cabinets to the customers. This was not so different from today's fibre to the cabinet proposals using VDSL technology which uses advanced modulation and coding techniques to exploit the legacy copper from the cabinet to the customer rather than installing new fibre all the way from the remote nodes to the customer site.

The first consideration of a PON approach for the access network was around 1982 when single mode fibre technology was being seen as a new way forward for optical communications. Single mode fibre offered many advantages compared to multimode fibre, much greater bandwidth being one of the more obvious. However, probably of equal importance for the future evolution of optical networking was the ability to make high-performance optical components. When working in a multimode environment it was very difficult to ensure that optical components would perform equally well for all the possible propagation modes and also it was very difficult to excite all these modes equally at all times. The upshot of this was that multimode components were difficult to manufacture, had variable performance and produced noise (modal noise) due to the unstable mode excitation in the incoming fibres. One area where multimode fibre did have a significant advantage over single mode was in the area of splicing and connectors. At this time it was much easier to get low-loss connectors and splices with multimode fibre using simple tools, and for this reason the data communications world continued with multimode fibre where distances are relatively short and bandwidths were moderate.

One component that became available fairly quickly with the arrival of single mode technology was the fused fibre directional coupler. This is a device whereby two fibres are brought into close proximity, usually by twisting together, heated close to melting point and then pulled such that the fibres with their guiding regions (the fibre cores) shrink in diameter. The effect of reducing the diameter of the fibre core is to reduce the confinement properties of the fibre and, contrary to simple intuition, the shrinking core causes the light in the core to spread out (the mode field expands). As it expands, it overlaps with the core of the adjacent fibre and coupling occurs. By controlling this coupling and the length of the interacting region any proportion of light in one fibre can be coupled to the other. If this coupling is stopped at 50 % we have an optical splitter. These splitters or couplers can be cascaded and any size of splitter or star coupler can be made. These optical splitters were the key component behind the PON concept. The same coupling principles can be applied to planar optical waveguide technologies and today both fused fibre and planar waveguide optical power splitters are readily available.

In the first half of the 1980s, the PON concept was centred on wavelength-switched networks. These used star couplers to interconnect network terminations and wavelength selection to route paths [2] across the network. At the same time ideas of using the couplers as simple passive splitters for broadcasting television signals was also being considered [3].

All this early thinking was technology led; it was thought that these networks would be used to deliver broadband services, with video being the obvious application. Little thought was given to revenues streams or the business environment. It was thought that if the bandwidth capability was delivered services and revenues would follow.

In the mid 1980s the operational arms of BT rather than just the research department, became interested in the possibilities offered by optical fibre access and this led to a refocusing of the PON approach. The operational units brought a much needed business focus to the research and challenged the research teams to develop a system that could be economic for telephony. This was a service with a known revenue structure as opposed to the unknown revenues from future broadband services. This approach became known as a 'telephony entry strategy' and led to the invention and development of the 'TPON' (Telephony over Passive Optical Network) system [4,5]. Several manufacturers became heavily involved in the development including Fujitsu, Alcatel, and Nortel. In the end BT deployed a Fujitsu system to several tens of thousands of customers.

TPON was TDM based and the early system had a limited bandwidth of 20 Mb/s – limited by the capability of low-cost consumer CMOS technology at that time – which was adequate for telephony and ISDN but not for broadband. The idea was that broadband would be added later, as an upgrade, by the addition of extra wavelengths. To facilitate this an optical blocking filter was added to the TPON ONUs, which only passed the original TPON wavelength and blocked all others, enabling additional wavelengths to be added to the PON at a later stage without disturbing the original telephony-only customers. However, because the system was never rolled out on any significant scale the upgrade system (called BPON at the time) was never developed into a commercial product.

In the latter part of the 1980s BT was developing an ATM version of Passive Optical Networks called APON. Also around this time, optical amplifiers were emerging as a viable network component. At the end of the 1980s and in the early 1990s, several experiments were performed at BT Laboratories that demonstrated the real potential of the passive optical networking approach and culminated in the publication of initially a 32 million-way split network delivering 12 wavelengths at 2.5 Gb/s each [6]. This was further enhanced a few months later with the publication of a 44 million-way split with 16 wavelengths at 2.5 Gb/s and 500 km of optical fibre, showing that the whole population of the UK could, in principle, be serviced from one PON [7]. Although no one was suggesting a real implementation of such a network, it did illustrate that the TPON and APON architectures and indeed the current ITU-T specifications for PON systems were only scratching the surface of the potential of this technology.

During the early 90s BT continued with the design and development of a more practical amplified PON architecture, and it became dubbed as SuperPON. This examined the design and implementation options of a PON that could service a split of up to 3000 and have a geographical range up to 100 km. The capacity was 1.2 Gb/s downstream (from network towards the customer) and 300 Mb/s upstream. At the time these bit rates were very ambitious for optical access and were considered to be the limits for low-cost consumer equipment.

Interestingly this solution was the lowest cost solution found for fibre to the home/business and even today there are as yet no FTTHome/business solutions being offered with lower projected costs.

The European ACTS PLANET project continued with the SuperPON concept up until 1999 [8]. Beyond this point there have been no further developments occurring in the industry.

No decisions were taken to deploy FTTH systems to the mass market and by the mid 1990s, it was realised that without international cooperation and standards there was not going to be a sufficiently large common market to drive down the costs of the base components necessary to enable such systems to become economically viable.

Around this time BT, Deutsche Telekom and NTT decided, with other operators, to set up a consortium to develop and standardise PON requirements and systems. This forum became FSAN. For a fuller account of the beginnings of the FSAN initiative see Reference [9]. In recent years PON systems have continued to be developed, largely along FSAN guidelines and mainly in the small/start-up company arena. More recently Japan, Korea, and the US has revitalised interest in the supply industry and PON access solutions are once more becoming the access solution of choice as FTTH deployment progresses into the 21st century.

8.3   STANDARD PON SYSTEMS

Broadband Passive Optical Networks (B-PON) are now reaching maturity in ITU-T and many vendors are offering compliance with the G.983.x series of standards. Capacity ranges from 155/155 Mb/s symmetrical to 620/1240 Mb/s asymmetrical transmission. G-PON (G984.x series) extends this capacity to 2.4/2.4 Gb/s and allows more efficient transmission of packet protocols such as Ethernet. These systems can operate up to 20 km (66 kft), and 32-way split optical distribution networks although power budget limitations of standard optical components will restrict both of these parameters being achieved simultaneously. The wavelength plan recommended by the ITU makes provision for additional channels to be added for applications such as a broadcast video service by using a wavelength in the erbium-doped fibre amplifier (EDFA) gain window.

The optical distribution network (ODN) shown in Figure 8.1 typically uses standard single mode fibre with minimum dispersion at 1310 nm, which is the nominal wavelength of

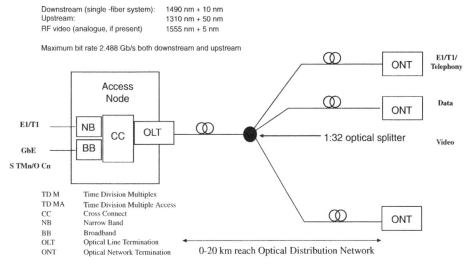

Figure 8.1   ITU-PON access system.

the upstream transmitter. Downstream transmission is at 1490 nm with an enhancement band around 1552 nm for services such as broadcast video transmitted as analogue RF signals. In a typical green-field deployment, the optical splitter may be a single monolithic waveguide device but can also be implemented as two or more (in principle) stages of split.

The ODN connects to an optical line termination (OLT) in the local exchange (Central Office, CO). Several of these can be provided on a single circuit board and plugged into a shelf. Several shelves can be included in an equipment rack, which also includes power supplies, protection switching, a set up control interface, and connections to the transmission equipment of the metro or core network. Some operators are asking for OLT cards, which can be plugged into a shelf with conventional ADSL (broadband) cards too. This will assist operators in making the transition from copper to fibre. At the customer end the ODN connects to an Optical Network Termination (ONT) unit which converts the light to electrical signals to connect to existing equipment such as a telephone, TV, PC, or home/office network hub.

8.3.1   WHERE IS INVESTMENT IN PON TAKING PLACE?

The largest deployments of B-PON have been made in Japan where competition is driving a speed war, and it is understood that company tax is reduced if an operator, telco or cable co, chooses to deploy fibre infrastructure.

In the USA, competition from multi-service (cable TV) operators with hybrid fibre coaxial (HFC) cable networks has been growing and this has led incumbent telcos to respond by upgrading their networks to enable triple-play bundled services. Fibre to the node (e.g. VDSL2) multi-pair ADSL2+ and fibre are the three technologies being considered to enable competitive high-definition (HD) and standard-definition TV services in addition to voice and data. Verizon has led the way in a push for fibre in a triple-play PON network. In their case investment has partly been justified by expected savings in operational costs, and the upgrade and integration of the management systems of two former Baby Bells.

8.3.2   THE ROLE OF THE ITU IN STANDARDIZATION

The International Telecommunication Union headquartered in Geneva, Switzerland, is an international organisation within the United Nations System where governments and the private sector coordinate global telecom networks and services. ITU-T Recommendations, such as G.982 (PON), G.983.x (Broadband-PON) and G.984.x (Gigabit-PON) are agreed by consensus and provide a framework for implementation. The procedure leading to standardisation includes contributing to an appropriate Study Group and Question, consensus building, agreeing final text, gaining consent from the Study Group chairman at a Plenary SG meeting, and a final public approval procedure. From start to finish a standard might take around a year to progress provided that there is sufficient buy-in and no objections.

From an operator's perspective, cost reduction is the key motivator for standards [10]. Interoperability and second sourcing are also important for derisking investment. From a vendor's perspective, it is the assurance that their products will satisfy the needs of a wide market.

8.3.3   HOW DOES FSAN RELATE TO STANDARDS?

As mentioned above FSAN was set up in the mid 1990s as collaboration between operators and vendors to bring about cost reduction of access networks and kick-start investment in a ubiquitous full services access network.

The Management Committee, which meets approximately yearly, gives overall steer and also reports on deployment.

The Optical Access Networks (OAN) Group meets approximately quarterly and focuses on systems, which will meet the service requirements of operators. Currently, there are 20 operators and 26 vendors in this group. Documents are agreed by consensus at meetings and by e-mail, and are used as input documents to the most appropriate standards body. This has most often been ITU-T/SG15/Q.2; however, there has also been strong input by members to the IEEE802.3 EFM group to gain agreement on key optical parameters such as wavelength, reach, transmission rate and split ratio. The OAN group includes three task groups. The Interoperability Task Group promotes standards conformance and interoperability among vendors. The Common Technical Specifications (CTS) Task Group brings together a (minimum) set of common requirements and standards into a single document to be used by operators when purchasing optical access systems. It is used as part of the documentation needed in a tendering exercise [10]. The third group is the Next Generation Access Task Group and is investigating future generations of optical access to give improvements such as increased capacity, reach and lower cost.

More information on FSAN can be found on the website http://www.fsanweb.org/cts.asp. The two major standardized systems BPON (G983.x) and GPON (G984.x) are described in more detail in Appendix 1.

8.3.4   INTEROPERABILITY

Interoperability is a key requirement for operators requiring independent supply of local exchange (Central Office) and customer-end equipment. Recommendations on the general characteristics (G.983.1), and the management and control interface (G.982.3), have been revised to take account of advances made since the original B-PON recommendation was

Table 8.1 B-PON interoperability events.

Where	When	Host	Functionally
Makuhari, Japan	March 9–11, 2004	NTT/FSAN meeting	Transmission Convergence (TC) layer with Ethernet
Geneva, Switzerland	June 2–4, 2004	ITU 'All Star Workshop'	TC Layer with Ethernet
San Ramon, CA, USA	September 28, 2004	SBC/FSAN meeting	TC Layer with Ethernet Voice and fax services via GR-303
Chicago, USA	June 7–9, 2005	TIA/ITU, SUPERCOMM	TC Layer with Ethernet Voice service via GR-303 H-D IPTV and optical RF Video

published in 1999. Interoperability for voice services has been demonstrated using the GR-303 interface which ensures transmission of TDM voice channels, for example, at the 1.5 Mb/s T1 (DS1) rate 24 channels are combined with signalling control and meet a low delay requirement. Demonstrations have been made possible by work done in both ITU and by the Full Services Access Networks (FSAN) initiative. A number of interoperability events have been staged, to assist vendors in achieving conformance with ITU-T Recommendations and interoperability with other vendors' products. Details are given in the Table 8.1.

The FSAN Interoperability Task Group is currently investigating the set-up of a PON test lab to validate both conformance (with ITU-T Recommendations) and interoperability (e.g. between an OLT from vendor A and ONT from vendor B). Currently, NTT-AT is the only source of a conformance tester, which enables the transmission convergence layer of a B-PON OLT or ONU to be checked for conformance with G.983.1.

At the June 2005 SUPERCOMM conference, a range of PON equipment was demonstrated according to the ITU-T G.983 and G.984 series of Recommendations. B-PON interoperability was demonstrated with equipment from six vendors. Three vendors exhibited Extended B-PON and two exhibited G-PON products. Two vendors showed optical distribution networks as recommended in G.982. These comprised 32-way splitters with 20 km (66 kft) of fibre. Each exhibitor was provided with fibre feed from the 'Central Office' area to his ONT at his kiosk, a set-top box and a High-Definition TV (HDTV) monitor (see Figure 8.2). Three vendors provided the high-definition and standard-definition video

Figure 8.2 Layout of the ITU-PON showcase at SUPERCOMM.

head-end equipment and set-top boxes for RF overlay or base-band IP transmission. During the event there were demonstrations of B-PON interoperability and conformance testing.

8.3.5 OUTLOOK

A key question facing investors in fibre is how the next 'step-up' in capacity will be achieved. Upgrades can be made either in-service, for example with the addition of wavelengths, or out-of-service, for example by replacement of OLT and ONU. PONs, unlike P2P technologies, do not lend themselves to out-of-service upgrades as many customers can be affected simultaneously. This leaves WDM or additional fibre as the two likely candidates. WDM suffers from the drawback of needing WDM filters at the outset to block the new wavelengths. These filters incur additional cost and at 'day one' generate no additional revenue. Additional spare fibre also incurs additional upfront cost, although often there will be spare fibre in the installed cables due to the step fibre count changes that occur as cable sizes increase and invariably operators will install a cable size the next incremental size above the initial planned demand.

Using spare fibre or wavelengths as upgrade strategies enables new generation of PON systems to be added in the future with greater capacities and capabilities than the original BPON or GPON systems.

8.3.6 STANDARDS SUMMARY

The B-PON and G-PON series of standards are largely complete in ITU-T. B-PON has reached maturity with many vendors demonstrating a degree of interoperability between OLT and ONU.

ITU-T Recommendations in the G.984.x series detail G-PON, the latest generation of standardised PON technology. G-PON maintains the same optical distribution network, wavelength plan and full-service network design principles of G.983.x series (B-PON) and, as well as allowing for increased network capacity, the GPON standard offers more efficient IP and Ethernet handling.

Different operators will have different market opportunities; they will want to focus deployment to match predicted service growth against their own local market conditions. The availability of a range of standardised systems will allow them to select those which best allow the required service sets to be transported that meet these local opportunities.

8.4 EMERGING DRIVERS FOR FTTH

What requirements are there for higher speed broadband services that could drive the need for fibre much closer to the customer, and what has changed in the last few years that is stimulating a resurgence of interest in fibre access?

8.4.1 THE INTERNET

It is generally agreed that the most significant phenomenon to impact the broadband debate is the emergence and popularity of the Internet, and in particular the World Wide Web. In the early days of service modelling, long lists of proposed 'services' would be generated with titles like 'home banking', 'home shopping', 'games', 'e-mail', 'VOD' etc. Many of these are now just applications and services available over the Internet.

Clever encoding and website design enables many of these services and applications to be available even with a 56 Kb/s modem connection while the 'broadband' connection ~256 Kb/s markedly increases the responsiveness and attractiveness of the experience. Apart from high-quality video, most of these services can be provided with an access speed of only a few hundred Kilobites per second. However, as the content becomes image and video rich and as the quality of that content increases in terms of resolution, frame rate or reduced latency (important for gaming and conversational services), the speed of the connection, for a satisfactory customer experience, will need to increase.

8.4.2 DIGITAL IMAGING AND VIDEO

Another area to consider is the rapid developments in consumer equipment, one example of which is digital video. Most digital video cameras use the DV format and now have a 'fire-wire' (IEEE1394) or USB 2 interface that will deliver a high-quality digital video signal at about 30 Mb/s. High-Definition Video (HDV) is also emerging and has been designed to use similar data rates by using real-time MPEG2 encoding rather than the cosine transform coding used by the DV standard. Currently, the only place that signal can be transmitted is to other CPE a few metres away. Multiple USB and IEEE1394 interfaces are now standard features on PCs and are capable of speeds up to 400 Mb/s (although admittedly the internal architecture and limitations of the PC, and its operating system will often limit this speed; however, file transfers at >100 Mb/s are readily achieved).

Digital photography is now rapidly displacing film photography and the ability to distribute these images via e-mail and file transfer is one of the major advantages of this media format. Modern cameras, even at the low end of the market, generate multi-mega pixel images and single image raw file sizes can easily be several Megabytes. The mobile phone camera market is also now taking off and people want to transmit the images captured to colleagues, friends and family. These camera phones can actually capture high-quality, high-resolution images but the transmitted quality is very poor due to the very heavy compression applied to the images prior to transmission over current narrow band mobile networks. Linking these mobile phones to other network devices, for example the PC via Bluetooth, could enable much higher quality and even the large raw, uncompressed, image files to be transmitted over the broadband fixed network at full resolution.

8.4.3 DEVELOPMENTS IN DIGITAL CPE

The processor speed of PCs has continued to follow Moore's Law and 3 GHz plus machines are now commonplace. In the relatively near future this is likely to increase to ~10 GHz. The increasing power of these machines with increasing speed of the interfaces will enable large files to be moved between different items of CPE at very high speeds, much higher than the few hundred Kilobites per second capability of DSL technology to the external network. This mismatch between internal and external networking capability will continue to increase until high-capacity optical fibre communications is brought much closer to the customer.

With increasing PC power, the popularity and rapid development of very high resolution imaging devices (including still image cameras with Mega-pixel images), DV cameras becoming commonplace, and HDV emerging, users in the consumer market place as well as business will be generating huge amounts of data. Storage of these files within the PC is already driving hard drive capacities to the hundreds of Giga-bytes sizes and will soon be in the Tera-bytes regime.

A major problem for owners of such storage is secure back up of valuable data. Storage area networking is a potentially valuable emerging market for the business sector, currently focussed on large business. But with the huge expansion of low-cost mass storage for the mass market, a future service could be protected on-line storage for valuable data or data that customers will want to access wherever they are. Such a service will only be useful and practical if upload and download speeds are sufficiently fast that typical transactions can occur in minutes rather than hours or even days.

8.4.4 EXAMPLES OF TRANSMISSION OVER FIBRE V XDSL

It can be seen from the above that very large data files are not just the domain of large businesses. Consumer equipment can now generate huge data files.

An 1 h DV tape can contain over 16 Gb of data, single layer DVD can hold ∼5 Gb even the humble CD holds ∼0.6 Gb of data. Transferring such large files electronic over current networks takes an unacceptable length of time, such that the usual method of 'large' file transfer is via physical movement of media, that is, the Postal Service. The limitations of access network technologies to move such large data files between distant locations (greater than a few metres) are illustrated in Figure 8.3., which shows the time taken to transfer large files of various sizes using the various access technology options that are available, note the log scales.

For peer-to-peer transfer the upstream rate is the limitation for transmit time with ADSL and VDSL technologies, but even for downstream transmission, only FTTH provides realistic transfer times for services such as purchase and network downloads, of electronic

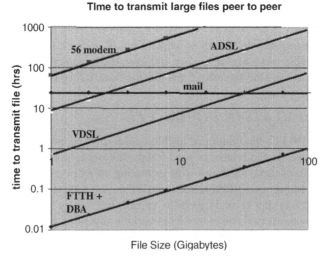

Figure 8.3 Time to transmit large files peer-to-peer.

media such as DVD files or even CDROM. The only practical option for say DVD purchase over the Internet or transferring a home movie at DV quality to friends and family is to use the postal service.

In the business community the need of tele-working is increasing steadily. Official estimates from a 2001 survey indicate that there were 2.2 million tele-workers in the UK. Tele-workers were defined as people who worked at least 1 day per week from home [11]. This definition excludes occasional tele-workers (people who have an office location but also work either irregularly or occasionally from home). If these were also included, the number of 'tele-workers' would increase significantly.

In many corporate companies employees have access to internal Intranets usually with 10 or 100 Mb/s Ethernet connections. In order to operate at the same service capability and efficiency as office-based workers, tele-workers ideally need a network connection speed comparable to that of the internal office LAN. Current, DSL-based broadband options go some way towards this but tele-working often needs broadband connections with much greater symmetry of transmission as well as higher data rates. VDSL technologies or FTTH would aid the tele-working environment enormously.

The conclusion from the above brief survey of the drivers for broadband is that DSL technologies can only be an interim solution and will not satisfy the demand generated by emerging consumer technology or future services. In Korea, the government is already suggesting that the future broadband convergence network (BcN) should operate at speeds up to 100 Mb/s to the end customer, $\sim$50 times faster than conventional broadband services.

8.4.5 FINANCIAL BARRIERS

The greatest barrier to large-scale penetration of fibre in the access network is of course the high investment costs required. Studies in the early 90s indicated a cost of $\sim$£15 billion to provide fibre to all the homes and businesses in the UK. Although some elements of the costs for fibre to the home/office (FTTH/O) have reduced over the intervening years, overall the costs, using current technology prices, have not changed significantly and the total cost will still be of the same order.

Early fibre systems targeted specific or niche markets. For example, many of the first generation of commercially available APON and BPON systems were aimed at data services for the medium- and small-enterprises market, and did not support POTS services. The problem with these solutions for mass deployment is that they cause a fragmentation of revenue streams with revenues from different service portfolios being required to support different platforms. For a mass-market solution this is not viable. All possible revenue streams will be needed to support the deployment and operation of a single platform. A FTTH/O solution needs to deliver all services: voice (including the regulated PSTN services), video and data to have any chance of being financially viable.

The other issue for broadband access is that historically, overall revenues from the customer base grow relatively slowly. Revenues for a particular service can grow quite fast, however, it is usually via revenue substitution from within the telecommunications business.

On a macro-economic scale this is perfectly reasonable – GDPs of countries, and indeed the world, only grow at a modest rate (typically $\sim$1–3 %). To have a revenue growth rate much greater than this over an extended period will mean significant underlying economic change, that is spending in telecoms will mean a down turn in some other sector of the economy. Although such changes do occur, they generally occur slowly. In the UK growth in telecoms spending has indeed exceeded growth in GDP for several years implying underlying changes in spending patterns. However, expecting significantly greater and sustainable changes in the spending ratios over extended periods of time is unrealistic.

The conclusion from the relative overall slow growth of revenues for telecommunications services, compared to the potentially very rapid increase in bandwidth demand from future broadband, is that the cost per unit of bandwidth must decline very rapidly, much more rapidly than traditional price declines of products in the electronics market.

To illustrate this, consider the three service scenarios outlined below for the UK telecoms market. These are potential growth scenarios, categorised as: Pragmatic Internet, Optimistic + Moderate Video and Very Optimistic & Video Centric. The main service and usage assumptions behind the scenarios are shown in Table 8.2. Note these are scenarios, not forecasts. The intent is merely to illustrate the impact different service growth and usage demand patterns could have on network capacity requirements, and what the possible network architectures and solutions could be to meet the demands economically.

The corresponding growth in network traffic for these scenarios is shown in Figure 8.4. Figure 8.4(a) can be considered a pragmatic Internet scenario. It only relies on ADSL and Cable Modem technology. Streamed video services are a minority product and most video entertainment is delivered through conventional systems, for example terrestrial, satellite or cable. Internet growth is assumed to be fairly robust but not particularly ambitious. Customers are assumed to be split between Cable Modem service operators and ADSL service operators.

Table 8.2 Assumptions for example traffic scenarios.

	Conservative internet		Optimistic + Moderate video		Video centric	
	2009/10	2014/15	2009/10	2014/15	2009/10	2014/15
Total broadband Internet customers (millions) (Includes cable Modem)	9.4	12.3	10.7	14.8	10.9	16.2
Number of VDSL/fibre customer (millions)	0	0	1.65	2.9	2.1	7.7
Video/VoD customers (millions)	0.14	0.54	1.5	3.5	2.0	8.7
Average internet session time/day (mins)	75	80	96	107	101	126
Average internet session bandwidth (kb/s)	77	114	320	1270	488	4780
Average video session time/day (mins)	24	24	58	70	94	128
Average video session bandwidth (kb/s)	2000	2000	7000	7500	7280	7400

Figure 8.4(b) is a more optimistic scenario but without stretching the bounds of plausibility for the types and usage of the services postulated. The main difference is the use of the network to deliver high-quality video entertainment services and fast file transfers for business and e-commerce. Even in this scenario video usage is only equivalent to about one feature film per day implying most entertainment video would still be delivered by

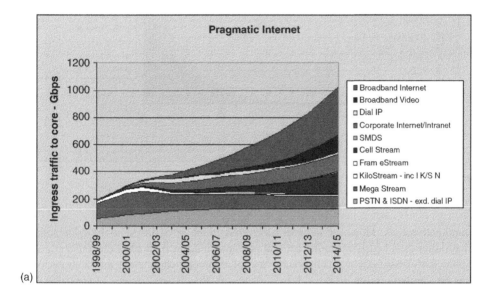

(a)

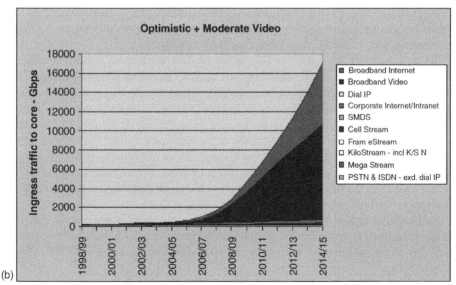

(b)

Figure 8.4 (a) Projected traffic growth for pragmatic Internet scenario; (b) Projected traffic growth optimistic and moderate video scenario; (c) Projected traffic growth for very optimistic and video-centric scenario.

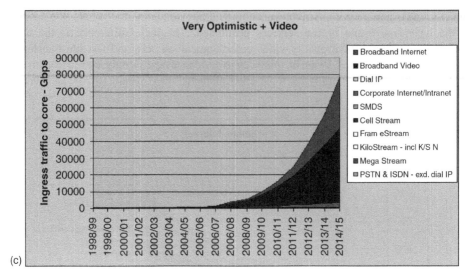

(c)

Figure 8.4 (*Continued*)

conventional means. However network delivered video is assumed to be a personalised service rather than broadcast/multicast and is transmitted to individual customers.

Figure 8.4(c) assumes a much greater use of content rich Internet applications, which pushes up average session bit rates. It implies that ADSL, where capable, will enable customers to operate up to ~2 Mb/s and that some VDSL and/or fibre solutions are also rolled out. The other major change from the 'pragmatic' scenario is that a large proportion of customers use their broadband connection for personalised video applications, although it is not the main entertainment video delivery system used by customers.

It can be seen from the results in Figure 8.4 that there could be big differences in the bandwidth demands that a future network may need to meet. An even bigger issue is that in the high growth scenarios, revenue growth will not match the cost of servicing the bandwidth growth. This will particularly be the case for the optimistic scenarios, unless new ways can be found of significantly increasing the price decline of the cost of bandwidth.

Ideally the industry would like to maintain the return on capital expenditure (ROCE) during a sustained growth period otherwise there is a risk of profitability suffering or even companies going out of business. The interesting question to ask of these scenarios described above is: what price decline or learning curve for bandwidth must be achieved, to enable the growths shown, whilst maintaining ROCE?

It can be shown that if all growths can be expressed as a compound annual growth rate (CAGR), then there is a simple relationship linking revenue growth, bandwidth growth and the learning curve for the price decline per unit of bandwidth to maintain ROCE, given by:

$$1 + G_R \geq (1 + G_B)^{1+L}$$

Where G_R is the CAGR for Revenues, G_B is the CAGR for Bandwidth, and $L = $ Log (L %/100)/Log(2)

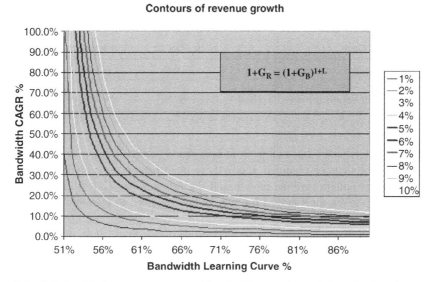

Figure 8.5 Relationship between revenue and bandwidth growths, and the price learning curve for unit bandwidth.

Where $L\%$ is the learning curve (expressed traditionally as a percentage).

A learning curve is defined as the percentage decline in price of a product as the product volume doubles, an 80 % learning curve will mean that the price of a product at a volume V will decline to 80 % of that price at volume 2xV. In this case the product volume is taken as bandwidth.

This is the macro-economic condition that needs to be met for a sustainable business as bandwidth grows in the broadband future.

This relationship is plotted in Figure 8.5 with contours of constant revenue growth. Typically, it is reasonable to expect time-averaged revenue growths in the range 4–7 % CAGR. Electronic systems have learning curves of ∼80 % (historically transmission systems have been following such a learning curve for the past decade at least), and in Figure 8.4, this price decline and revenue growth would suggest that overall network bandwidth growths of ∼10 % can be sustained. Before the advent of broadband, growths of this order or less were typical, and the whole economic system was internally consistent between price declines, revenue growth and bandwidth growth. Note this is a macro-economic argument applied to the total system size; it does not apply to individual products and services or sub-networks within the main network. These could grow at much faster rates by substitution of other products or markets, or they could be small niches and simply cause too small a perturbation to be noticeable in the bigger picture. In the scenarios illustrated above, we are looking at large scale changes to the capacity of the network, and though very simple, this model gives a useful guide to the bounds on growths that can be expected to be viable, given a set of revenue constraints.

The envelopes of bandwidth growth for the scenarios illustrated in Figure 8.4 do not exhibit constant CAGRs due to the many individual parameters and growth and service usage functions that influence the growth curves. However, if an approximation is fitted to

Table 8.3 Learning curves to meet given network bandwidth growths.

Scenario	Bandwidth CAGR fit	Bandwidth learining curve
Pragmatic internet	10 %	∼72 %
Optimistic and moderate video	50 %	∼54 %
Very optimistic and video centric	80 %	∼52 %

the regions that start where bandwidth from broadband begins to become significant, then approximate CAGRs can be obtained. Using these fitted bandwidth growth rates, assuming a nominal 5 % per annum revenue growth and the relationships shown in Figure 8.4, the required learning curves for bandwidth price decline required to maintain margins can be estimated; these are shown in Table 8.3.

The 'Pragmatic Internet' scenario is on the bounds of economic viability with electronic centric systems and network architectures not very different from today. So, the current strategies followed by operators deploying ADSL or Cable Modem for predominantly Internet surfing, e-mail, etc. with little video services, should be viable and will probably maintain ROCE, particularly if some additional revenue growth can be obtained from the richer service sets that can be offered. In addition, operators can control bandwidth growth to some extent by use of contention for backhaul and core bandwidth. From a user perspective this will appear as increased delay for upload and download as the network gets congested in busy periods, effectively reducing the average session bandwidth per user.

However, the case is very different as we go beyond this scenario to the higher bandwidth scenarios that assume greater than a few hundred kilobits per second session rates and greater use of personalised video services. It can be seen that the price declines required to meet the bandwidth growths in these scenarios, are much faster than the traditional 80 % learning curves typically arising in the electronics industry. In the case of the optimistic and video centric scenarios the required price declines are so fast that it can be safely claimed they will not be met. From this perspective, future network architectures that continue to use traditional electronic solutions will not be able to price decline sufficiently fast to be able to maintain operating margins and profitability.

8.5   LOWER COST ARCHITECTURES

The implications of this simple economic relationship are quite profound for the telecommunications industry. It implies that either growth of future bandwidth will have to slow to match that dictated by the price decline achievable in conventional electronic centric networks, or we need to develop and implement new alternative architectures that can enable much faster decline in the price of bandwidth. The interesting question addressed in this chapter is, can architectures that exploit a much greater use of optical networking technologies produce significantly faster price declines than conventional electronic centric networks?

It is not very likely that optical networking technology will price decline any faster than electronic technologies. Many of the manufacturing methodologies are similar and indeed are developed from the electronics world and are therefore likely to follow similar price decline trends. The potential strength of optical technology comes not from lower cost components, but from its ability to displace electronic sub systems and nodes within the networks, offering much higher capacities with a much smaller equipment inventory, that is reducing box and port count while enabling very much higher network bandwidth.

All optical networking in the core, including true optical or photonic switching, has had a long gestation period but practical systems are now emerging. The main outstanding issue to be resolved is the gaining of confidence by operators in the ability to be able to manage and operate systems employing such 'transparent' optical core networks. Transparent optical networks that are photonic end-to-end accumulate analogue impairments along the routes. In a dynamically reconfigurable, optically transparent network this is a real problem for an operational network. Network operators require optical networks to be capable of being managed with similar capability and features as today's networks based on conventional OEO (optical-electronic-optical) architectures.

The promise offered by all optical switching is lower cost transport across the network core, achieved by removing the large number of transponders required to terminate each end of every wavelength channel of the conventional OEO architecture. However, this advantage can only be realised if the operational costs do not increase as a result of the added complexity of managing the 'analogue' nature of the transmission paths of the all-optical proposition.

8.5.1   BY-PASSING THE OUTER-CORE/METRO NETWORK

The approach being evaluated in this chapter is to extend the 'transparent' all optical architecture concept to the access and metro networks with the same aim of removing electronic nodes and port cards. Traditionally access and outer-core/metro networks are two separate worlds, and the usual approach is to design access solutions and metro solutions that are largely independent of each other. This is not too unreasonable because usually there is an electronic access node (the local exchange or remote concentrator unit) placed between the access network and its technologies and the outer-core/metro network. Also historically this 'access' node has been placed where it is because of the physical limits of the copper network, indeed this node position has been called the 'copper anchor'.

Once optical access is considered this limitation of the first node position is no longer necessary, and one interesting architectural option is to use optical access to reach deep into the network and terminate on a core edge node (a Metro Node in the BT 21CN architecture), which becomes the place where traffic grooming, marshalling, and concentration etc. is performed. An option proposed for a *long-reach access network* is based on the passive optical network (PON) principles but with optical amplification to boost the power budget. This enables large increases in bandwidth, geographical range and optical split compared to a conventional PON. It can be pedantically argued that this is no longer a PON because of the use of an 'active' optical amplifier within the PON. However, the optical amplifier is replacing the electronic access node and the original access proportion of the network remains passive, and the term PON will continue to be used in this paper (for the purist the P in PON could become Photonic).

Potentially the long reach enabled by the amplifiers would enable bypass of the usual backhaul network which would conventionally be SDH possibly combined with Metro WDM systems for enhanced capacity. The long reach would also allow the possible bypass

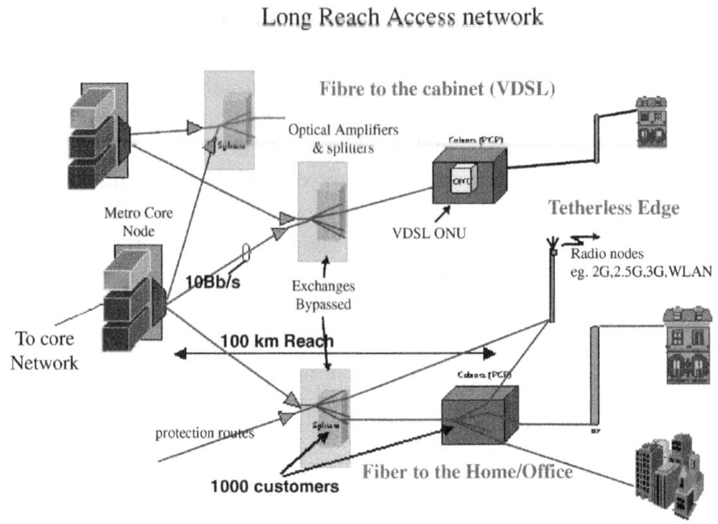

Long Reach Access network

Figure 8.6 Bypassing the outer core/metro with long-reach access PON.

and ultimate removal of the local exchange or remote concentrator site, and the increased capacity of the system coupled with the increased power budget enables a much greater optical split to be deployed. This allows more customers to be able to share the capacity and costs of the PON and the OLT, located in the 21CN Metro Node.

This combined access and backhaul network terminates on the 21CN Metro Node, which now provides the management and intelligence functions, and marshal and groom traffic onto a photonic inner core network. This photonic inner core interconnects these core edge nodes via simple wavelength channels. For the UK it is envisaged that there would be need to be about 100 of these 21CN Metro Nodes.

This architecture leads to a highly simplified network that has the potential for significantly reducing unit costs and enables scenarios with very high bandwidth growth. Such a future long-reach access network is illustrated in Figure 8.6. It operates at 2.5 or 10 Gb/s and could serve ~500 (possibly 1000) customer sites on each amplified PON [12]. The traffic from the long-reach PONs would be terminated onto the ~100 Metro Nodes, which are interconnected by an optical core network.

Although this network can now bypass the local exchange or concentrator, it would be many years before these node locations could be eliminated from the network. Legacy systems, particularly those still utilising the copper network, would still need to terminate on them. While these nodes exist they can provide a convenient place for locating and powering the optical amplifiers. However, the amplifiers and associated optical splitters are sufficiently small and low power consumption that street cabinets or even underground manholes/footway boxes could be used to house them. This could lead to the eventual elimination of those buildings.

This architecture would require a PON to be developed with higher split, longer reach and much higher capacity than the current generation of PON but if developed could offer many advantages and options for simplifying and cost reducing the end-to-end network.

8.5.2 ADVANTAGES OF LONG-REACH ACCESS

Analysis of the UK network suggests that an optical access network with 2.5–10 Gb/s, capacity, with up to 512 (or even 1024) way split and 100 km reach could be a very attractive option. The 100 km reach would be required because such a network requires protection paths and mechanisms to provide the outer core/metro network connectivity. Protection paths can be up to twice as long as the primary paths. The increased split would be required to minimise the cost per customer of the long reach, protected, portion of the network, and the increased capacity would be required to maintain the average bandwidth per customer, while serving many more customers. The high bandwidth would enable massively increased burst capability, exploiting the statistical multiplexing gain by providing a 'bandwidth reservoir' within the network. This would enhance the customer experience by providing a very responsive and high-speed network.

The deep-reach access network would effectively bypass the metro or outer-core collector network and could significantly reduce the cost of backhaul from the access node to the Metro Node. It can also provide a very flexible and highly functional traffic engineering capability. With the right functionality built into the transport protocol, bandwidth could be allocated to customers on demand, broadcast and multi-cast services could be offered with minimum bandwidth, and switching functionality required in the core. It could also act as a distributed concentrator and service switch, and simplify the switch and routing functions required in the Metro Node.

It would also be possible to incorporate sophisticated monitoring and diagnostic tools that could give early warnings of problems before they become traffic affecting and also provide terminal performance, identification and location information.

Security is always an issue with shared access media and needs to be addressed carefully to ensure adequate security from eavesdropping and malicious damage to the network integrity. There are many approaches that have been considered during the development of passive optical networks.

Other advantages that the long-reach access architecture can offer are:

All customer types and services supported: This includes residential customers, multi-dwelling units, small and medium enterprises (SMEs) and even large business sites. Existing larger business sites already served with point-to-point fibre systems would continue to use existing access fibre but could also be integrated into the long-reach architecture by exploiting the wavelength domain. Individual wavelength channels could be delivered to the largest customers and for the large customers not needing the capacity of dedicated wavelengths, sharing a wavelength over a number of business sites could be exploited.

Symmetrical capacity: Although asymmetrical capability could just as easily be offered, large-capacity PONs that serve all customer types will probably need to be symmetrical, the proportion of capacity allocated for broadcast services can be a relatively modest fraction of the total capacity and the bulk of the capacity may well need to be symmetrical for future service requirements.

Guaranteed QoS: The combined access and outer-core network that long-reach access architectures span could provide guaranteed QoS parameters by suitable choice of transport protocol and control system. Real time streamed services and PSTN quality voice (with all the associated delay constraints) together with common packet protocols such as Ethernet and IP could be supported.

Assuming high bandwidth demand and growth the high capacity of the system coupled with traffic management and dynamic bandwidth assignment produces the lowest cost per unit of bandwidth per customer.

Reduced capital expenditure: The major saving associated with this architecture is the large reduction in backhaul costs that become possible; the costs for the access portion remain the same if not slightly higher than conventional PONs because of the higher capacity of the optical network units. The saving in backhaul costs arise because the Metro WDM or SDH/SONET equipment is by passed with a simple node consisting of optical power splitters and a few optical amplifiers. This also means that these systems will also be viable in many more of the rural and low-density serving areas.

Reduced operational costs: The potential operational savings are also significant. The UK network could be reduced to approximately 100 nodes that will require staffing. In theory, the long-reach access network could by-pass the ~5500 local exchange nodes, replacing them with simple amplifier and splitter points that could be located in street cabinets or manholes/footway boxes. This could eliminate the cost of maintaining these local exchange nodes. If fibre was installed to the customer premises all service enhancements/changes could be implemented remote without the need for visits. This could be the ultimate in 'hands off access networks' and could mean that plant fault rates should be determined predominantly by third party dig-ups, producing significant reduction in plant maintenance costs.

Enhanced customer experience: The customer experience could also be dramatically enhanced by the very high burst speeds that could be offered. Essentially burst rates will be limited by the home/office network interface speeds. If these could be Gigabit Ethernet speeds in the future then customers would experience very low delays for most applications (click and it is there). Even for very large file transfers such as DVD downloads, the transfer times could be tens of seconds rather than hours.

8.6   AN END-TO-END VISION

Combining long-reach access with photonic core networks would lead to highly simplified networks that have potential for significantly reducing unit costs and enabling the very high bandwidth growths predicted by the more optimistic and even video centric broadband scenarios.

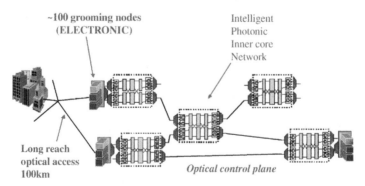

Figure 8.7   An end-to-end network vision using long-reach access with photonic core networks.

Such a future network is illustrated in Figure 8.7. The deep-reach access network serves a customer catchment area of up to 500 or 1000 sites. The traffic is terminated onto a multi-function node with functionality for IP, ATM and TDM traffic and services, if required. Note this does not imply that this functionality would be all in one box, although this would be one option, but that the functionality would be colocated at the same geographical node. Traffic from the long reach access networks terminating on the Metro Node would be marshalled into transport containers (e.g. this could be concatenated VC4-n containers over SDH or any other adequate transport protocol) and placed onto a wavelength channel destined for another Metro Nodes. The only circuit manipulation and routing that would take place in the core would be at wavelength granularity. This would avoid optical to electronic conversions followed by electronic processing and then electronic to optical conversion for onward transmission. This function would only be performed in the Metro Nodes. Note that in practice many of the optical nodes in the core would be geographically colocated with a Metro Node containing electronic switching and routing equipment, the optical switches providing through path routing.

This end-to-end architectural proposition could reduce a network the size of the UK to the order of 100 switching/routing nodes with $\sim$30 000 long-reach access PONs to serve the customer base. The long-reach PONs would have optical amplifying nodes rather than electronic equipment and these could be placed in manholes or street cabinets, and it is predicted they and would consume only 0.5 % of the electrical power of the equipment and buildings they would replace making this also a very energy efficient and environmentally friendly solution.

8.7   SUMMARY/CONCLUSIONS

A consequence of the collapse of the bubble in the telecommunications industry around the turn of the new millennium bought home the message that supplying bandwidth and network capacity will not *per se* lead to a corresponding increase in the growth of revenues. It is also true, however, that there is a latent demand for higher-bandwidth services and that the gap between the information processing capability of consumer equipment and the access capacity of telecommunications networks is growing apace.

To meet the challenge of these two opposing sides of the business environment – slow revenue growth but huge latent bandwidth demand – it is necessary to find network architectures that can radically reduce unit bandwidth costs. Optical networking may offer a solution. Certainly, it is clear from the analysis described in this paper that the price decline of network architectures based on today's electronic intensive solutions shows no historical trend to be able to reduce bandwidth costs fast enough.

In the core network, all optical networking may be the way forward, although there are still technical and management issues to be resolved. In the access network, migrating legacy copper networks to DSL architectures is only a step along the road to true broadband, and as service bandwidths continue to increase, the push to drive fibre closer to the customer will also increase. One way of impacting the cost of providing this greater bandwidth offered by a fibre-rich access environment is to integrate the outer-core or metro network with the access network and combining it with an all optical core. The long-reach optical access network using high-capacity amplified PON may be a solution. It can offer higher capacities, symmetrical access at much lower end-to-end unit bandwidth costs, than any electronic intensive solution.

This network architecture addresses the problem of the cost of bandwidth outstripping revenue growth by massively reducing the unit cost of bandwidth; at the same time by radically simplifying the network it also offers the prospect of significantly reducing network operational costs. Without some such architecture to produce much faster declines in the cost of bandwidth than conventional network solutions can produce, future bandwidth growth will necessarily be constrained, if networks are to be operated with reasonable returns on investment.

8.A1 APPENDIX 1

8.A1.1 B-PON GENERAL CHARACTERISTICS

The original 155/155 Mb/s symmetrical and 622/155 Mb/s asymmetrical broadband PON as defined in G.983.1 has been extended in capacity to include 622/622 Mb/s symmetrical and 1244/622 Mb/s asymmetrical variants via Amendments 1 and 2. These amendments arose to meet the need for additional capacity for businesses (Amend. 1), and the opportunity to use PON for VDSL backhaul (Amend. 2).

Dynamic bandwidth allocation G983.4 and G.983.7 give the B-PON a powerful conditional access mechanism which allows queues at the customer-ends of the PON to be served according to the priority assigned to the traffic flow ranging from circuit emulation through to best effort (using spare capacity). This mechanism also offers the benefit of 'concentration on the fly', with statistical gain for packet-based services [13]. Conditional access is likely to become increasingly important as users of IP begin to expect QoS-based services on congested networks.

The 'enhancement' band centred around 1552 nm was defined in G.983.3 principally to allow downstream video services to be delivered over the PON using a different multiplexing and modulation method from the cell-based PON [13]. Within this band, analogue of digital formats may be transmitted to conventional TVs or set-top decoders. In ITU-T/SG9/ Q.16, a PON-specific modulation scheme is defined which uses a wideband F.M. optical transmission technique pioneered by NTT.

Table 8.A1.1 illustrates how the B-PON standards can form part of a strategic business plan for deployment. Different operators will have different market opportunities and can

Table 8.A1.1 Business drivers for B-PON.

Business factors	B-PON Requirements	ITU-T Recommendation
Economical and reliable products	International standards compliances and multi-vendor interoperability	G.983.1 G.983.2
	Video signal overlay with 3-waves multiplexing	G.983.3 G.983.3amend1
Competition with CA TV providers	DBA (Dynamic Bandwidth Assignment) to improve transmission efficiency of upstream signals	G.983.4 G.983.7
Competition with CLEC's ADSL	Access line protection and survivability	G.983.5 G.983.6
Competition with Long Haul operators	Enhanced OMCI (ONT Management and Control Interface) for new services	G.983.2amemd2
	Adding 622Mb/s upstream	G.983.1amemd1
	Adding 1.2Gb/s downstream and security	G.983.1amemd2

focus deployment against predicted service growth and select standards, which best allow the service sets to be transported.

The G.983.2 Recommendation on 'ONT management and control interfaces' describes the control of additional services via the management interfaces. These include DSL, VLAN and WLAN. These edge technologies need the capacity of fibre for backhauling traffic and offer an additional level of distribution beyond that of the PON by aggregating traffic at the optical network termination (ONT).

Enhancements to B-PON standards?
November 2004
G. 983.1 Revised. 'Broadband Optical Access Systems Based On Passive Optical Networks (PON)'
This revision includes two previous Amendments, A Corrigendum and Implementers' guide.

G.983.2 Amendment 2, 'B-PON ONT Management and Control Interface (OMCI) support for Video Return Path',
B-PON wavelength overlay systems using G983.3 band-plan. This defines a method of translating the return path signals used in the video return path for interactive video services principally delivered through an RF channel on an extra wavelength. It facilitates the use of set-top boxes originally designed for cable networks.

May 2005
G.983.2 Revised 'B-PON ONT Management and Control Interface (OMCI)'.
All documents on OMCI have been merged into this revision, G.983.2 and G.983.6 through to G.983.10 plus the Amendments 1 and 2, and Implementers' guide. New functionality includes mechanized loop testing for telephony copper interfaces and 'last gasp' reporting. A reference to Ethernet MAC bridging was added. VoIP support will require a further Amendment possibly for early 2006.

G.983.3 Amendment 2, 'A broadband optical access system with increased service capability by wavelength allocation'
This establishes the industry best practice optical budgets for the B-PON system operating at 622 Mb/s downstream, 155 Mb/s upstream. Optical distribution networks having a maximum loss of 28 dB are supported with digital-only services. If an analogue video service is also provided the minimum loss is 27 dB at 1490 nm. Tighter specifications enable allow the downstream laser to be used for more accurate power measurements when commissioning the ODN.

G.983.1 Amendment 1 on Protocol Implementation Conformance Statements (PICS) for the OLT and ONT.
The Protocol Implementation Conformance Statement for OLT and ONU can provide evidence from vendors to purchasers that the devices conform to G.983.1 at the transmission convergence layer (e.g. synchronization, ranging and data transport are achievable). The NTT-AT conformance tester provides a reference device against which this assessment can be made. Further work is needed to verify performance at the optical layer.

8.A1.2   G-PON GENERAL CHARACTERISTICS

A Gb/s capable PON (G-PON) was conceived as a way of satisfying a Full Service Access Network's hunger for more capacity and allows more efficient transmission of services, including Ethernet, over the same optical distribution network as B-PON. Recent work on G-PON has focused on high-definition IPTV as an in-band service offering.

The first step towards standardisation was 'requirements capture'. Operators were asked to state their probable service requirements and network applications. These were input to the ITU-T and given consent in January 2003 as G.984.1 'General characteristics for Gigabit-capable Passive Optical Networks (G-PON)'. This Recommendation describes a flexible optical fibre access network capable of supporting the capacity requirements of business and residential services. It covers systems with nominal line rates of 1.2 and 2.4 Gb/s in the downstream direction and 155 Mb/s, 622 Mb/s, 1.2 Gb/s and 2.4 Gb/s in the upstream direction. Both symmetrical and asymmetrical (upstream/downstream) systems are described. User network interfaces include Ethernet (principally to support PSTN, IP traffic), ISDN (BRI), ISDN (PRI), T1, DS3, E1, E3 and ATM. A summary of some of the key service requirements is given in the Table A1.2 below. The logical reach of 60 km (37 miles) allows for possible longer reach requirements to be met by changing the optical path without changing the electronics.

Also consented at this time were the physical media-dependent requirements as G.984.2 'Gigabit-capable Passive Optical Networks (GPON): Physical Media Dependent (PMD) layer Specification'. Key physical layer specifications are summarized in Table 8.A1.3 below.

The ITU-T website- http://www.itu.int/newsroom/press_releases/2003/04.html contains a press release on these Recommendations, which includes the following text.

'These new standards build on the existing and widely adopted G.983 series Recommendations relating to Broadband PONs, by providing unprecedented network capacity. Increasing capacity to gigabit levels should more than satisfy foreseeable customer demands. G-PON maintains the same optical distribution network, wavelength plan and full-service network design principles of G.983. Besides allowing for increased network capacity, the new standard offers more efficient IP and Ethernet handling. B-PON and G-PON will allow service providers to deliver applications such as video-on-demand, streamed video, on-line games and voice over IP.

Table 8.A1.2   Service requirements for G-PON.

Items	Target descriptions
Service and QoS performances	Full Services (e.g. 10/100Base-T, voice, leased lines)
Bit rates	1.25 Gb/s symmetric and higher Asymmetric with 155 Mb/s and 622 Mb/s upstream
Physical reach	Max 20 Km and Max 10 Km
Logical reach	Max 60 Km (for ranging protocol)
Branches	Max 64 in physical layer Max 128 in TC layer
Wavelength allocation	Downstream: 1480–1500 nm (Video overlay is considered.) Upstream: 1260–1360 nm
ODN classes	Class A, B and C; same as B-PON requirements

Table 8.A1.3 Key physical layer specifications for G-PON.

Item	Specification
Bit rates	1.244 Gbit/s and 2.488 Gbit/s symmetric 155.52 Mbit/s and 622.04 Mbit/s only for upstream Error rate: Better than 1.0E-10
Dispersion and error correction	Up to 10 km: FP-LD without FEC
	UP to 20 km: DFB-LD or FP-LD with FEC
Optical device	LD*1 + PIN (APD*2 is available)
Upstream overhead	12 Bytes (1.244 Gbit/s), 24 Bytes (2.488 Gbit/s)

These new recommendations represent an evolutionary development of the basic PON-standard (G.983.1). They provide a very significant increase in speed whilst largely maintaining the basic, PON-based broadband optical access system requirements of G.983.1 to ensure maximum continuity with existing systems and optical fibre infrastructure.

The G-PON Transmission Convergence layer and ranging protocol for G-PON are described in G.984.3. The approach being taken is to consider TDM, ATM and Ethernet mappings into physical layer frames and to define a new flexible technique, which is named 'GEM: G-PON Encapsulation Method' with good efficiency for all three traffic types. This approach will allow a wide range of service delivery mechanisms with QoS capability using dynamic bandwidth assignment. OMCI documents relating to G-PON are also being drafted.

Enhancements to G-PON standards?
May 2005

G.984.3 Amendment 1 to G-PON Transmission Convergence Layer. Peak information rate and sustained information rate parameters are now included and are analogous to ATM for alternative cell lengths such as Ethernet packets. Multicast services may now be supported over GEM (e.g. IPTV). (GEM is the generic encapsulation mode use at in the transmission convergence layer.) The mandatory method uses a single Port-ID (identifier) for all streams, while the optional method uses multiple Port-IDs.

G.984.4 Amendment 1 'Gigabit-capable Passive Optical Networks (G-PON): ONT Management and Control Interface specification'. This proposes management features on G-PON in support of Ethernet and IPTV service such as the IEEE802.1p priority mapper, GEM traffic descriptor, and support of multicast connection.

REFERENCES

1. Payne DB, Davey RP. The future of fibre access systems? *BT Technology Journal* 2002; **20**(4): 104–114.
2. Payne DB, Stern JR. Wavelength switched passively coupled, single mode optical networks, ECOC, Venice, October 1985.
3. Stern JR. Optical wideband subscriber loops and local area networks in the UK, ICC, Amsterdam, May 1984.
4. Stern, et al. Passive optical networks for telephony applications and beyond. *TPON Electronic Letters* 1987; **23**(24): 1255–1257.
5. Stern JR, Hoppitt CE, Payne DB, Reeve MH, Oakley K. TPON – A Passive optical network For telephony ECOC 88 Brighton, UK, September 1988, proceedings, 203.

6. Hill AM, et al. 39.5 million-way WDM broadcast network employing two stages of erbium-doped fibre amplifiers. *Electronic Letters* 1990; **26**(22): 1882–1884.

7. Forrester DW et al. 39.81 Gb/s, 43.8 million-way WDM broadcast network with 527 km range. *Electronic Letters* 1991; **27**(22): 2051–2053.

8. ACTS PLANET project final report – ftp://ftp.cordis.lu/pub/infowin/docs/fr-050.pdf

9. Faulkner DW, et al. The full service access network initiative. *IEEE Communication Magazine* 1997, 58–68.

10. Ueda H, et al. Deployment status and common technical specifications for a B-PON System. *IEEE Communication Magzine* 2001; **39**(12):134.

11. http://www.statistics.gov.uk/articles/labour_market_trends/Teleworking_jun2002.pdf

12. Nesset D, et al. Demonstration of 100 km reach amplified PONs with upstream bitrates of 2.5 Gb/s and 10 Gb/s. ECOC 2004, Stockholm, September 2004, paper We 2.6.3

13. Effenberger F, et al. Advances in Passive Optical Networking Technologies. *IEEE Communication Magzine* 2001; **39**(12): 118.

9

An Evolutionary Fibre-to-the-Home Network and System Technologies: Migration From HFC to FTTH Networks

James O. Farmer

Wave7 Optics Inc., Atlanta, Georgia, USA

9.1 INTRODUCTION

Fiber-to-the-Home (FTTH) networks offer the highest quality, widest bandwidth, and lowest maintenance of any available technology for providing service to homes. They can be used for broadcast video (analog and digital), IP video (IPTV), data, and voice, all using but a single fiber to one or several homes. In order for a new technology such as FTTH to succeed, it must be amenable to installation alongside current-generation telecommunications systems, with minimal disruption to current operations. This is a good definition of compatibility. It must be easy for people to understand, and it must utilize similar interfaces on either end of the communications system. It must accomplish this while being not significantly more expensive to install, and must look much the same to both system operators and the end subscriber. This chapter describes an architecture for FTTH that is very compatible with cable TV systems and yet allows an operator who deploys it to be competitive with any next-generation network, while maintaining backwards compatibility with legacy hybrid fiber-coax (HFC – cable TV) networks.

9.2 ELEMENTS OF COMPATIBILITY

For a new system to be compatible with an old one, several conditions should be met, including

- Similar architectures that can be understood easily by those working in the field.
- Compatible with existing equipment at the headend (central office to a telephone-oriented individual).

- Compatibility with existing subscriber equipment.
- The cost to connect the first subscriber should not be excessive, compared with connecting to the legacy network.

This chapter will show how an FTTH system can be made compatible with HFC, or hybrid fiber-coax systems. Cable operators have installed HFC architectures since the late 80's, and that architecture itself was evolutionary from the tree-and-branch architectures of an earlier generation of cable television systems. We shall first deal with the physical compatibility of HFC and FTTH systems, and then we shall move on to the higher levels.

9.3   THE STATE OF HFC NETWORKS

Cable TV in North America is accepted today as part of the larger communications infrastructure. Cable systems in North America were mostly rebuilt from about 1995 to 2003, in order to add two-way services and more bandwidth in the downstream direction. There are exceptions, but the dominant architecture for North American cable systems today is downstream transmission from 54 to 750 MHz, though a few systems go to a maximum of 860 or 870 MHz.[1] Typically the spectrum from 54 to 550 MHz is filled with analog (NTSC) transmissions, with the spectrum above 550 MHz usually reserved for digital video, shared with data and perhaps voice downstream. About 78 analog channels are currently supplied, along with perhaps 150–200 digital programs, which can be packed about 5 to 12 to a 6 MHz channel, depending on program content, desired compression quality, and the type of modulation used.

These digital programs include an expanded broadcast tier comprising mostly newer networks that have formed about the same time that digital transmission became a commonplace (circa 2000). The programs also included expanded premiums sold as subscription, pay-per-view, and video-on-demand services. Pay-per-view is commonly presented with start times every 10–30 min, so that the subscriber can start watching it at the most convenient time. Traditional subscription channels such as HBO and Showtime have also been moved to the digital tier in most systems.

Video-on-demand (VOD) is a movie (or other program) service provided by a video file server at the headend, to one subscriber at a time. The subscriber has 'VCR-like' control over starting and stopping, rewinding, and fast-forwarding the program. VCR-like control implies that the one subscriber must have exclusive use of that one program channel for the duration of the program. It also implies rapid communications from the set top terminal (STT) to the headend. VOD is expected to be a big revenue generator in the future, but it is a heavy bandwidth user. Typically systems size their VOD capacity to serve some 5–10 % of their subscribers simultaneously. We shall see that VOD is a service that works much better in FTTH systems.

High definition programs (HDTV) are becoming standard fare on cable systems, both from off-air programmers and from cable programmers. Whereas up to maybe 12 standard definition channels can fit in one 6 MHz channel, only about 2 or maybe 3 HDTV signals can fit. These new services, along with demand for data and voice, are stressing even the widest-bandwidth HFC networks. However, since the systems have been recently rebuilt, it is

---

[1]As of this writing there is some movement to use the spectrum up to 1 GHz or even higher.

financially impossible to rebuild them for FTTH at this time. Systems in smaller towns have generally not kept up with modern practice, and these systems are being overbuilt by alternate suppliers using FTTH.

9.4 COMPARING THE TECHNOLOGIES

Figure 9.1 compares HFC systems to FTTH systems. The HFC network today consists of a node, which does two-way conversion between optical and electrical (RF) signals, followed in each of several directions by a cascade of 3–6 amplifiers. This makes it an active network, with a number of power-consuming devices located in the field. FTTH networks tend to be all-passive in the field. The headend element of the FTTH system is generically called an Optical Line Termination (OLT). Normally it is located in the headend, but in some cases may be located in the field, connected to the headend using standard gigabit Ethernet and broadcast optics. The OLT would be located in the field in the interest of achieving a longer distance between the headend and the subscriber. The passive configuration is called a Passive Optical Network, or PON.

9.4.1 BROADCAST SERVICE

The HFC industry has long wanted the ability to control service on/off from the headend. This significantly reduces the cost of disconnecting and reconnecting subscribers, either due to a move or due to nonpayment of the bill. In HFC networks, there is no natural place to provide on/off control because the only subscriber-specific device is a passive tap. Active, addressable taps have been built for the purpose of effecting service on/off, but have not achieved widespread usage due to cost. FTTH networks, on the other hand, terminate in an active device on the side of the house (in the house in some cases), and this device can and is configured to provide the on/off function via the element management system (EMS).

In terms of specifications, the ones that the distribution network most often impact negatively are carrier-to-noise ratio (C/N) and distortion. Poor C/N results in a snowy TV

	HFC	FTTH
Passive Network?	No	Yes (from OLT)
Broadcast on/off	Generally no	Individual control
Typ EOL C/N	48 dB	50 dB
Typ EOL distortion	53 dB	57 dB
Data Rate	37 Mb/s per many subs (down), Usually 8 Mb/s per many subs (up)	500 Mb/s per 16 gateways symmetrical, managed QoS
Data Encryption	Yes	Yes (but EFM standard does not specify)
Voice	POTS (NCS)	POTS (MGCP/NCS or SIP)
T1/E1	External	Business Gateway
Construction Cost	Benchmark	Competitive with HFC in new-build
Maintenance Cost	Benchmark	Much, much lower
Network Management	Add-on	Built in
Subscriber hardware	Inside	Outside except STT (inside available, limited features)

Figure 9.1 Comparison of HFC and FTTH systems.

picture, and poor distortion results in 'busyness' in the picture. Modern cable TV systems typically achieve an end-of-line (EOL) C/N of about 48 dB, which produces generally undetectable snow under common viewing conditions, though the snow is visible under good conditions. Properly designed FTTH networks typically deliver at least 2 dB better performance. Similarly, the distortion number for FTTH can be several decibels better (e.g., -57 dB instead of -53 dB).

9.4.2 DATA SERVICE

Data performance is an area in which FTTH networks provide much, much better performance than do HFC Networks. HFC networks almost always use DOCSIS-based modems for data communications. In 2004 North American DSL providers got much more aggressive in their pricing. HFC operators responded, not by moving their prices down, but by increasing their bandwidth offerings. The speeds demanded are stressing the common DOCSIS platforms. Three versions of DOCSIS (1.0, 1.1, and 2.0) are deployed today, with DOCSIS 1.1 being by far the most common.

In HFC networks, a single DOCSIS channel can supply at most 43 Mb/s of downstream data, with a net payload (after DOCSIS overhead) of about 37 Mb/s. Downstream speed is the same for all three versions of DOCSIS in use today. This data is shared over the number of users the operator has built his system to accommodate for a given optical node. Frequently up to a few hundred subscribers will share one DOCSIS downstream channel. The average downstream bandwidth per subscriber is a fraction of 1 Mb/s.

The example FTTH system provides up to 500 Mb/s shared over only 16 subscribers, resulting in an average of 31.25 Mb/s per subscriber. Of course, both DOCSIS and FTTH take advantage of the fact that not every subscriber will use bandwidth at exactly the same time, so both technologies can offer higher speeds, depending on statistics to keep them from overload.

The FTTH system offers symmetrical bandwidth: the same bandwidth upstream as downstream. DOCSIS was designed as an asymmetrical service, with more downstream than upstream bandwidth. In DOCSIS 1.1 systems, the upstream wire rate is 10 Mb/s, with the usable bandwidth being about 8 Mb/s shared over some large number of subscribers. This asymmetrical bandwidth works well for web surfing, and somewhat less well for email. It works very poorly for peer-to-peer services, and makes it essentially impossible to host a web server in one's home. FTTH, with its symmetrical bandwidth capability, does not have these drawbacks.

Data can be encrypted in both HFC and FTTH configurations. In HFC, DOCSIS specifies a 56-bit encryption key. Recently-released standards include encryption, though earlier standards did not, forcing manufacturers to develop their own. Encryption keys up to 128 bits long are in use.

As we showed above, data bandwidth is a significant differentiation between HFC and FTTH, with FTTH offering many, many times the bandwidth that HFC systems can offer. Currently there is movement toward a DOCSIS 3.0 specification that includes higher bandwidth from bonding several DOCSIS channels together. This could help DOCSIS approach the bandwidth capability of FTTH for one customer, but it is not clear how it is going to help close the gap when a lot of customers are served from a bandwidth pool. Higher density modulation technology is not actively being considered, so the greater bandwidth gained by bonding channels ('channel bonding') means more RF spectrum taken for data, and hence less available for video.

For example, suppose 10 RF channels were bonded to improve downstream bandwidth. This would yield on the order of 370 Mb/s net bandwidth (after DOCSIS overhead). The bandwidth required would reduce the number of RF channels available by about 100. And the result would be 370 Mb/s shared over the subscribers on that optical node – usually a few hundred today. Compare this against the 500 Mb/s shared over 16 subscribers of the example FTTH system.

9.4.3 VOICE SERVICE

Any system can support VoIP to an IP phone, but the most common voice solution is to present analog POTS lines to the user, while the underlying voice protocol is one of several VoIP protocols. For DOCSIS, a variant on MGCP, the TGCP/NCS protocol is defined. In the next version of the PacketCable specifications, SIP will also be supported. The example FTTH system can support MGCP, TGCP/NCS, or SIP, ensuring compatibility with HFC systems that adhere to the PacketCable standards. There are some differences in how quality of service (QoS) is implemented, but any system that supports HFC can also support FTTH. This is a key element of compatibility: use of the same protocols regardless of the plant technology.

DS-1/E1 transport is useful for business applications, primarily to connect legacy switchboards. Such transport can be added to either DOCSIS for HFC. Business terminals in FTTH systems have DS-1/E1 transport built in.

9.4.4 ELEMENT MANAGEMENT SYSTEMS

For historical reasons, most HFC systems do not have integral element (or network) management systems (EMS), though third-party systems exist and can be added to most plants. Typically sensors and transponders must be added to each element that is to be managed, and this must be done by the operator. FTTH systems, by virtue of having been developed more recently, have integral EMS.

9.4.5 EQUIPMENT LOCATION

Partly for historical reasons and partly due to technical complexity, essentially all HFC termination equipment is located inside a subscriber's home. Operators have long asked for outside equipment, so that they can get to it without the subscriber having to be home, and where it presumably is less susceptible to damage or theft. With FTTH systems, the most common configuration is to mount everything outside the home, with the possible exception of a power supply and a set top terminal (STT).

9.5 INTRODUCTION TO THE ARCHITECTURES OF HFC AND FTTH NETWORKS

In analyzing the issues in developing an FTTH system that is compatible with HFC, we should start by showing the reference HFC-compatible FTTH architecture.

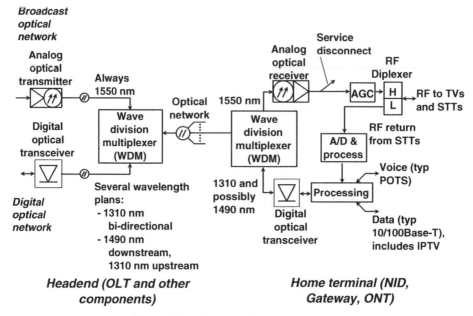

Figure 9.2   Elements of an FTTH network.

9.5.1   ELEMENTS OF AN FTTH NETWORK

Figure 9.2 depicts the elements of a generic FTTH network that carries broadcast video as well as data, and which uses the familiar (to HFC engineers) tapped bus architecture. The video headend is identical to that for HFC, and terminates in a high-quality analog optical transmitter of the type currently used for longer distance reach in HFC networks. Fifteen hundred fifty nanometer is always used for broadcast TV in FTTH networks, as it is necessary to amplify the optical signal in the network. Headend and field hardened EDFAs[2] are available as needed.

9.5.2   DATA LAYER

The data layer is separate from the broadcast layer, and does not use any of the broadcast frequency spectrum. This can free up spectrum for one or more additional video channels. Rather, in FTTH, data (including voice) is transmitted on a different wavelength, preferably on the same fiber as the broadcast video. Two wavelength plans are used for data. One plan, used by the example system, uses 1310 nm in both directions for data. This is possible because of the isolation that can be achieved in the optical network, and the fact that all 1310 nm signals are digital.[3]

---

[2]Erbium-doped fiber amplifiers – a practical, cost-effective technology for amplifying optical signals in the 1550 nm window.

[3]Digital signals do not need nearly as much signal-to-noise ratio as do analog signals, so a level of crosstalk between upstream and downstream will not make the difference it would if analog signals were carried.

Other systems use 1490 nm for downstream data and 1310 nm for upstream data. The problem with this plan is that stimulated Raman scattering (SRS) creates interference between the downstream data at 1490 nm and broadcast video at 1550 nm. This is particularly true on the lower RF frequencies, so it affects the low-band channels the worst. It may be possible to put digital video on those lowest channels, but usually they are the channels most favored for analog video for a number of reasons. The only way to control this problem is to carefully control optical signal levels, the number of splits in the systems, and the distances covered. Two-wavelength systems, which use 1310 nm bi-directional transmission for data, do not have this problem and have been carrying analog (and digital) video for several years, with no problems.

The digital optical transceiver shown in Figure 9.2 connects to the headend data structure, usually using standard Gigabit Ethernet connections. In common terminology, this transceiver, along with some data processing, would comprise the OLT. The same routing equipment used to connect CMTSs[4] and other data elements can be used, simply by inserting Gigabit Ethernet blades. CMTS operation is not affected in any way. Thus, FTTH interfaces in the headend are quite compatible with HFC interfaces.

9.5.3 OPTICAL NETWORK

As shown in the middle of Figure 9.2, the optical network can be a tapped architecture not unlike a tapped HFC architecture. The only difference is that only a few taps configurations and values are needed. Typically, one fiber from the headend will serve a limited number of subscribers, and since the fiber loss is so much less than coax loss, various tap values are not needed. In the example system, only three taps are normally used: an 8-way terminating tap, a 4-way through tap, and a 4-way terminating tap. There are no different tap values, easing maintenance issues.

9.5.4 HOME TERMINAL

The home receiving equipment is variously known as an ONT (optical network termination), a gateway, or an NID (network interface device). In the example system, a single fiber carries broadcast video at 1550 nm and bi-directional data at 1310 nm. A wave division multiplexer (WDM) in the front end of the gateway splits the optical signal into the two wavelengths. The 1550 nm wavelength is supplied to an analog optical receiver similar to that in an HFC node. AGC is used to take out any RF level variations due to variations in received optical levels. An RF diplexer is used to support RF return from STTs in the home. The RF output is identical to that from an HFC network.

The 1310 nm data signals are routed to and from a digital optical transceiver. Processing following the transceiver supports the FTTH protocol, and converts the signals into, typically, one or more POTS[5] lines and 10/100Base-T connections, which can be supplied to data networks in the home.

In order to support RF return from common cable TV STTs, a diplexer routes the RF upstream signal to an analog-to-digital (A/D) converter and appropriate processing. North

---

[4]Cable Modem Termination Systems: the headend termination for cable modems.
[5]Plain Old Telephone Service: standard analog phone lines.

Migrating from HFC to FTTH Networks

HFC Network

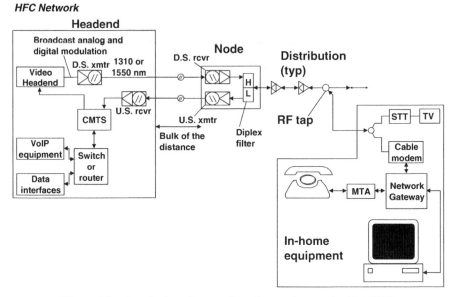

Figure 9.3 A typical HFC network, to be supplemented with FTTH.

American practice is to use one of two SCTE specifications for control communications between the headend and the STT. In order to support SCTE 55-1 communications systems, the RF is digitized, put into an IP packet, and transmitted to a box at the headend, which converts the signal back to RF to be fed into the RF return receiver in the headend. The STT control system does not know that the FTTH system is in use. For SCTE 55-2 systems, the RF signal is demodulated at the home and returned to the headend as an IP packet. Modified software in the STT and headend control system allows compatible operation with STTs installed in HFC plant.

Figure 9.3 shows a high-level view of an HFC network. We shall use this drawing to show how a FTTH system is compatible with an HFC system. On the left is the headend, with the video headend supplying signals to a downstream optical transmitter. A CMTS handles network interface for data and voice, interfacing to some sort of switch or router infrastructure. An upstream optical receiver supplies return signals to the CMTS and other upstream receivers.[6] An HFC node is shown, followed by an amplifier cascade. In the home the signal is split to go to the STT for video and a cable modem. The output of the modem may be supplied to a network gateway or a simple switch, to supply data to multiple computers and to an MTA (media terminal adapter) for voice applications.

Contrast Figure 9.3 with Figure 9.4, which shows the corresponding FTTH network, with similar interfaces. The video headend is identical to that in the HFC network of Figure 9.3.

---

[6]Other common upstream services include the STT upstream control receiver, status monitoring (element management) systems, and sometimes switched circuit voice systems. Sometimes the upstream direction will be used to backhaul video to the headend.

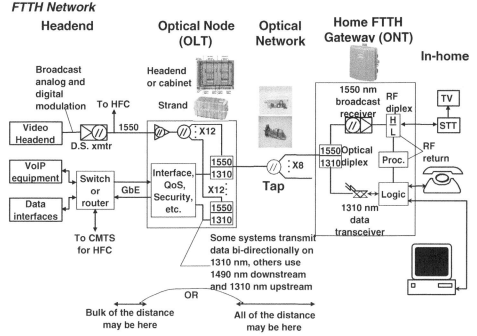

Figure 9.4 FTTH network corresponding to Figure 9.3.

It terminates on a downstream optical transmitter, which is always operating at 1550 nm. External modulation is recommended for good quality. The switch or router is the same as that used in the HFC portion of the network, as is the VoIP and data structure on the network side of the switch. Interfaces to the CMTS are shown, but use of a CMTS is not necessary in the FTTH portion of the network. Rather, standard Gigabit Ethernet interfaces from the router go directly to the OLT. The OLT conditions the data for a point-to-multipoint environment, adds security and QoS, and modulates the data onto a 1310 nm optical carrier. Of course, it also recovers upstream data.

Depending on network configuration, the OLT may be located in the headend, or may be located remote from the headend. It would be located remotely if distances to the subscriber were longer. When the OLT is remotely located, it is powered from the same 60–90 Vac power supplies as are used to power HFC line gear. The RF portion of the OLT normally comprises an EDFA. It amplifies the 1550 nm video optical signal as required. The optical signal is split to a number of outputs, which are wave division multiplexed (WDM) with the data signals, so that both the video and data tiers are delivered to sets of subscribers on a single fiber optic cable.

As mentioned above, some systems employ a 1490 nm wavelength to transmit downstream data and 1310 nm to transmit upstream data. Use of 1490 nm for downstream data transmission is an unfortunate choice of the standards bodies, which complicates video transmission on such so-called three-wavelength systems. SRS interference causes the data signal to interfere with the video signal on lower RF channels. In Gigabit Ethernet systems such as the Ethernet in the First Mile system standardized by the IEEE, the problem is exacerbated for some channels (such as channel 3) due to strong frequency components from the Ethernet idle frame. The result is that three-wavelength systems must use recently-developed special techniques in order to carry analog video. Two-wavelength systems, which use 1310 nm for bi-directional data communication, do not suffer from this problem because the wavelength separation between 1310 nm and 1550 nm is sufficient to prevent problems. The example system is such a two-wavelength system, so it does not have this problem.

In the example system, the OLT may optionally be mounted in the headend or in the field. If mounted in the field, either cabinet or strand mounting may be used. If the OLT is mounted in the headend, then the network is all-passive (a PON, or *passive optical network*) – no powered equipment is used between the headend and the home. If the OLT is field-mounted, then one active device is located in the field for each group of subscribers. The reason you might want to mount the OLT in the field is to achieve a longer distance from the headend to the home. PONs are generally limited to about 18–20 km. While this covers a lot of cases, there are some situations in which the subscriber is located too far from the headend to allow an all-passive network. In this case, the OLT can be mounted in the field, with a distance from the headend to the OLT of up to 70 km, limited by the range of long-distance Gigabit Ethernet optics. Thus, with a field-mounted OLT, you can cover up to about 90 km from the headend to the home, with only one active serving a group of subscribers.

The optical network between the OLT and the home is all-passive, and consists of only dielectric fiber and optical taps. Because there is no metallic conductor, no sheath currents can exist and corrosion is not a problem. Obviously there is no possibility for RF leakage or ingress, either. An optical tap works just like an RF tap except that it works on optical signals. It separates some signal to go to each drop, and passes the remaining signal to other drops or downstream to another tap. Because the loss of the fiber is so small and the number of taps on one optical strand is low, there is no need for a lot of tap values. It is usually sufficient to use three total taps: 8-way terminating, 4-way terminating, and 4-way through. There are no tap values, greatly simplifying design.

At the home, a FTTH gateway (ONT) is usually located on the side of the home, though some are designed for inside mounting. The optical wavelengths are separated into different 1550 and 1310 nm paths (for a two-wavelength system), using an optical diplexer. The 1550 nm path goes to a 1550 nm broadcast receiver. Since there is some variation in light level arriving at the receiver, AGC is employed to maintain a constant RF output level.

In order to support standard set top terminals (STTs), an RF diplexer is used to separate the upstream RF signals coming from the STTs in the home. The lower frequency upstream signals are routed to suitable processing. Depending on whether the RF communications system meets the SCTE 55-1 or -2 standards, processing is slightly different. For SCTE 55-1 support, the RF signal can be digitized, some fast data reduction performed, and the resultant data packet sent to the headend. At the headend, a special circuit reconstructs the RF signal, which is then supplied to the normal RF receiver in the STT control system. For SCTE 55-2 systems, which use a tight timing loop between downstream and upstream, a different technique is used. Here the processing demodulates the RF return signal, packetizes it, and returns the packet to the headend. It is sent directly to the STT control system, bypassing the RF subsystem. The STT and the headend control system both run modified software to support this application.

Returning to a description of the circuitry in the ONT, bi-directional data signals interface with a 1310 nm data transceiver, and from there to suitable logic. From the data signals, data is converted to one or more standard 10/100Base-T Ethernet interfaces, and voice is converted to standard analog POTS lines. Note that the functions of a Cable Modem and voice MTAs have been subsumed into the ONT. The ONT is usually located on the side of the house at the utility entrance. Typically, a coax entry will already exist (or would have to be installed in any event), and phone entrances will also exist at that location. Only the data input, on standard Category 5 cable, will need to be added. There are several groups working on putting Ethernet over coax cable plant. Of course, other options for getting data signals into the home include wireless (802.11a, b, g, n), HPNA, and HomePlug.

9.6   ELEMENTS OF COMPATIBILITY

From the above comparison, we can identify a number of elements making the FTTH system compatible with HFC systems, easing the job of transitioning from HFC to FTTH. First, a broadcast layer that uses the same headend as is used for HFC is important. The only possible addition is a 1550 nm optical transmitter. Some HFC systems now use 1550 nm transmitters, but others use 1310 nm.

9.6.1   POWERING

Powering of equipment must be compatible with other equipment in use. The OLT, or optical node, is either strand mounted in a casting, as with HFC nodes and amplifiers, or it is rack mounted. The strand mounted configuration is a good fit with other cable TV equipment, and is often powered by the same 60–90 Vac power supplies as are used for HFC line gear. Optionally, the optical node may be powered from either 48 or 130 Vdc, common in headends and Central Offices. The rack mounted version is powered from 48 Vdc, a common voltage in use at both cable TV headends and hubs, and in Central Offices.

9.6.2   BANDWIDTH COMPATIBILITY

The HFC system will need to tie up some of its downstream spectrum for data and voice whereas the FTTH system can use this spectrum for more video if desired. In many cases, the data and voice signals are simply left on the FTTH plant: they do not do any good in FTTH, but they do not cause any problems either. With some re-cabling the operator can use the freed spectrum for video, but in many cases this is not being done.

9.6.3   SET TOP TERMINAL SUPPORT

Support for the same STTs as are used in the HFC plant is critical. Without this, the operator would have to provide for different STTs, usually including two different control systems, and different interfaces into the billing system. Downstream support is no problem: whatever frequency is used for STT data transport on the HFC plant is good enough for the FTTH plant. Upstream must be accommodated differently as described above, since there is no inherent RF path from the home to the headend in FTTH plants.

9.6.4   DATA INTERFACES

Since the data interfaces to the OLT are standard Gigabit Ethernet, all that is usually needed to support data are low-cost Gigabit Ethernet interfaces from whatever router is used to connect CMTSs to other data services.

The most common cable modem interface is 10/100Base-T Ethernet, which is also the same interface as is common on FTTH termination equipment. Features and protocols supported are also similar.

The most noticeable difference between the two at the subscriber's home relate to the location of the Ethernet port(s). Cable modems are intended to go inside the home, and a coaxial cable must be run to the location of the modem. It is not always true that the modem is used where a TV is used, so often new coax cable must be run.

In-home wiring may have to be upgraded to accommodate the reverse path from the cable modem, and the HFC operator may want to install a filter that stops all return frequencies other than that used by the cable modem, from getting back to the plant. It is known that the biggest source of problems with interference in the return band are in-home wiring. Sometimes the in-home wiring may be cleaned up during modem installation, but the subscriber has the right to install his own extra outlets, and this is often a source of interference that develops long after the modem is installed. In order to protect the plant, many HFC operators install filters that block all return path frequencies other than that needed for the cable modem, and for the STT if used. Since data is not transmitted in the RF spectrum in FTTH, the same in-home interference issues do not exist with FTTH.

HFC systems need to run new coax cable to the location of the cable modem. FTTH systems usually have the Ethernet interface on the side of the home, so the task is to run a Cat 5 cable from there to the computer or network interface. The same network interfaces work for both HFC and FTTH networks.

All of the cable modems we have seen have a single 10/100Base-T interface. FTTH interfaces intended for single-family dwellings have up to four 10/100Base-T interfaces, so in some cases an external router is not needed.

9.6.5   VOICE PROTOCOL

HFC networks often use a voice control protocol specified by the PacketCable standards, and called NCS/TGCP.[7] This is a profile of the MGCP protocol, which is defined by the Internet Engineering Task Force RFC 3435. It is possible to operate systems that are compatible with both NCS/TGCP and MGCP – the two are quite similar though not identical.

Concern has been expressed that DOCSIS, which is used as the transport for HFC voice implementations, uses a different QoS mechanism than do many FTTH systems. DOCSIS defines a particular QoS paradigm called DQoS, or Dynamic Quality of Service.[8] DQoS establishes a *service flow*, or reserved bandwidth, in each direction through the DOCSIS subsystem. A service flow must be established in each direction for every voice call (and for

---

[7]*PacketCable*[TM] *Network-Based Call Signaling Protocol Specification*, PKT-SP-EC-MGCP-I10-040402, available at www.packetcable.com.
[8]*PacketCable*[TM] *Dynamic Quality-of-Service Specification*, PKT-SP-DQOS-I11-040723, available at www.packetcable.com.

every other service that uses DQoS). The service flow is established during call set-up, and is cancelled during call tear-down.[9] The service flow establishment is initiated by the softswitch, which controls the call set-up.

Contrast this with FTTH systems, in which the QoS mechanism is often based on IETF recommended practices, and consists of identifying the type of packet (voice bearer packets in this case) and assigning a suitable priority to them. When two packets present themselves for transmission at the same time, higher priority packets (such as those containing voice information) get to go first. This prioritized QoS mechanism does not need to be initiated or torn down with each phone call; it is there all the time, and comes into play when a voice packet is presented. For instance, the example system has the ability to assign one of four different priorities to downstream packets. The highest priority is usually reserved for system maintenance messages. The next highest priority is assigned to voice and DS-1 traffic, and the lowest priority is assigned to web surfing and email. The intermediate priority might be used for video, or could be reserved for other services.

Softswitches that are compatible with NCS are usually configurable on an individual line basis, as to whether they will attempt to initiate a DOCSIS service flow or not. Thus, they can be configured to simultaneously work with both HFC and FTTH.

PacketCable specifies a required encoder for voice, and also some optional encoders.[10] The required G.711 encoder produces a 64 kb/s voice channel (before IP overhead) without compression. Since upstream bandwidth is severely limited in most DOCSIS systems, PacketCable also specifies alternate encoders that include compression. FTTH systems do not specify compression, and usually use G.711 encoding. They have so much more bandwidth available that compression is not particularly useful, and compression always causes some level of problems. It adds undesirable delay, and can also interfere with operation of modem-based services such as fax.

Another protocol, SIP, is gaining a lot of traction today for Voice over IP. Some FTTH systems already support SIP as well as MGCP/NCS/TGCP, though usually not at the same time. CableLabs has indicated that a future release of the PacketCable specification may also support SIP.

9.6.6   QUALITY OF SERVICE (QoS)

There are differences in the QoS mechanisms of HFC plants and FTTH plants. We mentioned some of the differences above in connection with the discussion on voice. HFC QoS mechanisms are based on DQoS, which must be set up for each instance of a service that wants to take advantage of QoS mechanisms. If there is no mechanism to request a service flow (reserved bandwidth for a particular application), then every service is a best-effort service, limited by the bandwidth the subscriber is allowed.

FTTH QoS mechanisms are frequently based on IETF recommended practices, in which each packet is examined to determine what type of data it contains (but the contents are not

---

[9]Cicora *et al.*, *Modern Cable Television Technology: Video, Voice, and Data Communications*, San Francisco: Elsevier, 2004, Chapter 5.
[10]*PacketCable*[TM] *Audio/Video Codecs Specification*, PKT-SP-CODEC-I05-040113040723, available at www. packetcable.com.

examined). Packets can be identified as belonging to a particular subscriber, and can also be identified as being part of a particular application. Based on the application and the operator's policies, the packets are put through a rate-limiting mechanism that meters out data at a set rate. The packets are then prioritized, and higher priority packets that arrive simultaneously with lower priority packets will go first. For example, assume that in a shared network two packets present themselves for transmission simultaneously. One is part of a voice call, which must get through with minimum delay, and the other is part of an email, which is not at all time critical. The voice packet will be put ahead of the email packet, so that it goes first when the channel is available.

In addition, Ethernet quality markings (often called *P/q markings* from the Ethernet standard sections defining them) are respected. From the above, it can be seen that the QoS mechanisms in FTTH systems are quite versatile and do not require set-up as do cable modem QoS mechanisms. Rather, the operator sets policy and as packets present themselves to the FTTH system, they are subjected to those pre-set policies.

The IETF-based QoS mechanisms are sufficiently versatile to allow, for example, an IP video session at 4 Mb/s, while simultaneously limiting the same subscriber to 250 kb/s or whatever he has purchased, for data. Or a subscriber might be limited to 1 Mb/s of data, if that is what he has bought, but when he is on a VoIP telephone call, the roughly 100 kb/s extra bandwidth required for that does not necessarily have to come from his data bandwidth.

9.7 VIDEO ISSUES

There are two technologies used to distribute video to subscribers today. Broadcast is the older technology by far, having been used for audio for a century and for video for half that time. The other technology is TV distributed on IP, or IPTV. HFC operators have talked about using IPTV, but the arguments for it on HFC are less than compelling. However, broadcast works very well on HFC. Operators using any form of DSL cannot use broadcast, and are restricted to IPTV. FTTH systems, on the other hand, can use either or both broadcast and IPTV, so an HFC operator is capable of taking advantage of the best of both worlds. We believe that in the future we will use broadcast for what it is best at, sending the same program to all subscribers at the same time, with the intention that they will either watch it then, or will record it for future viewing. At the same time, IPTV will be used for what it is best at, which is to send video to one subscriber at a time. This is the model for video-on-demand (VOD), in which the subscriber orders a program from a central video file server. The program is played out only to that subscriber, who has 'VCR-like' control of pausing, rewinding, and fast-forwarding.

There is plenty of bandwidth available in most FTTH systems to allow video on IP, IPTV, as well as broadcast video transmission if desired. While broadcast video is ideal for broadcast applications, it is not ideal for point-to-point video. The most common example of point-to-point video is VOD, which is intended for consumption by one subscriber only. In order to reach that one subscriber using broadcast techniques (either in HFC or the 1550 nm tier in FTTH), one must consume bandwidth going to every home in the node. But with IPTV distributed on the 1310 nm data tier, the signal only goes into 16 homes in order to reach one, and in other homes, the same bandwidth can be reused to reach other subscribers.

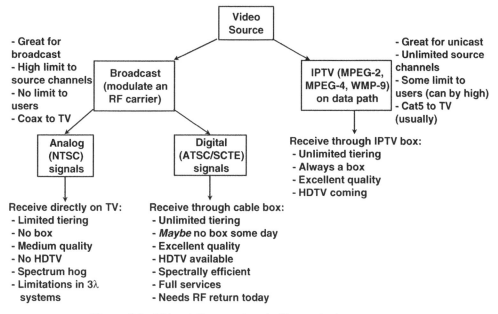

Figure 9.5 Video delivery options in fiber-to-the-home systems.

9.7.1 COMPARING BROADCAST TO IPTV

Figure 9.5 illustrates the options in video delivery, which were described in a paper by this author in the 2003 FTTH Conference.[11] Video may be sent via broadcast. This is the method that has been used by off-air television stations, cable television systems, and satellite broadcasters since the beginning of television. A TV signal is modulated, or impressed, on an RF carrier at a particular frequency and sent to all subscribers simultaneously. Broadcasting video is great for, well, broadcast, where you expect a large number of subscribers to watch the same program at the same time. There is a limit to the number of channels you can transmit, though the limit is quite high with FTTH systems. The only limit on the number of users is the number of subscribers you have.

Within the category of broadcast television, you have two choices today, analog and digital. Analog is the traditional TV transmission method, but it is in the process of being supplanted by digital. We expect the complete transition to digital to take many, many years, due to the large number of analog TVs in the hands of consumers. A TV lasts approximately 15 years, and it is unacceptable to obsolete something a consumer has bought in good faith that it will serve him for that long. Analog transmissions can be of quite high quality, but the effort required to keep the quality high is substantial. Fortunately, FTTH systems can deliver higher quality signals than can any delivery alternative.

The other choice is digital broadcast, the new technology that is gaining ground quickly in cable TV systems and, to may be a lesser extent, in over-the-air broadcasting. Digital

---

[11]James Farmer, *The Basics of Video*, 2003 FTTH Conference.

broadcast is also modulated onto carriers, but the signal that goes onto the carrier is quite different from analog. The signal is digitized and a lot of processing is used to *compress*, or eliminate unnecessary ('redundant') information in order to get the bandwidth down. The compression method being used today is called MPEG-2, developed by an international group known as the *Motion Picture Experts Group*. The MPEG-2 compression technology is used in off-air television, known in the US as ATSC, for the *Advanced Television System Committee*, the body that developed the standard. MPEG-2 is also used elsewhere, often within the DVB, or digital video broadcasting, standard. Both standard definition and high definition programming is transmitted using MPEG-2 compression. The system used in cable TV is almost identical except for the modulation technique used, and for a few other differences which are being resolved now. Satellite television uses similar compression, but there are some proprietary twists, particularly for DirecTV, which launched their system ahead of the standards.

There are two newer compression techniques that offer better compression than that offered by MPEG-2. MPEG-4 was originally developed for Internet moving pictures, and did not offer the quality demanded for entertainment. In the last few years, a new version of MPEG-4, known as advanced video coding, AVC, or Part 10, has been developed for entertainment-grade video. More recently Microsoft has proposed their Windows Media Player 9, WMP-9,[12] as an alternative to MPEG-4 AVC. The competition between AVC and WMP-9 is ongoing. Neither is being used for cable TV, off-air, or satellite transmissions today (AVC is being used in certain professional TV broadcast products, but is not being transmitted to subscribers. Some DSL providers are close to using it. Both MPEG-4 and WMP-9 are used for video on the Internet).

There is an agreement in place between cable and the consumer electronics industry to enable TVs to receive cable digital broadcasts directly. However, the standard being used to enable direct reception of cable signals without a STT only works for services that do not need a return data path, so acceptance is likely to be minimal until two-way standards are available.

What works on cable TV works on FTTH, so you can use cable TV headend equipment and STTs on FTTH systems, too. There are two caveats that go with that statement, though. One is that many cable TV services today demand a return path to allow signals to be transmitted from the STT in the home back to the headend. The most common application is for pay-per-view (PPV) a service in which the subscriber orders a program through his STT and is billed for it at the end of the month. In order for the operator to know to bill him (and to pay the program supplier), data must be transmitted upstream, from the STT to the headend. Not all FTTH systems are compatible with all upstream transmission systems.

The other precaution is that in three-wavelength FTTH systems that use 1490 nm for downstream data transmission, the data can cause cross-talk into the video system by way of a fiber nonlinearity called Stimulated Raman Scattering, SRS. The interference from SRS can cause unacceptable quality degradation if analog signals are transmitted on lower channels (approximately below 100–150 MHz). Unfortunately, for several reasons, the lower channels are the preferred spectrum on which to transmit analog signals. Solutions to this problem are needed.

---

[12]The Society of Motion Picture and Television Engineers, SMPTE, is in the process of establishing a standard based on WMP-9. The standard is known as VC-1.

All broadcast transmission in FTTH systems uses optical transmission in the 1550 nm region, as it is feasible to use EDFAs to optically amplify the signals, a practical necessity. The transmitter and receiver must be analog, as a large number of RF carriers are used, and if digital components were used, the cross-talk between the carriers would be completely unacceptable.

The other way to transmit video shown in Figure 9.5 is IPTV, Internet Protocol Television. IPTV does not use broadcast carriers. Rather, it puts the same MPEG (or WMP-9) data into IP packets, and transmits them in the data layer of the FTTH system. IPTV is a much newer way to deliver video, having been used in the past by a few DSL carriers who are delivering video.

IPTV is great for programs intended for use by one subscriber only, as a minimum amount of the network is tied up to serve that need. In contrast to broadcast video, IPTV has no inherent limitation in the number of channels that can be offered for transmission. There is a high limit to the number that can be carried to subscribers, depending on the transmission capacity of the network and how much of that capacity is devoted to IPTV. The same data transmission capacity is used for all other data traffic. In most cases, it is necessary for the FTTH operator to install category 5 (Cat 5) cable to the TV, a significant expense in existing homes. To answer this problem, several systems for transmitting Ethernet on coax cable are coming on the market. An STT is always used to convert the IPTV signal to something the TV can use, and we know of no movements to incorporate IPTV reception into TV sets. HDTV is possible, though we are not aware of anyone who is commercially transmitting HDTV on IP yet. The bandwidth required for HD-IPTV is much greater than that required for standard definition, making the commercialization of advanced compression more important.

So far MPEG-2 has been used for most IPTV transmissions, though MPEG-4 and WMP-9 are candidates for future service. Since IPTV is not as mature as is broadcast digital transmission, there is less equipment in the hands of consumers that would be stranded if an operator used MPEG-4 AVC or WMP-9. And a number of IPTV STTs tend to use software-based decoders, which can be downloaded with new decoder software. (This capability could be one reason that IPTV STTs tend to be expensive compared with cable TV boxes, though.)

9.7.2 HFC VIDEO OPPORTUNITIES

HFC systems were originally designed to carry broadcast information exclusively, and they do an extremely good job of it. It is theoretically possible to carry IPTV over a DOCSIS modem system, but it is of questionable usefulness to do so. There is talk in the industry now of doing just this, with the primary intent being to reduce the number of networks the cable operator must deal with, to just one: an IP network. This does have merits in the internal plant, but as you get to the edge of the network (where the HFC plant is located) the arguments weaken. Putting video into IP over the HFC plant means taking today's MPEG transport stream, putting additional IP protocol around it, and transmitting it on a DOCSIS modem, which has about 37 Mb/s of downstream bandwidth.

9.8 CONCLUSION

FTTH systems are really complementary to HFC systems and not competitive with them. In order to deploy your first FTTH systems, it is not necessary to change anything you are doing in the headend, so cable TV operators can continue to operate HFC plant while adding FTTH for business or new-build residential applications in an evolutionary manner. FTTH systems offer superior quality and reliability, and operational costs are a small fraction of what they are with HFC.

REFERENCES

1. Ciciora W, Farmer J, Large D and Adams M. *Modern Cable Television Technology: Video, Voice, and Data Communiations* (2nd ed), San Francisco: Elsevier; 2004.
2. Society of *Telecommunications Engineers, Data-Over-Cable Systems Radio Frequency Interface Specification* 1.1, available at www.cablemodem.com. Other documents in the DOCSIS series will also be useful.
3. *PacketCable*^{TM} *Network-Based Call Signaling Protocol Specification*, PKT-SP-EC-MGCP-I10-040402, available at www.packetcable.com. Other documents in the PacketCable series will also be useful.
4. LeBrun G, Sr. *Alternatives for Delivering Voice in FTTP*, available at http://www.ftthcouncil.org/documents/856118.pdf
5. Farmer JO. *Optimizing Video Delivery*, available at http://www.ftthcouncil.org/documents/959776.pdf. A large number of other useful documents can be found at www.ftthcouncil.org. Papers from the most recent conference on available to members only for a year, then are made available to anyone.

10

FTTH Systems, Strategies, and Deployment Plans in China

Jianli Wang and Zishen Zhao

Wuhan Research Institute of Posts & Telecom, Wuhan, China

10.1 CURRENT STATUS OF BROADBAND ACCESS

10.1.1 CHINA'S BROADBAND USERS GROWING RAPIDLY [1,2]

In the last few years, the number of Internet users in China grew very fast. By June 2005, China had 103 million Internet users, an 18 % increase over June 2004. Among them, 53 million are broadband users, a 71 % increase in a year.

In spite of the rapid growth of broadband users in the last few years, by June 2005, China's broadband penetration rate is only 3.54 %, much lower than that in some developed countries (e.g., ~25 % in South Korea). China, therefore, has a great potential for broadband growth. It is forecasted that China's broadband users will reach 144 million by 2007.

10.1.2 ACCESS SERVICE REQUIREMENTS [12]

In China, the access users can be divided into three major categories according to their service requirements.

- Large corporations: This category includes headquarters of large corporations, large government departments, universities, financial institutes, media, hotels, business buildings, etc. This type of users usually has a large-sized private network and has high requirement for network availability, security, and quality of services. For this type of users, the major access services they need are various leased line services. Among those leased line services, DDN and SDH with line speed of 2 Mb/s and below accounts for 70 % of current users. Others include frame relay leased line and ATM leased line.

Broadband Optical Access Networks and Fiber-to-the-Home: Systems Technologies and Deployment Strategies
Chinlon Lin © 2006 John Wiley & Sons, Ltd

- Medium- and small-sized enterprises: This type of users has a relatively wide range in access network capacity. The services they need include DSL, DDN, FTTx + LAN, dial-up and ISDN. The data rate provided to an end-user includes a few 10 Kb/s, 512 Kb/s, 1 Mb/s, 2 Mb/s, and 10/100 Mb/s.
- Residential users: For this type of users, the major services are voice and Internet access. These users are in general more concerned about simplicity, plug, and play and cost. The major current access approaches are dial-up, ADSL, and LAN.

10.1.3 ACCESS TECHNOLOGIES AND KEY PLAYERS

According to the subscriber categorization described above, the major broadband access technologies used for small-sized enterprises and residential users include ADSL, LAN, broadband fixed wireless (BFW), etc. In the following, we describe the main features of major access technologies and their key players in China.

10.1.3.1 ADSL (Asymmetrical Digital Subscriber Line)

In China, ADSL is the major broadband access technology deployed today. It plays a major role in family and small business Internet access.

The line speed of ADSL is from 512 Kb/s to 2 Mb/s. But, most subscribers are using 512 Kb/s and 1 Mb/s. Although the line speed provided to the subscribers is claimed to be 1 Mb/s or 512 Kb/s, the actual speed the subscriber can enjoy is much less than the claimed speed due to the congestion of the metro network and/or the low-response speed of the website server people visit. In terms of equipment price, it has been dropped dramatically in the last few years. Now the central office (CO) equipment price is around $40 USD per subscriber. The monthly fee is about $12 USD per subscriber.

In China, the deployed DSL so far are almost all ADSL. The line rate is from 512 K to 2 M. In late 2004, ADSL2+ products became mature and were tested by service providers. The massive deployment of ADSL2+ is expected to start in 2006. China already has mature VDSL products, but we have not seen urgent deployment requirement because ADSL2+ is able to provide high bandwidth and will therefore delay VDSL deployment or even make VDSL deployment unnecessary.

Major service providers: ADSL is a technology based on current copper line, so it is straightforward that the two major local service providers, *China Telecom* and *China Netcom* became major ADSL service providers. China Netcom has all the Northern 10 provinces' ADSL market, and China Telecom has almost all the rest of ADSL market. Other than the above two ISPs, China Railcom also has certain amount of copper lines where it is able to provide ADSL services. But, because its copper resource is very limited, it is hard to compete with the other two ISPs.

Major equipment vendors: For central office equipment, Huawei is the biggest ADSL equipment vendor in domestic market with about 30 % market share, while Alcatel, ZTE, and UTStar are in the second, third, and fourth places, respectively. In addition to those four big vendors, China also has a number of small equipment players including Fiberhome, Nokia, and Harbor. For end-user equipment, the vendors include Daya, Tongwei, Shida, ZTE, and Alcatel.

10.1.3.2 LAN

With this approach, fiber from central office reaches a point close to a residential area or a building, called optical drop. From this optical drop a LAN leads to each family or office. With this approach, each family or office usually has about 100 Mb/s Internet interface.

Although a large number of residential communities have been passed by LAN, the take-rate is low. The major reason is its cost, which is still high in comparison with that of ADSL. Another reason is that only 13 cities in China are open to competitor service providers, which are the key LAN access promoters. The monthly fee is about $18 USD per subscriber.

The LAN ISPs include China Telecom, China Netcom, China CATV, China Unicom, and Changcheng Broadband.

10.1.3.3 BFW (Broadband Fixed Wireless)

In China, there are three types of fixed wireless access solutions deployed. WLAN based on 2.4 GHz is the most promising fixed wireless technology in China. In the past few years, it has grown very fast due to a number of advantages including easy deployment, simple network architecture, and smooth combination with mobile communications. With the support to mobile in 'hot' spots, it will be more used in enterprises and homes. 3.5 GHz technology has also grown fast in the last couple of years, while 5.8 GHz deployment has just started.

WLAN: As for frequency spectrum, according to 'China Wireless Management Bureau,' 2.4 GHz is mainly assigned for WLAN and Bluetooth with a range of 100 m. It is mainly used for offices and homes. Another usage of WLAN is in the public area, called PWLAN. It can be used in hotels, airports, conference centers, and libraries. The major obstacle for WLAN use is its security issue. As for the equipment vendors, Cisco, D-Link, Netgear, and Huawei are on the top of the list. Those four vendors own 68% of China's market. ISPs include China Telecom, China Netcom, China Mobile, and China Railcom.

3.5 G: In 2003, China saw a rapid growth in 3.5 GHz. Two bids for frequency bandwidth issued licenses in 37 cities in China. Since then, however, no further deployment has happened. The major service providers are China Unicom, China Netcom, China Mobile, and China Telecom. The major equipment vendors include ZTE (China), Alvarion (Isral), Hongxin (China), and VYYO (US).

5.8 G: This is an open frequency band. China has just started the use of this system. There are a few field trials by service providers. But the deployment of 5.8 GHz systems has just started. No standard and therefore the interoperability issue are the major obstacles.

10.1.3.4 Others

In addition to broadband access technologies described above, the following technologies also have certain market deployment.

Cable Modem: There is only one type of ISPs, that is Broadcasting and Television companies. It is the only ISP that owns the cable TV delivery system infrastructure (HFC, hybrid fiber coax networks).

FTTx: So far the optical access is mainly FTTB and FTTC. In China almost 90% of business buildings have fiber drops. Some communities also have fiber drops. But almost no office or residential house has fiber reached. See next few sessions for more discussion about FTTO and FTTH.

PLC power line communications: There is only one PLC ISP, China Feidian. PLC has a very small market in China.

10.2   DRIVING FORCES OF FTTH [3,4]

10.2.1   INCREASED BANDWIDTH DEMAND

The most important driving force for FTTO and FTTH is more and more broadband applications needing more bandwidth. In the last few years, China's broadband services grew very rapidly. In addition to the old applications such as web surfing, e-mail, and file download, many new applications and services have become popular. Examples include online gaming, IPVOD, videophone, etc. Those applications need much more bandwidth.

Another important application is broadcasting TV. China has launched a national TV digitalization plan. The goal is to discontinue analog TV programs nation-wide by year 2015. Currently, the TV programs are delivered by a separate network operated by CATV service providers; however, the nation is promoting convergence of CATV network, telecom network, and Internet, especially for the access portion, to realize triple play. In that case, the bandwidth need for a typical family is estimated to be around 60 Mbps. With the introduction of HDTV, the bandwidth requirement is even higher. With that amount of bandwidth, none of the current available broadband access technologies is able to meet the needs, so FTTH is needed.

10.2.2   REDUCED SYSTEM COST

Another important driving force for FTTH deployment is that the cost of optical-electronic components has been reduced dramatically thanks to the advance of technology. In the past few years, almost all the optical components used for FTTH have seen a great drop in cost. Fiber is now about $16 USD per kilometers and optical transceiver at 1 Gbps is about $80 USD per unit. The cost of Ethernet card, passive optical components, etc. keeps dropping every month. Reduced system cost makes FTTH and FTTO economically acceptable.

10.2.3   COMPETITION

Competition is a driving force for the application of any new technologies and there is no exception for FTTH. In China, FTTH is widely accepted as the dominating broadband access technology in the future. Although there is no urgent immediate deployment needs, most carriers and service providers are actively promoting this technology because they realized that sooner or later FTTH will take off and they want to occupy the territory ahead of time. This competition can be seen in the following aspects:

- For CLECs to compete with ILECs: China Netcom owns local network resources in 10 provinces in Northern China. China Telecom owns local network resource in the rest of the country, mainly in Southern China. For the 10 provinces in Northern China, China Telecom needs to find another approach to provide local access services to compete with ADSL provided by local carriers. Likewise, China Netcom needs to do the same thing in

Southern China. It is also true for other carriers including China Unicom that has almost no local network resource.

- For local carriers to bypass residential service providers: In 2001, China started a program to open access network operation to residential service providers. With this trial, a new model, known as 'Tailong Model' in China, emerged. With this model, a new type of service provider, residential service provider, owns and operates the residential network which can be connected to any local carrier's network according to residences' interests. After this trial, so far Ministry of Information Industry (MII) has not found any serious problems about this new operation/business model. It is expected that this new model will be stated in 'China's Telecommunication Law' which is expected to be issued late 2005 or early 2006. In that case, the local service providers will have their access network reach at some point of a community and then it is up to the residential service provider to provide the last mile access to their subscribers. In order to bypass the residential network, local service provider prefers to have their fiber reached each family directly. FTTH is a solution for that purpose.

- For telecos and CATV providers to compete for data access and other services: Currently, CATV providers are not allowed to provide telephone services and Telecos is not allowed to provide CATV services while both are allowed to provide Internet services. Currently, the government is working on the regulations to converge CATV network and telecom network. Now it is very important that whoever owns this access facilities will likely own the subscribers. Although the final goal is for the subscriber to freely access to any service provider they want, for a certain period of time, the access network and the local service will be in monopoly. He who owns the access network owns the subscriber.

10.2.4   MARKET [7]

FTTH has a huge market. It includes not only FTTH systems and equipments but also FTTH related devices, components, accessories, deployment, services, etc. Assuming 100 million FTTH subscribers within 5 years, it is estimated that China's average annual FTTH market is between $30–40 billion USD.

10.2.5   REGULATORY

In many cases, new technology promotion needs encouragement from government's regulation. So far, Chinese regulation is not favorable for network convergence and therefore not for FTTH. But some local government is considering of setting rules to promote FTTH deployment. For example, new-built house or apartment must have fiber deployed. There is also a sign that Chinese government is promoting convergence of three networks and triple play, it is said that Chinese government will likely open residential networks to allow residential service providers to build and operate residential networks. All of those will help to promote FTTH deployment in China.

10.3   LATEST FTTH INITIATIVES

In Asia Pacific Optical Communications Conference (APOC) 2003 held in Wuhan, many telecom experts in China, led by Fiberhome Technologies Group, discussed FTTH in China for the first time in the last few years. The topics included opportunities, challenges, FTTH applications, cost analysis, network architectures, etc. That forum attracted a lot of attention of the China's telecom community.

10.3.1 NATIONAL FTTH RESEARCH PLAN

Starting 1999, China launched FTTH research and development program to study key technologies of FTTH and real FTTH systems for field trials. The first research program was a national '863' hi-tech program called APON, which was completed in 2001. Another similar state '863' hi-tech program on EPON started in 2001 and was completed in 2003. A number of organizations, with Fiberhome Telecom as the leader, took part in these two programs, and the results are very positive. In China's 'eleventh five years plan' period, many proposals have suggested to put FTTH as one of the major areas for government financial support.

10.3.2 FTTH PRODUCTS

The very first FTTH system in China was developed in 2001 by Fiberhome Technologies Group in Wuhan as part of the effort of the national 863 APON plan. Now there are a number of FTTH equipment vendors. Fiberhome is still the leading FTTH equipment vendor with most advanced technologies and richest functionalities. Other vendors include UTStar, Salira, Greenwill, Firstmile, Optical Solutions, and Fohope.

Most vendors focus on EPON system and point-to-point system. GPON system is under development by some vendors, but none has mature products yet. There is only one WEPON vendor, Fohope. The FTTH products can be basically divided into three categories according to their capability of supported services. First category is the product that can only support data service, such as Fiberhome's 100 Mb/s EPON, UTStar's 1000 Mb/s EPON, etc. This type of FTTH system can only provide Internet access. The second category can support data interface and CATV interface by using a separate wavelength for CATV. Those products include Fiberhome's point-to-point FTTH system and Fohope WEPON system. The third category can provide real triple play, using three wavelengths to support all data, TDM/voice and analog or digital CATV services. Fiberhome's 1000 Mb/s EPON (called GEPON) is a typical example. All above products are mature and have been deployed in a number of field trials and commercial FTTH networks.

10.3.3 EARLY FTTH FIELD TRIALS

Since 2002, there have been a few field trials in China. Here, we describe two of earliest field trials: one is in Wuhan and the other in Chengdu.

Wuhan FTTH project: The purpose of this project is to demonstrate the feasibility of various FTTH systems. The project consists of 3 buildings and 87 homes. The equipment vendors include Fiberhome, Optical Solutions, Fohopee Networks, Salira, and Wave 7. The FTTH systems deployed include 100M EPON, 100M P2P, and WEPON. The services provided include Internet access and CATV. The service provider is Wuhan Telecom. This project concludes that (1) FTTH products are mature to be commercially deployed; (2) the cost of the FTTH is still very high in comparison with ADSL.

Chengdu FTTH project: This project is to build a residential FTTH network. The network is deployed and operated by Chengdu Tailong, the only residential service provider in China. The project is designed to provide triple play through a single fiber. The total number of homes covered by this project initially is only a few tens of homes. The equipment vendors include Fiberhome, etc. The FTTH systems deployed include 100 M EPON. The service provided is Internet access only. Currently, the service providers connected to this residential FTTH network include China Telecom and China Railcom.

More field trials: After the above two FTTH field trials, the technologies for FTTH has been proven to be mature. Next step of field trials is to study the business model. For that purpose, more than 30 field trials have been carried out all over the country. Among those field trials, the number of subscribers is between 400 and 1000. The cities for the field trials include Wuhan, Beijing, Shanghai, Hangzhou, etc.

10.3.4 THE FIRST FTTH EQUIPMENT TESTING BY CARRIER

In December 2004, China Netcom held the very first FTTH equipment testing in Beijing. This is the first equipment testing organized and carried out by carriers in China. That means the Chinese carrier is seriously considering FTTH commercial deployment. There were five FTTH equipment vendors participating the test: Fiberhome, UTStar, ZTE, Huanwei, and Gaohong. The equipment tested is GEPON and the result is positive. Among the five vendors, Fiberhome's solution is much more advanced in system functionally and performance, especially in multi-service support. The result shows that the FTTH technology and products in China are mature for commercial deployment.

10.3.5 COMMERCIAL FTTH NETWORKS

So far, there are a number of commercial FTTH networks in operation. In the following, we describe three of them.

Wuhan Zisong FTTH network [5]: This is the very first FTTH network deployed and commercially operated by leading service providers in China. This network is deployed by (Wuhan) China Telecom in a brown field for old multi-dwelling apartment buildings. The purpose of this project is to obtain experience in FTTH deployment and installation in existing apartment buildings, and to study multi-service or triple-play capability of EPON systems and the business model for triple play with FTTH. The project consists of about 480 subscribers. It uses GEPON equipment with 1 : 32 split ratio from Fiberhome Technologies Group. The services it provides include POTS phone, VOIP phone, CATV, IPTV, FAX, Internet access, and E1 TDM service. It uses single fiber with three wavelengths WDM. The network deployment completed in January 2005 and started operation right after. The network structure is illustrated in Figure 10.1.

Wuhan South Lake FTTH network: This commercial FTTH network was deployed and commercially operated by China Network in Wuhan. Similar to the Zisong FTTH project above, the GEPON equipments and cable used in this project were provided by Fiberhome. This project has about 700 subscribers. The purpose of this project is similar to the one above but for different service providers. The services provided are the same as the one above. The only difference is that this project uses two fibers with one fiber exclusively used for CATV services. The project completed and started operation in August 2005.

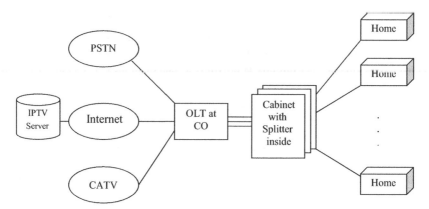

Figure 10.1   Wuhan Zisong FTTH network.

Beijing broadband house FTTH network: This commercial FTTH network was deployed and commercially operated by China Network in Beijing. It also uses GEPON system to provide triple play. It has about 1000 subscribers. The project was completed and started commercial operation in July 2005.

A bigger FTTH network is planned: A very big FTTH/FTTO network has been planned in Wuhan for a combined residential and business district. The district occupies about 7 km with a population of 160 000. The district will use FTTH and FTTO for all the home and business access.

In Wuhan, the government has been considering speeding up FTTH deployment with a goal of about more than 200 000 subscribers within 2 years.

10.4   FTTH TECHNOLOGY CONSIDERATIONS [9,10,11]

Currently, the three major FTTP technologies commonly considered in China are EPON, GPON, and P2P. APON will not be deployed due to its complexity and high equipment cost. Those three technologies are complementary rather than conflicting. EPON is mainly used for FTTH in the area with high population density; GPON will be mainly used for enterprises or businesses that need TDM leased line services. P2P can be used for both business cases and residential cases. For business case, P2P will be used when the required bandwidth is more than that a PON can provide. For residential case, P2P is preferred to EPON when the residential homes are scarcely located. When there is a high requirement in scalability in terms of the number of subscribers per system or the amount of bandwidth per subscriber, P2P is also preferred.

10.4.1   EPON WILL DOMINATE RESIDENTIAL FTTH MARKET [3]

In China, most of time people use FTTH to actually mean Fiber-to-the-Premise (FTTP), including both FTTH and Fiber-to-the-Office (FTTO). For homes and small offices, EPON is preferred to GPON. The considerations are as follows:

- Technology wise, EPON is more mature than GPON at the present time. Most Chinese FTTH vendors can provide EPON systems, although they are different from each other in multi-service support. GPON is a relatively new technology and there is no mature GPON equipment in China yet.
- EPON can meet all the current service requirements for residential uses. For family home usage, the services need to be supported are POTS telephone (including fax), Internet access (file download, IPTV, games, etc), and CATV. Fiberhome's EPON equipment can meet all the above requirements. It can provide telephone service by using either VOIP with embedded Integrated Access Device (IAD) function in ONU or through TDM (E1) over IP; it can provide Internet access in a straightforward way because of its Ethernet-based transport nature; it provides CATV service through a third wavelength which is dedicated for analog CATV transport. The fax service can be either through telephone line over IAD or via VOIP T.38.
- EPON is easy for the next-generation network (NGN) evolution. NGN is the future of the network. One of the key features of the NGN is that the transport network of NGN will be packet-based IP network. Ethernet-based EPON is very easy for NGN evolution because Ethernet is the best layer 2 technology for IP support.
- EPON equipment is cheaper. Ethernet is the widely used layer two technology, and more than 80 % of the IP traffic is now carried by Ethernet. Ethernet protocol is very mature and many types of Ethernet cards have become a commodity. Because of all the above, EPON equipment is cheaper.
- For home or small business, it usually does not need TDM E1/T1. In some cases, a small business may need a couple of E1/T1s, which EPON can easily support.

10.4.2   GPON WILL BE MOSTLY USED FOR FTTO [3]

GPON is a new PON standard proposed by ITU-T. The biggest advantage of the GPON is its capability of multiple service support, especially the TDM service. While some EPON products can also support TDM service, GPON can provide large amount of TDM services in an efficient way. For some small- or medium-sized business, TDM leased line service is required. Although EPON can also provide TDM service, when the number of TDM T1/E1 is larger, it is not easy or efficient for EPON to provide it. So in this case, GPON is a preferred solution. In some corporation or business cases, other leased line services such as ATM and frame relay are also required. GPON is able to carry out those services.

10.4.3   P2P WILL NOT BE WIDELY DEPLOYED [3]

Currently, the biggest advantage of P2P type of FTTH system is its flexibility in terms of bandwidth expansion. Because each user has a dedicated fiber exclusively used for itself, the bandwidth can be easily expanded without affecting other users. Therefore, P2P is a very good solution for those with large bandwidth need and fast bandwidth increase. However, P2P type of FTTH system has a problem, that is, lower central office equipment density. In comparison with PON system, it has more ports (one for each end user) in central office, and therefore more central office space, more power consumption, more likeliness of system problem, and more complicated network management. This problem becomes severe especially in big cities where central office space is very limited. Another problem with

P2P is that it needs more fiber. Although fiber cost keeps dropping, fiber/cable installation and deployment is still very expensive, especially when new pipe needs to be placed. So, in China P2P system will not be widely deployed.

10.4.4 MULTI-SERVICE SUPPORT IS GENERALLY REQUIRED

In China, one of the driving forces for FTTH is service integration: one fiber serves for three networks. This is quite different from that in Japan, where the FTTH is only used for Internet access. In China, especially for the new buildings, the services that need to be provided by FTTH include POTS telephone, CATV, and Internet access. The most common approach to provide multiple services is via WDM. Just like that specified in ITU-T BPON standards, two wavelengths are used for upstream and downstream voice and data, and the third wavelength used for downstream CATV only. The reason for multiple-services support is that people have well realized that the integration of the access networks is the best way to avoid multiple network construction. To realize 'one fiber three networks,' the biggest barrier is state regulations, CATV and voice services are provided, and operated by different parts of the government, and they each prefer to have their own separate networks. The good news is that the proposal of 'CATV network and TV program separation' is now under government's consideration. Also, the 'Telecommunication Law' to be issued soon will very likely relax the restrictions on telecom service operation, to allow third party to run telecom services. This third party is usually called virtual carriers. All of those are good signs that the government encourages the integration of the three networks at the access part.

10.4.5 FTTH FUNCTIONALITY EXPANSION

Currently, most FTTH systems only function as transport systems. A typical FTTH system consisting of an OLT and multiple ONUs. The service provided by the FTTH system is either Internet access only or triple play. The vendors are now considering of expanding the functionality of the FTTH system in the following aspects:

- Multiple OLT functions for smooth evolution: Currently, FTTH has not been deployed in China. In the near future FTTH will not completely replace existing access solutions, especially DSL solutions. Even in a small area, when FTTH starts to be deployed, DSL and FTTH will coexist for a long time. In a residential area, very likely it will be that some families prefer FTTH while others prefer DSL. So, an integrated function OLT is a good solution: a single chassis consisting of both ADSL and FTTH (e.g., EPON) cards and those different cards can be configured by any combination according to the customer's requirements. For the uplink interfaces, the equipment can provide interfaces to different networks, including PSTN, Internet, and CATV network.
- Integrated ONU for home network: A typical FTTH ONU only performs optical termination and provides different interfaces for family usage. The type of interfaces is dependent on the type of services provided. Typical interfaces are RJ11, RJ45, and CATV cable interface. Recently, local area and home wireless technologies such as WLAN are mature and accepted by the industry. An FTTH and WLAN combined solution might be useful in some cases. With this solution, FTTH has the optical drop at the door and then voice and Internet access is provided by WLAN, and video service is connected to TV set

through coax cable. For this solution, the ONU of the FTTH system has to provide the WLAN AP (Access Point) function too.

Another case for function integration is to integrate the home gateway function into the FTTH ONU. The extra function may include various interfaces for home security, hydro and power meter reading, and TV STB (set-top box) can also be integrated into the FTTH ONU. With all those functions integrated, the ONU actually becomes a versatile home network gateway. With this solution, each family has only one box at home performing communication, entertainment, security, and control functions.

10.5 MAJOR FTTH PLAYERS AND PRODUCTS [3]

10.5.1 MAJOR FTTH SERVICE PROVIDERS

Currently, China Netcom and China Telecom are the two most active FTTH promoters. They view FTTH as part of their 'Broadband' strategy. China Railcom is also considering the building of its FTTH networks. Another new service provider called Tailong is very much involved in promoting FTTH by encouraging government to open its residential networks.

For China Netcom and China Telcom, in the cities where they have local network resource, they want to use FTTH for new residential buildings, in the cities where they do not have local network resource, they try to use FTTH to compete with their competitors for access networks. For Tailong, it wants to build its own residential networks and do the network operation and maintenance while leave the service to service providers. The driving force for them to build FTTH network is to expand its residential networks.

10.5.2 MAJOR FTTH VENDORS

Since 2001, a number of equipment vendors started to develop commercial FTTH systems. Those vendors include Fiberhome Technologies Group, UTStarcom, Greenwill, and Fohope Networks. In addition to the domestic companies, there are also a few foreign companies participating FTTH filed trial in China, including Wave 7, Optical Solutions. Most of those companies have their FTTH products deployed in various field trail or commercial FTTH networks.

10.5.3 FTTH PRODUCTS [6]

There are a number of FTTH equipment vendors in China. The major FTTH systems include EPON system and point-to-point system. There are also one vendor developing WEPON system and some considering GPON system. In the following, we briefly describe EPON and point-to-point products.

10.5.3.1 EPON Products

From its application, current EPON products in China market can be divided into three types: (1) for data services only; (2) for both data services and TDM/voice services; (3) for data, voice/TDM, and along CATV services.

EPON system: there are a number of EPON vendors in China. Their products are mostly follow IEEE 802.11 standards. The main EPON vendors include Fiberhome Technologies, UTStar, Greenwill, and Fohope Networks. Fiberhome is accepted as the best in terms of product maturity, functionality richness, and current deployment in China. In the following, we will describe the functionalities of EPON products with Fiberhome's as an example.

Fiberhome's EPON products focus on two types of applications: FTTH and FTTO. The main difference is that for FTTO usage it supports TDM and leased line services, while for FTTH usage, it does not have to support TDM services.

Fiberhome's EPON system has line speed of 1000 Mb/s optical Ethernet interface. It adopts triplex to transmit three wavelengths to realize triple play on a single fiber. For the downstream TDM and data, it uses 1.31 and 1.49 µm for upstream and downstream TDM and data, respectively, and 1.55 µm for downstream video. TDM service is provided by TDM over IP. Both TDM user's data and the synchronization signals are encapsulated on RTP on top of UDP/IP. Each PON can support 32 or 64 end-users. Downstream bandwidth of 1000 Mb/s is shared by all the 32 users; upstream bandwidth is dynamically adjustable with a granularity of 512 Kb/s within the range of 0~1000 Mb/s. System is protected. It provides traffic control, priority control, DBA (dynamic bandwidth allocation), and many operation, administration, and maintenance (OAM) functionalities.

The central office equipment is a 7U multiple service equipment in which one shelf can have two chassises, and each chassis can have mixed DSLAM and EPON OLT cards. One chassis can have a maximum of 32 EPON configuration. So, the total capacity of the OLT is 1024 ONUs. OLT has 100 Mb/s, Gigabit Ethernet (GE), E1, and CATV interfaces to the network.

There are two typical ONUs (1) BP5002 is a small-scale box, designed for home usage. It can be put on the top of the desk, or to hang on the wall, and it provides a separate power. It has one PON interface, two FE interfaces, one CATV coax cable interface, and one CATV optical interface; (2) BP5006 is a 19" 1U box, designed for office or small business usage. It can provide two PON interfaces and supports PON ODN 1:1 protection. It provides two E1(G.703) to support TDM services and eight POTS to interface to the user's PBX.

Figure 10.2 gives a few examples of the commercial EPON-based FTTH products, including both OLT equipment and ONU equipment.

10.5.3.2 P2P Products

The P2P system is to use one fiber for each ONU. Fibehome's P2P FTTP system is also based on Ethernet. It has two types of products: one is used for FTTO and the other is for FTTH usage. The major difference of those two types is that the equipment for FTTO supports TDM leased line services while that for FTTH does not.

For either case, the system consists of central office unit and user end unit. The bandwidth for each end-user is 100 Mb/s. The system provides traffic control, priority control, VLAN capability, and OAM capability.

One OLT box at the central office can provide access for 48 end-users. One OLT provides 2 GE interfaces, 24 E1 interfaces, and 48 POTS interfaces on the network side.

The ONU box at family home has one 100 Mb/s optical Ethernet uplink interface. For the end-user services, it has four 100 Mb/s Ethernet interface, two POTS interfaces, and one CATV interface.

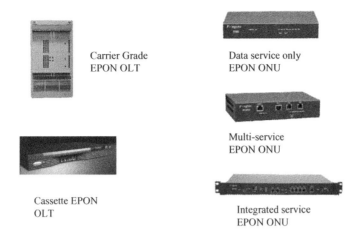

Carrier Grade
EPON OLT

Data service only
EPON ONU

Multi-service
EPON ONU

Cassette EPON
OLT

Integrated service
EPON ONU

Figure 10.2   Examples of EPON-based FTTH products.

10.6   MARKET BARRIERS [4,8]

Since November 2003, FTTH has become a very hot topic in China's access networks; there have been a number of forums, technical papers, state-funded research programs, and a number of field trials and commercial deployments. But massive commercial deployment and operation of FTTH network in China seems to have a number of barriers to overcome. These barriers include high system cost, lack of killer applications, lack of encouraging regulations, and competition from other technologies. Those barriers are not standing along, but interdependent. For example, high cost and lack of killer application are correlated to each other. If there are some applications that need bandwidth which only FTTH is able to provide, then people will not compare FTTH cost with ADSL cost because FTTH can solve problems that ADSL cannot.

10.6.1   HIGH SYSTEM COST

The biggest barrier for massive FTTH deployment in China is that the cost of FTTH system is still very high. In comparison with today's dominating access solution ADSL, FTTH system's cost is much higher. Today, ADSL equipment cost is about $50 USD per subscriber, while that for FTTH is about $180 USD per sub for Internet access only and $500 USD per sub for triple play. For brown field where the existing telephone service and CATV service are already available, FTTH is not attractive. For ADSL, the equipment cost is very low and it typically does not need installation of copper or fiber. And the bandwidth provided by ADSL is enough for today's usage. For green field, FTTH does not have immediate cost advantage, but realizing that FTTH is the future direction of the access network, for the long term's point of view, people will consider FTTH, especially for the luxury houses. People do not like to have copper, CATV cable for now and replaced with fiber in a few years.

Overall, the cost of the system is still the major problem for massive FTTH deployments. That brings a challenge to FTTH equipment vendors.

10.6.2   LACK OF APPLICATIONS

Another problem with FTTH deployment is that there are no killer applications for FTTH. Although the overall bandwidth requirements are increasing, there is no one killer application that makes FTTH a must. Currently, most Internet applications are low bandwidth applications. CTAV is the number one bandwidth consumer; however, currently in China, CATV programs are not allowed to be carried by networks other than CATV network.

New applications include IPTV and on-line gaming. But none of them need more than 5 Mb/s bandwidth. Although in recent months, IPTV has been a very hot topic in broadband applications, China Netcom, and many other telecos start to provide VOD through the Internet. However, people are not used to pay for each TV program in China because TV program has been free or almost free since it was first delivered to the people.

Another new broadband application in China is the digital TV. Currently, TV programs are carried by cable and HFC systems where TV signals need to be converted between analog and digital for a few times, which not only makes equipment complicated but also seriously derogates the quality of the TV signals. The ideal solution is to carry digital TV signals at baseband without modulation and demodulation. That means more bandwidth is needed. In comparison with 32QAM modulation, for example, the baseband transport needs three times more bandwidth. So, digital TV is a potential killer application. However, the popularization of digital TV is happening much slower than expected. Just take Wuhan as an example, in 2004, the actual number of digital TV subscribers is only about 10 % of that expected.

So, lack of high-bandwith killer applications is another big barrier to FTTH.

10.6.3   REGULATIONS

For most new technologies, the early-stage application needs encouragement from government's regulation. The FTTH in Japan and broadband in South Korea are good examples. The unbundling rule issued by FCC of the United States for FTTH is another good example. But so far the Chinese government has not taken any measures to promote FTTH. The government regulations that need to be modified to promote FTTH have to cover the following aspects:

- The government allows other service providers to operate CATV services with only the source of the TV programs under government's control while leaving the network operation open for competition.
- Currently, the access network is still in monopoly to certain extent. Chinese carriers have a geographically partition of the access network. To bring real competition into access and residential network is to be regulated. The broadband development experiences in Japan show that competition plays a very important role in new technology application at the introduction phase. In this aspect, so-called 'Tailong Model' as we described previously is a good attempt.

10.6.4   COMPETITION FROM OTHER TECHNOLOGIES

Another factor which prevents FTTH from massive deployment now is that existing access technologies are good enough for various current usages. Without TV programs to be carried by telecos' access network, current ADSL is good for most service and applications, The ADSL deployed in China today can provide bandwidth of 2 Mb/s for each subscriber, which is good for most family usage. The new versions of ADSL, ADSL2+ can provide much more bandwidth (25 Mb/s) in many cases. That is enough for today's and most of the near-future's applications. ADSL2+ has been tested by a number of carriers in China and will start to be deployed soon.

In addition to ADSL, HFC, and LAN are also competing with FTTH. For broadcasting TV, HFC is the most preferred solution for now due to its existing network infrastructure and regulation restrictions on TV program broadcasting. LAN is also a good solution for green field because it is straightforward for Internet access and for other IP-based services (e.g., VOIP). Both HFC and LAN are able to provide triple play.

Fixed wireless technologies are also competing with FTTH. Those technologies include WLAN and newly developed WiMAX. Those technologies can provide enough bandwidth when the distance between family home and the central office is within a certain range.

10.7 MARKET OPPORTUNITIES AND DEPLOYMENT STRATEGIES

As we discussed in the previous sections, we have many difficulties to overcome for FTTH deployment in China mainly due to its high cost and lack of killer applications. But that does not mean there are no immediate opportunities at all. In the near future, the following are some immediate FTTH deployment opportunities in China [4,8]:

• FTTO: When we discuss FTTH, it actually includes two types of application environment: one is FTTHome and the other is FTTOffice. For FTTH case, one of the major issues is its high cost. But that issue is not that serious for FTTO case. For FTTO, the customers are more concerned about the services and stability of the system, while the system cost is not their number one concern. In comparison with today's access approach, FTTO is not that costly considering its TDM leased line services. A case study shows that for small business office usage, payback period for teleco operator for FTTO is about 2 years with an assumption that the take rate of FTTO will reach 70 % at the end of the 2 years. So, one of the strategies for promoting FTTH in China at present time is for business office use.

• FTTH for new residential district: Although FTTH is not really needed and not economically realistic at the present time, especially for the houses that already have copper line, the industry accepted that FTTH is the future of broadband access, so for new buildings, people prefer to have fiber to be deployed to home at house construction time rather than to have copper for now and fiber to replace copper a couple of years later. What is more, the cost for copper deployment and the cost for fiber deployment are compatible for the new builds.

• FTTH for CLEC to compete against ILEC: For CLECs, even in the brown field where copper already reached family home, FTTH is a good solution for CLECs to compete with ILECs.

• FTTH for telcos to compete with CATV service providers: Three networks' convergence is an ideal network infrastructure to avoid duplicate network construction. That is more important for access networks. Although this cannot be realized in the near future, people realize that it is the trend, sooner or later the three networks will be converged into one. With that belief, the CATV providers and telcos are competing for FTTH territory. That will also drive FTTH deployment.

- HDTV needs FTTH: Although HDTV is not yet popular in China now, people accept that it will happen in the near future, especially for 2008 Olympic Games. When customers started to demand HDTV and more dynamic programming, such as on-demand programming, the issue of bandwidth emerged. Satellite could deliver HDTV thanks to its nondependence on a physical plant. However, CATV may not adequately support HDTV programming due to bandwidth limitations. The best solution for support HDTV is to use FTTH.

Given above opportunities for FTTH, the strategies for carriers and equipment vendors to take are suggested as follows:

- Starting with FTTO: At the present time, the biggest barrier to FTTH is the cost of the system, especially for family usage where people are very sensitive about the cost. For FTTO, customers are more concerned about services capability and the quality of the services. They are less sensitive to the cost of the system. So massive FTTH deployment can start from FTTO.
- To work together with house builders: One way to promote FTTH is for carriers to work together with house builders. FTTH as a communication system is expensive, but in comparison with a house, it is nothing. In China, most people have their new houses luxuriously furnished. For new houses, people prefer to have their wire (or cable, fiber, etc) to be deployed at the construction time rather than after the house is furnished. When people believe that FTTH is a trend, they can accept current cost. So, FTTH deployment can be bundled together with new house construction.
- To work together with service providers to create more applications and services: The most important thing to promote FTTH is to create more broadband applications. With those applications, the customers can enjoy more services and the service providers can make more money. So, the carriers and equipment vendors need to work together with service providers to create more broadband applications and services.
- Flexible solutions for different environment: The cost of FTTH system is very much dependent on the service it provides. The system cost for triple play is much higher than that for Internet access only. So, we need to differentiate different application environments. For example, for the brown field where people already have phone service and CATV service, FTTH system only needs to provide Internet-based services; the system can be much simpler and the cost can be much lower.
- Take the best advantage of competition: Competition among carriers and service providers is one of the important driving forces for any new technologies. As government, it should make the best usage of the competition to promote FTTH deployment.

REFERENCES

1. CNNIC, China's 16th Internet report, http://it.sohu.com/s2005/netreport16.shtml
2. Zhang X. The current telecommunication status and China's information industry strategies, Eleventh Forum on New Telecomm Technologies and Applications, Beijing, September 2005.

3. Wang J. FTTH in China: Current status and future trends, 2005 Asia-pacific FTTH forum, Taibei, July 2005.

4. Wang J. FTTH in China: Challenges and opportunities, FTTH forum at APOC 2004, Beijing, November 2004.

5. Fiberhome Communications, Zisong FTTH network, Fiberhome's internal report, 2005.

6. Xiong Z. Fiberhome's understanding about China's FTTH, FTTH forum, Shenzhen, September 2005.

7. Leping Wei, "On broadband access networks," Internal technical report, 2005.

8. Jianli Wang, "FTTH in China," China Communications, Vol. 2, No 6, Dec 2005.

9. ITU-T G.984.1, "General characteristics for Gigabit-capable Passive Optical Networks (GPON)," 2003.

10. ITU-T G.984.2, "Gigabit-capable passive optical networks (GPON): Physical media dependent (PMD) layer specification," 2003.

11. IEEE 802.3ah, "Media Access Control Parameters, Physical Layers and Management Parameters for subscriber access networks," 2003.

12. Fiberhome and Beijing Communication FTTP Working Group, "Service Requirement for FTTP," Internal technical report, 2004.

11

Integrated Broadband Optical Fibre/Wireless LAN Access Networks

Ton Koonen and Anthony Ng'oma

COBRA Institute - Technical University of Eindhoven, The Netherlands

11.1   INTRODUCTION

Wireless communication is more and more becoming the preferred first link for people into the worldwide information society. The penetration of mobile phones has already surpassed the one of the fixed telephones. New mobile telephony standards such as UMTS are being introduced into the market in order to extend the communication bandwidth while preserving mobility, the key virtue of wireless communication. In the future, the presence of wireless communication links everywhere in conjunction with mobility support functions will enable a roaming person-bound communication network, in which a person experiences his own specific communication environment everywhere. This 'personal network concept' requires myriads of radio access points maintaining the wireless links and supporting mobility. The evolution towards the wireless-everywhere scenario hence inevitably asks for cheap and easy-to-maintain radio access points, in which as much as possible the signal processing is simplified and where the radio network functions are consolidated in a central site.

Next to their prime feature, mobility, wireless communication links are also offering growing bandwidths to the end-users. This entails an increase of the radio carrier frequencies, which leads to smaller radio cell coverage due to the increased propagation losses and line-of-sight needs. Wireless LANs in the 2.4 GHz range according to the IEEE 802.11b standard carry up to 11 Mb/s, evolving up to 54 Mb/s in the IEEE 802.11g standard. The IEEE 802.11a and the HIPERLAN/2 standard provide up to 54 Mb/s in the 5.4 GHz range. Research is ongoing in systems that may deliver more than 100 Mb/s in the radio frequency range well above 10 GHz (e.g., LMDS at 28 GHz, HyperAccess at 17 GHz and 42 GHz, MVDS at 40 GHz, MBS at 60 GHz, etc.). Due to the shrinkage of radio cells at higher radio frequencies, ever more antenna sites are needed to cover a certain area such as the rooms in an office building, a hospital, the departure lounges of an airport, etc.

Broadband Optical Access Networks and Fiber-to-the-Home: Systems Technologies and Deployment Strategies
Chinlon Lin   © 2006 John Wiley & Sons, Ltd

Hence, for reasons of extending mobility support as well as of extending broadband communication capacity, in wireless networks it becomes increasingly important to simplify the antenna stations and to consolidate the signal processing in a centralised site. Thus, the capital expenditures for installing network equipment can be minimised, as well as the operational expenditures for maintaining and running the network. Carrying radio signals over fibre is an interesting solution to achieve this. It exploits the basic virtues of optical fibre, namely its large bandwidth and low loss, in supporting cheap ubiquitous broadband mobility.

Several techniques have been explored for the transport of radio signals over fibre. These techniques can be divided into two categories: those putting the microwave radio signals on the optical carrier by direct intensity modulation of an optical source, and those using optical frequency conversion for generating the microwave carrier at the antenna station.

11.2   DIRECTLY MODULATED RADIO-OVER-FIBRE SYSTEMS

Basically, three methods may be discerned for transporting radio signals over fibre by deploying direct intensity modulation of the optical source:

(a) Baseband transmission.
(b) Transmission in an intermediate (IF) frequency band.
(c) Transmission in a radio frequency (RF) band.

In baseband transmission, the data are transported in digital format to the base station, where they are modulated on the microwave carrier in an appropriate format (QPSK, multi-level QAM, OFDM, . . .), amplified and radiated by the antenna to the user terminals. As the transport along the fibre is in digital format in the baseband, the fibre line termination equipment is relatively simple, without strong linearity requirements. However, the base station needs to be equipped with signal modulation circuitry which is specific for the wireless communication standard adopted. In case an upgrade is made to another standard, the equipment at all base stations needs to be upgraded.

Following the second method, the data is modulated on a carrier in the IF band. This modulation takes place in the central site, according to a dedicated standard. At the base station, the signal is up-converted to the desired RF band, which requires circuitry of modest complexity. The bandwidth requirements on the line termination equipment are also modest. The linearity requirements may be high, depending on the modulation format chosen. Upgrading the system to another RF frequency band requires an upgrade of the up-conversion circuitry at all base stations.

According to the third method, the data is modulated on the RF carrier directly at the central site. At the base station, the signal just needs to be converted into the electrical domain, and after some amplification it can be radiated by the antenna. If the amplifier and the antenna are designed to handle a broad range of radio frequencies, an upgrade of the system to another RF band does not imply that the base station equipment needs to be replaced. The system also can support multiple RF signals, and thus may readily provide multi-standard operation. However, the bandwidth and linearity requirements on the line termination equipment and the fibre are high, and may restrict the application to the low-to-medium RF range. In multimode fibre links, due to multimodal dispersion the frequency characteristics may show bandpass lobes at higher frequencies beyond baseband, which may be exploited for a higher RF range, provided that the light launching conditions and fibre conditions are stable (see Subsection 11.4.2). In single-mode fibre links, the operation range is restricted by severe dips in the frequency characteristics which occur at certain fibre lengths, caused by fading due to chromatic dispersion (see Subsection 11.4.3.1). Commercial systems are available which support GSM and UMTS wireless communication.

11.3 RADIO-OVER-FIBRE SYSTEMS DEPLOYING OPTICAL FREQUENCY CONVERSION

Carrying multi-GHz analog signals over fibre by direct RF intensity modulation of the optical source requires very-high-frequency analog optical transmitters and receivers, including carefully optimised fibre dispersion compensation techniques. Also, depending on the modulation format, the requirements on the linearity of this modulation process are high. These burdens may be relieved by using optical techniques to generate the microwave carrier. In this section, two methods are outlined, based on optical heterodyning and on a novel principle termed optical frequency multiplication, respectively.

11.3.1 HETERODYNING SYSTEMS

An attractive alternative avoiding the transport of multi-GHz intensity-modulated signals through the fibre is to apply heterodyning of two optical signals of which the difference in optical frequency (wavelength) corresponds to the microwave frequency. The principle is illustrated in Figure 11.1. When one of these signals is intensity-modulated with the baseband data to be transported, and the other one is unmodulated, by optical heterodyning at the photodiode in the receiver, the electrical microwave difference frequency signal is generated, amplitude-modulated with the data signal. This modulated microwave signal can via a simple amplifier be radiated by an antenna; thus, a very simple low-cost radio access point can be realised, while the complicated signal processing is consolidated at the headend station.

The spectral width of the microwave is given by the sum of the linewidths of the two heterodyning light sources, which may exceed the allowable spectral width for adequate detection of the RF modulation format chosen. Hence, this approach requires two light sources with narrow spectral linewidth and carefully stabilised difference in optical emission frequency. Furthermore, the heterodyning requires alignment of the polarisation states of the two received optical signals, and that may be disrupted during transmission in the fibre link.

An alternative approach requiring only a single optical source is shown in Figure 11.2. The optical intensity-modulated signal from a laser diode is subsequently intensity-modulated

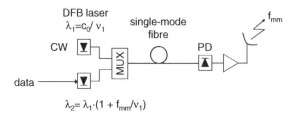

Figure 11.1 Optical heterodyning with two sources.

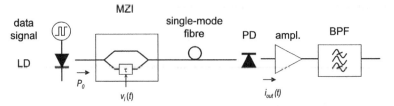

Figure 11.2   Optical heterodyning with a single source.

by an external Mach Zehnder Interferometer (MZI) which is biased at its inflexion point of the modulation characteristic and driven by a sinusoidal signal at half the microwave frequency. At the MZI's output port a two-tone optical signal emerges, with a tone spacing equal to the microwave frequency. After heterodyning in a photodiode, the desired amplitude-modulated microwave signal is generated.

When assuming that the laser diode has a central optical frequency ω_0 and an average output power P_0, and that the MZI's delay τ is modulated by the external signal $v_i(t)$ sinusoidally with an amplitude τ_m and a frequency ω_m, the output current $i_{\text{out}}(t)$ of the photodiode having a responsivity R_d can be shown to be

$$i_{\text{out}}(t) = R_d \sqrt{P_0} \left\{ 1 - J_0(2\omega_0\tau_m) - 2 \sum_{n=1}^{\infty} J_{2n}(2\omega_0\tau_m)\cos 2n\,\omega_m t \right\}$$

The signal $i_{\text{out}}(t)$ thus contains even harmonics of the sweep frequency ω_m, and the $2n$-th harmonic has an amplitude $2R_d\sqrt{P_0}J_{2n}(2\omega_0\tau_m)$. It can also be shown that the phase noise of the optical source is effectively suppressed, provided that the optical source linewidth is much smaller than the MZI's free spectral range. Hence, this method can easily generate pure microwave carriers by generating harmonics of the relatively low modulating sweep frequency.

The system is quite tolerant to dispersion in the single mode fibre link, as the heterodyned sidebands are located relatively close to each other in the optical spectrum.

The transmitter site may use multiple laser diodes as shown in Figure 11.3 [1]. A multiwavelength radio-over-fibre system can be realised with a (tunable) WDM filter to select the

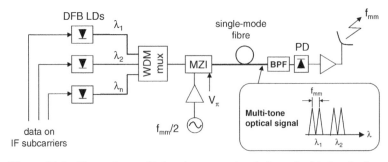

Figure 11.3   Generating multiple microwave signals by optical heterodyning.

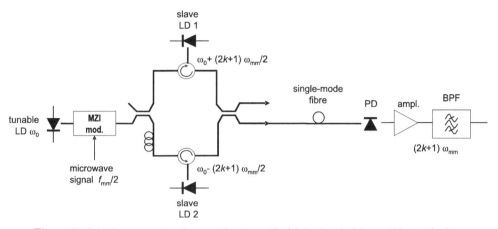

Figure 11.4 Microwave signal generation by optical injection locking and heterodyning.

desired wavelength radio channel at the antenna site. The data signals are modulated on IF subcarriers, which are intensity-modulated on the respective DFB lasers at different wavelengths. The wavelength multiplexed channels subsequently are converted into multi-tone optical signals at different wavelengths by the MZI, of which one wavelength channel is selected at the antenna station and by heterodyning converted into a microwave signal carrying the respective data. The system is tolerant to fibre dispersion, and also the laser linewidth is not critical as laser phase noise is largely eliminated in the heterodyning detection process.

   A very clean microwave signal can be obtained by injection locking two lasers on specific sidebands selected out of the many ones generated by the MZI [2]. As illustrated in Figure 11.4, each slave laser diode is phase-locked to a specific sideband by tuning the laser close to this tone. The two tones at which the slave laser diodes are lasing are separated by the desired microwave frequency. The resulting microwave carrier after heterodyning in the photodiode at the receiver is of high purity and high power.

11.4 OPTICAL FREQUENCY MULTIPLYING SYSTEM

The basic Optical Frequency Multiplying (OFM) concept proposed is shown in Figure 11.5 [3]. In the centralised headend station, the wavelength of a tunable optical source is periodically swept over a range $\Delta\lambda_0$. At the radio access point (RAP), a periodic optical bandpass filter is deployed, of which the bandpass transmission peaks are spaced by $\Delta\lambda_{FSR}$. Linearly sweeping the source wavelength back and forth with a sweep frequency f_m across N bandpass transmission peaks, yields light intensity bursts on the photodiode (PD) and thus a microwave signal with the fundamental frequency $2N \cdot f_m$ and also higher harmonics. The electrical bandpass filter (BPF) selects the particular harmonic for radiation by the antenna. The data signal is chirp-free intensity-modulated on the frequency-swept optical carrier by means of a differentially driven Mach-Zehnder modulator. This intensity-modulated signal is

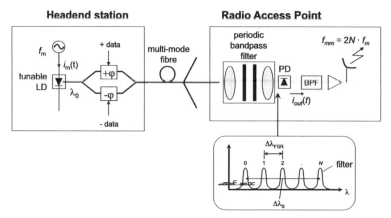

Figure 11.5 Remote generation of microwave signals using Optical Frequency Multiplying.

not affected by the OFM process, and appears as an envelope of the microwave signal at the photodiode output.

When multimode fibre is used, modal dispersion occurs due to the different propagation delays of the many individual guided modes. Assuming a linear behaviour of this fibre, the sequence of transmission of the modes by the fibre followed by processing of the modes by the periodic filter at the receiver may also be reversed, without affecting the OFM process. Thus, the periodic bandpass filter may equally well be positioned at the headend station instead of at the antenna station, yielding the basic setup shown in Figure 11.6 (where an MZI acts as the periodic filter). When the fibre network is passively split, the filter can be shared by a number of RAPs and thus the system's costs are lowered. The periodic filter may be equipped with single-mode fibre pigtails, and may even be integrated within the tunable source module. Furthermore, the periodic filter can be locally controlled and tuned to the source's sweep behaviour, which eases system operation and upgrades.

11.4.1 OFM SYSTEM ANALYSIS

Basically, there are various options to realise the optical periodic bandpass filter, such as a fibre Fabry-Perot filter, a fibre Bragg grating filter or a Mach-Zehnder Interferometer (MZI)

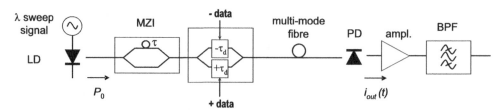

Figure 11.6 Point-to-point OFM system using a Mach-Zehnder Interferometer periodic filter, and multimode fibre (MMF).

filter. Using the latter filter, the point-to-point OFM system can be modelled as shown in Figure 11.6.

Assuming that the optical frequency of the laser diode is harmonically swept at a radian sweep frequency ω_m over an optical radian frequency range $2\beta \cdot \omega_m$ around a central optical frequency ω_0, that the laser light has a phase noise $\varphi(t)$, and that τ is the time delay difference between the two MZI arms (so its bandpass maxima are spaced at its free spectral range $\Delta\Omega_{FSR} = 2\pi/\tau$), it can be derived that neglecting fibre dispersion the output current signal $i_{out}(t)$ of the photodiode is

$$i_{out}(t) = \frac{1}{2}R_d\sqrt{P_0} \cdot \{2 + 2 \cdot \cos[\beta \sin \omega_m t - \beta \sin \omega_m(t - \tau) + \omega_0\tau + \dot{\varphi}\tau]\}$$

The signal $i_{out}(t)$ thus contains even harmonic frequency components $2n \cdot \omega_m$ with relative amplitude $R_d|E_0|^2\cos(\omega_0\tau + \dot{\varphi}\tau) \cdot J_{2n}(2\beta \cdot \sin(\frac{1}{2}\omega_m\tau))$, and odd harmonic frequency components $(2n-1) \cdot \omega_m$ with amplitude $R_d|E_0|^2\sin(\omega_0\tau + \dot{\varphi}\tau) \cdot J_{2n-1}(2\beta \cdot \sin(\frac{1}{2}\omega_m\tau))$.

The relative power of the harmonic components as a function of the optical FM modulation index β is shown in Figure 11.7, assuming a sweep frequency $f_{sw} = \omega_{sw}/2\pi = 2\,\text{GHz}$, and an MZI free spectral range $\Delta\nu_{FSR} = \Delta\Omega_{FSR}/2\pi = 10\,\text{GHz}$. Obviously, a higher FM index β is needed to achieve the maximum signal power at higher harmonic frequencies. For example, $\beta \approx 6.3$ is needed to achieve the maximum power generated in the 6th harmonic of the sweep frequency (i.e., at 12 GHz).

The contribution of the laser phase noise (represented by $\dot{\varphi}$) has negligible impact if $\dot{\varphi}\tau \ll 2\pi$. So if the laser linewidth is much smaller than the MZI's Free Spectral Range $\Delta\omega_{FSR} = 2\pi/\tau$, the OFM process effectively suppresses the impact of the laser phase noise.

Thus, using the OFM method microwave signals with very low phase noise can be generated without requiring a narrow laser linewidth. Experiments have shown extremely narrow microwave spectral linewidths, below the measurement resolution; see Subsection 11.4.4. These pure microwave signals allow comprehensive signal constellations, and therewith high-capacity wireless transmission.

Comprehensive data modulation formats such as QPSK or multi-level QAM can be transported by putting these first on a subcarrier, which is subsequently fed to the Mach-Zehnder

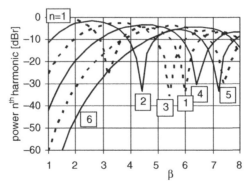

Figure 11.7 Impact of the FM modulation index β on the power of the microwave frequency components (nth harmonics of the sweep frequency) generated by OFM using an MZI periodic filter.

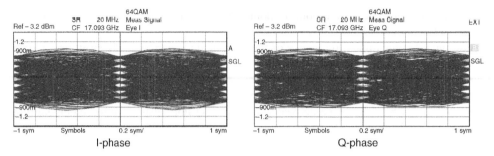

Figure 11.8   Eye patterns of 64-QAM demodulated data at 120 Mbit/s.

intensity modulator. Simulations have shown the feasibility of carrying high-capacity data at 56 Mb/s in 16-QAM on a subcarrier of 225 MHz and a microwave carrier of 5.4 GHz [3,4]. As shown in Figure 11.8, open eye patterns were achieved experimentally after transmission of a 64-QAM signal over 4.4 km graded-index multimode fibre, at a microwave carrier frequency of 17.2 GHz with a symbol rate of 20 MBaud, which implies a data rate of 120 Mb/s (cf. [5]). The error vector magnitude measured was 4.8 %, clearly below the 5.6 % required according to the standards. This indicates the potential for successful data transmission at very low bit error rate.

11.4.2   IMPACT OF DISPERSION IN MULTIMODE FIBRE SYSTEMS

The dispersion in multimode fibre (MMF) is composed of three components: waveguide dispersion, material dispersion and modal dispersion. The first two components also occur in single-mode fibre. Modal dispersion is caused by the difference in propagation delays of the various modes in an MMF.

Usually, modal dispersion strongly dominates the first two components. When the fibre has negligible mode coupling, the impulse response of the MMF then may show multiple impulses, corresponding to the delays in the individual mode groups. Hence, also the MMF's frequency response $H_{IM}(\omega)$ may show various lobes beyond its baseband transfer function, as illustrated in Figure 11.9.

It can be shown that the output current signal $i_{out}(t)$ of the photodiode is

$$i_{out}(t) = \frac{1}{2}R_d|E_0|^2 \cdot \sum_{i=1}^{M} c_i^2 \left\{ 1 + \cos \omega_0 \tau \cdot J_0 \left(2\beta \cdot \sin \left(\frac{\omega_m \tau}{2} \right) \right) \right\} + R_d|E_0|^2 \cdot \sum_{n=1}^{\infty} (-1)^n$$

$$\times \left\{ \sum_{i=1}^{M} c_i^2 \left\{ \begin{array}{l} \cos \omega_0 \tau \cdot J_{2n} \left(2\beta \cdot \sin \left(\frac{\omega_m \tau}{2} \right) \right) \cdot \cos \left[2n \left(\omega_m (t - \tau_i) - \frac{\omega_m \tau}{2} \right) \right] \\ + \sin \omega_0 \tau \cdot J_{2n-1} \left(2\beta \cdot \sin \left(\frac{\omega_m \tau}{2} \right) \right) \cdot \cos \left[(2n-1) \left(\omega_m (t - \tau_i) - \frac{\omega_m \tau}{2} \right) \right] \end{array} \right\} \right\}$$

where c_i is the strength of the ith mode group. This implies that $i_{out}(t)$ contains

- even harmonic frequency components $2n \cdot \omega_m$ with amplitude

$$R_d|E_0|^2 \cdot \cos \omega_0 \tau \cdot J_{2n}(2\beta \cdot \sin(\omega_m \tau/2)) \cdot |H_{IM}(2n \cdot \omega_m)|, \text{ and}$$

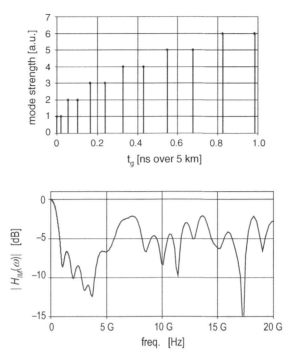

Figure 11.9 Transfer characteristics of parabolic graded index fibre, neglecting chromatic dispersion and mode mixing (numerical aperture NA = 0.2, core diameter $2a = 50\,\mu\text{m}$, fibre length $L = 5\,\text{km}$, wavelength = $1.3\,\mu\text{m}$, full-NA central launching) *left*: impulse response; *right*: frequency response.

- odd harmonic frequency components $(2n - 1) \cdot \omega_m$ with amplitude

$$R_d|E_0|^2 \cdot \sin \omega_0\tau \cdot J_{2n-1}(2\beta \cdot \sin(\omega_m\tau/2)) \cdot |H_{IM}((2n - 1) \cdot \omega_m)|.$$

Thus, due to modal dispersion in an MMF, the amplitudes of the harmonics generated by the optical frequency multiplying process are scaled linearly with the amplitude of the MMF's intensity modulation transfer function.

11.4.3 IMPACT OF DISPERSION IN SINGLE-MODE FIBRE SYSTEMS

The dispersion in single-mode fibre (SMF) is composed of waveguide dispersion and material dispersion, and is caused by its nonlinear phase characteristic. It has a larger impact on the transmission of microwave signals when using RF intensity modulation than when using the Optical Frequency Multiplying method, which is discussed in the following section.

11.4.3.1 RF Intensity Modulated System

In an RF intensity modulated system with direct detection at the receiver (RF IM-DD), when intensity modulation with a modulation depth m is done by a single tone ω_m, the input field to the fibre shows two sidebands, one at each side of the laser central frequency ω_0. These sidebands are spaced by ω_m from this central frequency, and will experience different phase delays with respect to the carrier because of the nonlinear phase characteristic of the dispersive single-mode fibre. The output current signal $i_{out}(t)$ of the photodiode in the receiver can be shown to be

$$i_{out}(t) = 2R_d\sqrt{P_0}\left\{1 + \tfrac{1}{4}m^2\cos^2\omega_m(t - \beta_0'L) + m\cos(\tfrac{1}{2}\beta_0''L\omega_m^2)\cos^2\omega_m(t - \beta_0'L)\right\}$$

where β_0' and β_0'' are the first and the second derivative of the fibre's phase characteristic $\beta(\omega)$ at the laser frequency ω_0, respectively. The second derivative can be related to the fibre's dispersion D by $\beta_0'' = -D\lambda_0^2/(2\pi c_0)$.

For small m, the amplitude of the output signal is thus proportional to $|\cos(\tfrac{1}{2}\beta_0''L\omega_m^2)|$. Hence, for an RF modulation frequency ω_m the transfer of RF IM-DD signals shows severe dips at fibre lengths $L = (2k + 1)\pi/(\omega_m^2 \cdot \beta_0'')$, with k integer. For example, when the modulation frequency is 30 GHz, the fibre dispersion $D = 16$ ps/nm·km, and the light's wavelength $\lambda_0 = 1.55$ μm, $L = (2k + 1) \cdot 4.34$ km. This has been confirmed by simulations; see Figure 11.10.

Therefore the transport of microwave signals by means of the IM-DD method through dispersive single-mode fibre may suffer from serious fading problems at certain fibre lengths.

11.4.3.2   Optical Frequency Multiplying system

In case of an OFM system, the input field to the fibre is actually frequency modulated by the sweep signal, and therefore contains much more frequency components. After transmission through the dispersive single-mode fibre, having a length L and assuming negligible attenuation, the output current signal $i_{out}(t)$ of the photodiode can be shown to be

$$i_{out}(t) = i_{out,0} + R_d|E_0|^2\sum_{n=1}^{\infty}\sqrt{A_n^2 + B_n^2}\,\cos\left[n\omega_m(t - \beta_0'L - \tfrac{1}{2}\tau) + \tfrac{1}{2}\beta_0''Ln^2\omega_m^2 - \arctan\frac{B_n}{A_n}\right]$$

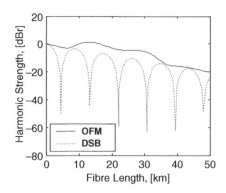

Figure 11.10   The impact of chromatic dispersion on the OFM system and on the RF IM-DD system (DSB), in case of delivery of a 30 GHz signal at 1550 nm wavelength in an SMF link.

where

$$i_{\text{out},0} = \tfrac{1}{2}R_d|E_0|^2 \sum_{k=-\infty}^{\infty} J_k^2(\beta)[1 + \cos(\omega_0 + k\omega_m)\tau]$$

$$A_n = \sum_{k=-\infty}^{\infty} J_k(\beta)J_{k-n}(\beta)\cos(\beta_0''L\omega_m^2 nk)\left[\cos\big((\omega_0 + k\omega_m)\tau - \tfrac{1}{2}n\omega_m\tau\big) + \cos\big(\tfrac{1}{2}n\omega_m\tau\big)\right]$$

$$B_n = \sum_{k=-\infty}^{\infty} J_k(\beta)J_{k-n}(\beta)\sin(\beta_0''L\omega_m^2 nk)\left[\cos\big((\omega_0 + k\omega_m)\tau - \tfrac{1}{2}n\omega_m\tau\big) + \cos\big(\tfrac{1}{2}n\omega_m\tau\big)\right]$$

The output signal $i_{\text{out}}(t)$ thus contains the higher harmonics $n \cdot \omega_m$ of the sweep frequency, of which the amplitude $\sqrt{A_n^2 + B_n^2}$ is affected by the fibre dispersion through $\beta_0''L$.

11.4.3.3 Comparison of impact of chromatic dispersion on an RF IM-DD and on an OFM system

Simulations have shown that the OFM system is quite robust against dispersion in an SMF link, and in particular more robust than the RF IM-DD system. As illustrated in Figure 11.10, the strength of the 5th harmonic at 30 GHz generated by OFM is much less dependent on the fibre dispersion than the amplitude of a 30 GHz signal delivered by the RF IM-DD transmission technique (double sideband, DSB, technique). The transmission wavelength assumed is 1550 nm, and the fibre dispersion $D = 16 \, \text{ps/nm.km}$. The IM-DD signal strength shows the dips corresponding with the nulls in the transfer function derived before. The increase in the OFM harmonic strength after 10 km is attributed to the extra FM-to-IM conversion in the fibre.

11.4.4 EXPERIMENTAL RESULTS

Experiments have confirmed the feasibility of the OFM process, and its inherent laser phase noise suppression characteristics as mentioned in Subsection 11.4.1 [6]. Figure 11.11

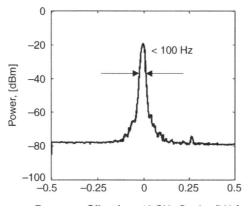

Figure 11.11 Generated microwave carrier at 16 GHz.

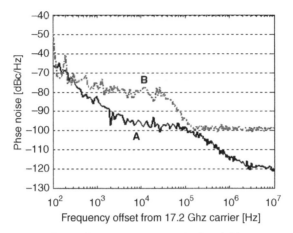

Figure 11.12 Phase noise of 17.2 GHz signal measured after 4.4 km MMF; (A) using the OFM concept, (B) using direct intensity modulation.

shows that a 16 GHz microwave carrier can be generated having a FWHM spectral width of well below 100 Hz whereas the laser linewidth was 1 MHz. A very pure carrier at 17.2 GHz was found after traversing 4.4 km of silica 50 μm core multimode fibre, using a sweep frequency of 2.87 GHz. Its phase noise was found to be considerably lower than that of a 17.2 GHz microwave delivered through the same fibre link by a commercial high-quality signal generator which directly intensity-modulated the laser diode; see Figure 11.12.

Also clearly open eye patterns were measured using the OFM process, when transmitting 100 Mbit/s PRBS data in on/off keying format on a carrier of 17.2 GHz.

11.5 BI-DIRECTIONAL MULTIPLE-ACCESS SYSTEM

In order to support delivery of interactive services to multiple user sites, the OFM system concept can be extended to the bi-directional point-to-multipoint system exemplified in Figure 11.13. Two wavelengths λ_0 and λ_1 are used to separate down- and upstream signals, respectively. For upstream, the microwave carrier generated at the Radio Access Point (RAP) plus a frequency shifter is used to down-convert the microwave signal received from the mobile terminal, and the resulting IF signal is sent by direct intensity modulation of the RAP's laser at the upstream wavelength λ_1. A medium access control mechanism is needed to avoid collision of the upstream data sent by the respective RAPs; for example, an SCMA scheme where a different subcarrier per RAP is used by means of frequency-shifting the microwave carrier with an adjustable local oscillator.

The analogue nature of the radio-over-fibre systems and the propagation delays added by the fibre have to be taken into account for enabling bi-directional wireless systems. Thus, systems with distributed controlled MAC schemes (such as the current IEEE 802.11 wireless LAN systems) put stringent requirements to the maximum fibre lengths that can be deployed without degrading dramatically the system performance in terms of spectrum efficiency and mean data throughput. Systems with centrally scheduled MAC schemes (such as HIPERLAN/2 and the fixed-wireless access IEEE 802.16 systems) relax fibre length

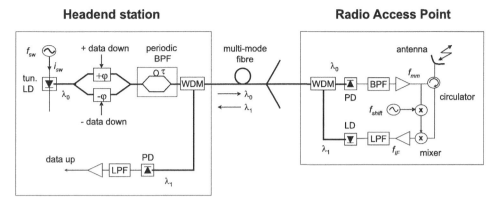

Figure 11.13   Bi-directional point-to multipoint system.

constraints and allow more flexibility for dynamically controlling the propagation delay added by the fibre.

11.6   INSTALLATION ASPECTS OF IN-BUILDING RADIO-OVER-FIBRE SYSTEMS

Single-mode fibre has highly suitable characteristics for distribution of microwave signals, for transporting directly intensity modulated RF signals as well as for remotely generating microwave carriers by optical heterodyning [7]. However, its tiny core necessitates delicate handling in installation, requiring highly skilled personnel, which results in relatively high installation costs.

In highly cost-sensitive areas, such as in-building networks, multimode fibre is an interesting alternative. Its large core facilitates splicing and easier light injection at the source, and avoids nonlinearities due to reduced light intensity. Polymer multimode optical fibre (POF) is even easier to install, due to its flexibility and ductility. It allows connector-isation by just cramping a metal ferrule on the fibre, without cracks as otherwise would occur with silica fibre. Moreover, multimode fibre is already widely accepted for short-range data communications in broadband LANs, benefiting from low-cost transceiver modules and the installation easiness. Several Ethernet standards have been established using multimode fibre: 100 Mb/s Fast Ethernet IEEE 802.3u standard 100BASE-SX for up to 2 km multi-mode fibre at 850 nm wavelength, and 100BASE-FX up to 2 km at 1310 nm; Gigabit Ethernet (line rate 1.25 Gb/s) IEEE 802.3z standard 1000BASE-SX up to 550 m multimode silica fibre at 850 nm, and 1000BASE-LX up to 550 m at 1310 nm; even 10 Gigabit Ethernet (line rate 10.31 Gb/s) up to 300 m multi-mode fibre, at 850 nm wavelength.

Striving for convergence of in-building networks for reasons of service integration, upgradability and economy of installation and maintenance, an attractive scenario would be to build radio-over-fibre systems on top of (already installed) multimode fibre data networks such as the Ethernet-based ones mentioned above. Such a multimode fibre-based integrated-services in-building network is exemplified in Figure 11.14. Both high-speed fixed-wired devices (e.g. desktop PCs with Gigabit Ethernet ports) and mobile devices (e.g. equipped with wireless LAN access) can be hosted in a single multimode fibre-based infrastructure.

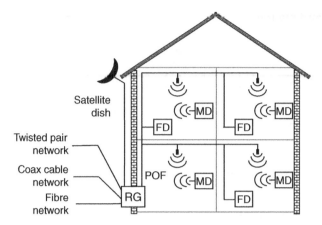

Figure 11.14 Integrated-services network using multimode (Polymer) Optical Fibre in a Residential/ Office/Public building. POF, Polymer Optical Fibre; FD, fixed device; MD, mobile device; RG, residential gateway.

11.7 DYNAMICALLY ALLOCATING RADIO CAPACITY

Wireless networks typically show considerable dynamics in traffic load of the Radio Access Points (RAPs), due to the fluctuating density of mobile users in the radio cells. By allocating the network's communication resources to the RAPs according to their instantaneous traffic load, these resources can be more efficiently deployed, thus generating higher revenues for the network operator.

Figure 11.15 illustrates how the OFM technique in combination with wavelength routing can yield such a dynamic capacity allocation [8]. At the central site, a number of frequency-swept sources each operating at a specific central wavelength are intensity-modulated by

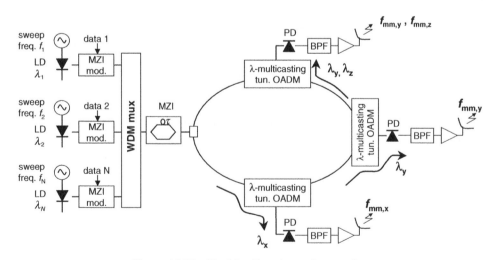

Figure 11.15 Flexibly allocating radio capacity.

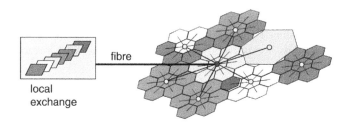

Figure 11.16 Flexible capacity allocation among network cells.

their respective downstream data streams. After wavelength multiplexing, a single periodic bandpass filter (i.e. an MZI) performs the OFM processing. The signals are injected into the ring network, where multicasting optical add-drop multiplexers (OADMs) can be tuned to drop one or more wavelength signals to a RAP.

The RAPs are wavelength-agnostic; a RAP thus emits all the microwave signals carried by each of the wavelength signals dropped to it. Besides the electrical bandpass filter which has to select the desired microwave frequency, a RAP is also fairly frequency-independent. Therefore, this setup allows multi-standard operation, in which the same RAP can handle a single microwave signal as well as simultaneously multiple signals following different standards.

By allocating the radio capacity dynamically among the cells in a network, capacity can be provided on demand. For example, when a hot spot with high traffic load emerges, the respective RAP can provide an extra microwave carrier as soon as another wavelength channel is directed to it (see Figure 11.16). A network operator can thus readily re-allocate the wireless communication capacity in his network.

11.8 CONCLUDING REMARKS

Radio-over-fibre techniques enable simplified antenna stations and centralisation of radio signal processing functions, and thus may reduce both the capital expenditure and the operational costs of broadband wireless communication networks. Optical frequency conversion techniques such as heterodyning and optical frequency multiplying (OFM) relax the requirements on the fibre link bandwidth. With the OFM technique, microwave signals can be delivered to simplified Radio Access Points through single-mode as well as multimode fibre networks; this enables the integration with fixed services such as Gigabit Ethernet in in-house multimode fibre data networks. In comparison to conventional RF intensity-modulated direct-detection (IM-DD) systems, the radio frequency generating electronic modules operate at much lower frequency, and the OFM system is more robust against fibre dispersion both in single-mode and in multimode fibre networks. Furthermore, the OFM process yields microwave signals with a strongly reduced phase noise, which allows complex signal constellations (e.g. higher-level QAM) and thus high transmission capacities.

By means of wavelength multiplexing, OFM radio-over-fibre wireless services can be readily integrated with wired services (such as Gigabit Ethernet) in a single network. As a RAP can handle several wavelength channels simultaneously, the OFM technique supports operation at multiple radio standards and multiple radio carrier frequencies. Dynamic wavelength routing combined with OFM enables flexible allocation of wireless communication capacity, and thus improved efficiency of using the network's resources.

ACKNOWLEDGEMENT

Part of the work described has been funded by the Dutch Ministry of Economic Affairs in the B4 Broadband Radio@Hand project and in the IOP GenCom programme, as well as by the European Commission in the 6th Framework Programme project MUSE – Multi Service Access Everywhere. The authors thank Maria García Larrodé for valuable discussions.

REFERENCES

1. Griffin RA, Lane PM, O'Reilly JJ. Radio-over-fibre distribution using an optical millimeter-wave/DWDM overlay. Proceedings of OFC '99, San Diego, February 22–25, 1999, paper WD6.
2. Ismail T, Liu C-P, Mitchell J, Seeds A, Qian X, Wonfor A, Penty R, White I. Bidirectional transmission of broadband wireless signals using a millimeter-wave over fibre CWDM ring architecture. Proceedings of NEFERTITI Symposium, Brussels, January 17–19, 2005.
3. Koonen T, Ng'oma A, Smulders P, van den Boom H, Tafur Monroy I, van Bennekom P, Khoe G-D. In-house networks using Polymer Optical Fibre for broadband wireless applications. Proceedings of ISSLS 2002 (XIVth International Symposium on Services and Local Access), Seoul, April 14–18, 2002, paper 9–3, pp. 285–294.
4. Koonen T, Ng'oma A, Smulders P, van den Boom H, Tafur Monroy I, Khoe G-D. In-house networks using multimode polymer optical fiber for broadband wireless services. *Photonic Network Communications* 2003; **5**(2): 177–187, ISSN 1387-974x.
5. Ng'oma A, Koonen AMJ, Tafur-Momroy I, Boom HPAv.d., Khoe GD. Using Optical Frequency Multiplication to Deliver a 17 GHz 64 QAM modulated signal to a Simplified Radio Access Unit Fed by Multimode Fiber. Proceedings of OFC 2005, Anaheim, March 6–11, 2005, paper OWB2.
6. Ng'oma A, Tafur Monroy I, Koonen T, van den Boom H, Smulders P, Khoe G-D. High-frequency carrier delivery to graded index polymer optical fibre fed next generation wireless LAN radio access points. Proceedings of ECOC 2003, Rimini, September 21–25, 2003.
7. Mitchell JE, Attard JC. WDM overlay fibre-radio access networks. Proceedings of NOC 2003, Vienna, July 1–3, 2003, pp. 212–219.
8. Koonen T, Ng'oma A, Larrode MG, Huijskens, F, Tafur Monroy I, Khoe G-D, GD 2004. Novel cost-efficient techniques for microwave signal delivery in fibre-wireless networks. Proceedings ECOC 2004, Stockholm, September 5–9, 2004, pp. 120–123.

12

Broadband Optical Access, FTTH, and Home Networks – the Broadband Future

Chinlon Lin
Center for Advanced Research in Photonics, Chinese University of Hong Kong, Hong Kong, China

12.1 INTRODUCTION – A HISTORICAL PERSPECTIVE

The landscape of telecommunications is now undergoing a major change. Over the last decade the world has seen a great transformation in telecommunications, and its impacts on the human civilization in digital information transmission and storage, media and entertainment. This is in many ways due to the advances in computer networking and the emergence of global Internet, which were made possible by the various advances in the high-speed, high-capacity broadband telecommunications technologies. While satellite and wireless microwave transmissions played important roles, the invention of lasers and optical fiber communication systems has made the ubiquitous global telecom infrastructure a reality. Without this global telecom network interconnecting millions of Internet sites over many continents worldwide, such a transformation would have been impossible.

Since the invention of lasers by Prof. Charles H. Townes and the first demonstration by Dr. Ted Maiman in 1960, many laser scientists and engineers have explored the idea of using laser light for high-bandwidth optical communications, but the early free-space laser communication systems or gas-lens-guided laser systems were not suitable for transmission over long distances due to the uncontrolled nature of the atmospheric conditions and the laser beam diffraction in an unguided medium or unstably guided medium. It was not until after 1970 when the first low-loss silica glass optical fiber was developed at Corning and Bell Labs, based on Dr Charles Kao's earlier measurement, analysis, and predictions, that long distance-guided laser communications based on optical fibers waveguides became a reality. Over the last ~35 years, we witnessed the tremendous progress in photonic technologies for optical fiber communications, thanks to the great effort of a few pioneers and many scientists and engineers in the major telecom R and D labs of the world [1].

Broadband Optical Access Networks and Fiber-to-the-Home: Systems Technologies and Deployment Strategies
Chinlon Lin   © 2006 John Wiley & Sons, Ltd

Today the world has a well-connected global optical fiber network, consisting of hundreds of long-haul trans-Atlantic and trans-Pacific undersea optical fiber links and systems, as well as thousands of major terrestrial optical fiber routes. The development of dense-wavelength-division-multiplexing (DWDM) and erbium-doped optical fiber amplifier (EDFA) technologies have greatly helped the successful progress toward establishing this global telecom infrastructure, which interconnects 100s of countries in the world and has changed the world in a very fundamental way. The world's basic information infrastructure, it can be said, relies critically on these global optical fiber backbone networks [2].

The successful establishment of these global high-capacity long-haul networks is essential, but this alone does not mean that residential customers worldwide will automatically have broadband access to high-speed data and multimedia services. The idea of Fiber-to-the-Home (FTTH) was started back in the mid-1980s when telecom R and D labs [3–6] began to consider the possibilities of bringing optical fibers and 'information superhighway' all the way to the residential homes and home offices, making possible almost unlimited bandwidth to the residential customers. Some of the previous chapters in this book described the early efforts in the mid-1980s to early 1990s in the research and development of FTTH, Fiber-in-the-Loop (FITL), and Passive Optical Network (PON) in Bellcore (on behalf of seven RBOCs), British Telecom, and NTT, among others. The early development of FTTH led to several well-known and successful technology field trials in, for example, Bell South, AT&T, Bell Atlantic, British Telecom, France Telecom, Deutsche Telecom, NTT, etc. [7]. The subsequent analyses showed that the cost for FTTH with fiber to every customer's home was too high and no high-bandwidth killer applications or services existed (long before the Internet is here) to justify deployment of any significant scale at that time. Some fiber- and optoelectronics-sharing architectures such as FTTC/FTTN, C for Curb, N for Node, or Neighborhood, with copper plant from the neighborhood optical node to multiple customers, were considered a lower cost alternative in order to bring broadband access near the homes. The need for a lower-cost FTTH system was then envisioned, and Full Services Access Network (FSAN) consortium was established in 1995, mainly by the major carriers around the world, later expanded to include most major PTTs and Telcos worldwide, and major equipment vendors. The idea of FSAN was to define common technical requirements and standardize some FTTH systems such that potential worldwide deployment of standard-based FTTH systems would mean lower cost and good interoperability [8]. FSAN has since developed several PON standards specifications that were proposed to and adopted by ITU-T [9]. For the history of FSAN and many PON-related technology standards development, refer to Chapters 2, 7, and 8. In its operation FSAN was successful in the end result that the first installation cost of FTTH system was reduced to about $1500 USD in 2000 (costing about $5000 USD per home passed in the late 1980s), making FTTH deployment within economic reality. In fact, FSAN-related ATM-PON test trials at NTT and Bell South were started in the 1998–1999 time frames, the fore-runners of today's FTTH systems. For more detailed discussions on the various PON systems (A-PON, B-PON, G-PON, GE-PON, etc.) refer to the earlier chapters in this book (e.g., Chapters 7, 8, and Chapter 2). Note point-to-multipoint (P2MP) PON architectures are popular in North America and part of Asia but nonPON-based P2P Ethernet architectures are currently more popular in Europe, as discussed in Chapter 3.

In the year 2003–2004 timeframe, NTT started a large-scale deployment of PON-based systems with RF video overlay for their FTTH and FTTB residential customers across Japan (see Chapter 1), partly due to NTT's new vision on "optical generation" but mainly in response to emerging competitions within Japan. Such competitions in Japan are important FTTH drivers and will be discussed in more details in a later section in this chapter. Today NTT is leading the world in the large-scale FTTH deployment, with close to 3 million FTTH customers and growing rapidly. New subscribers for FTTH-based services (more than 100 000 new homes added every month) has exceeded that for ADSL-based services in NTT for the first time in the first quarter of 2005, even though VDSL would allow a higher bandwidth (up to 30 Mb/s) than the old ADSL services [10]. Even more significantly, NTT has established a target of moving 30 million customers to FTTH-based services by year 2010 [11]. This requires a significant commitment in their CAPEX investment in the next 5 years and an aggressive time schedule, with total CAPEX for FTTH/NGN to reach 5 trillion JPY in 2005–2010, and the expected cost savings from the OPEX reduction of 800 billion JPY by 2010. See Chapter 1 for details on FTTH design and deployment considerations of various PON systems (BPON, GPON, GE-PON, and WDM PON) for FTTH-based broadband services in NTT.

In the US, in 2004, SBC (now new AT&T) and Verizon also announced serious efforts in FTTP trials and deployment planning, and started massive FTTP deployment in 2005. It is called FTTP (P for premises) to include various customer premises including residential homes, apartments, etc., but FTTH is now often used interchangeably. 'Verizon is committed to making FTTH cost effective and is actively changing the way, it does business in order to ensure that FTTP is a success. *Verizon is beyond the point of return on FTTP*' [10]. As declared in Chapter 2 of this book by the SBC team, '*There is no doubt that the age of Fiber-to-the-Home (FTTH) has arrived.*' Indeed, the era of nearly unlimited-bandwidth broadband optical access with FTTH has arrived, finally, almost 20 years after the very early FTTH trials in the US in the mid 1980s by the telecom companies like Bell South and Bell Atlantic, and AT&T.

12.2 THE BROADBAND ACCESS TECHNOLOGY OPTIONS – xDSL VERSUS CABLE MODEM, HFC VERSUS FTTH, PON VERSUS P2P ETHERNET

It is now well known that there are various technology options for broadband access providing high-speed Internet access and triple-play services including data, voice, and video. Most well-established broadband access platform now is the xDSL service for the Telcos, and the Cable Modem-based services for the Cable TV companies (often called Cable MSOs, for multi-system operators). It is in general agreed that FTTH provides the ultimate in bandwidth and flexibility in upgrades when considering really high-speed broadband access, especially with data rate of 100 to 1000 Mb/s or above.

Over the last 10 years the development of high-speed xDSL (digital subscriber line) technologies started in the US using advanced signal processing and modulation techniques has allowed the short-distance transmission of 1.5–8 Mb/s and even 20–50 Mb/s data over the traditional copper wires designed for the traditional voice services (or POTS, for plain-old-telephone service). Majority of ADSL services today have data rate of 1.5–6 Mb/s, though VDSL services of up to 50 Mb/s or more seem quite possible, if the copper distance is sufficiently short. However, as the speed of xDSL goes up, the distance of transmission becomes shorter and shorter, making it necessary to have a fiber-deep architecture such as FTTN or fiber-to-the-node or curb/cabinet (FTTC) deployed to support such higher speed xDSL services. Certainly one can continue to use even more bandwidth efficient spectral coding and modulation techniques for xDSL service beyond 100 Mb/s or even Gb/s, but it will reach a point of diminishing return in terms of gained advantage. While it is a good economic idea to reuse the existing copper plant as much as possible, it may reach a point when the upgrade to higher-speed xDSL services is more expensive, when OAM cost is considered in addition to the first installation cost, than a fiber-all-the-way to the customers (FTTH or FTTP, P for premises) deployment. Compared with FTTC/xDSL combination for the existing household or brownfield broadband installations, FTTH is now the preferred choice in new ('greenfield') installations. In some 'brown field' installation or upgrades where FTTC/xDSL may or may not prove to be less expensive, FTTH is also considered. Replacing copper plant with an all-fiber plant has indeed started, as stated by SBC, but it may take a relatively long time for the complete change-over, depending on the local economic considerations and customers' service needs as well as service providers' individual optimization criteria.

South Korea is a well-known leader in the broadband xDSL services. Broadband access with 8–10 Mbps provided by xDSL platforms now reaches 70–80 % of the homes and businesses in South Korea. There is a burgeoning demand for better and faster data communications possible with FTTH. South Korea is ready to move to the next level of high-speed service with nationwide 50–100 Mbps FTTH and mobile access to a single IP-based network capable of offering all multimedia-based broadband services. In South Korea, there is already a well-established demand for faster, higher-speed Internet access for various types of broadband services, so the migration from xDSL to FTTH seems to be a natural evolution and will happen with or without new additional FTTH services drivers. See Chapter 4 for details of broadband access in Korea.

For the Cable TV industry in the US, it is a different situation compared with Telcos. The Cable TV industry started in the US in the late 1970s as a CATV, a community antenna TV network for better reception in rural areas where the reception of the over-the-air broadcast TV was poor, so the community set up a larger antenna to have better reception, and subsequent distribution of the signals over coaxial cables in a tree and bus architecture to multiple community residential users. The coaxial-cable (Coax)-based transmission and distribution system, working with analog RF subcarrier-multiplexed multiple-channel TV signals, is not an advanced signal transmission system as the RF amplifiers and coax cables are prone to signal degradations and distortions. However, over the last 30 + years many significant changes have taken place [12,13]. First, the development of sophisticated Headend (HE, to be compared with the CO, central office of the Telephone companies) multi-channel video systems by companies such as Scientific Atlanta (SA, recently acquired by Cisco) and General Instruments (GI, acquired by Motorola and now part of Motorola Broadband) have significantly changed the nature of the CATV from a community master antenna concept to that of a popular video services provider. When the Cable TV companies merged and became bigger, they interconnected the consolidated HEs with digital optical fiber transmission technologies developed mainly by the R and D labs of the Telecom industries, significantly improving their video trunking quality and capability. Secondly the use of lasers and optical fiber transmission technologies even for analog RF subcarrier-multiplexed multi-channel video signal transmission was made possible by R and D engineers at major Telecom labs, such as Bellcore, AT&T Labs, GTE Labs, and others [see, e.g., 14–16]. This was followed by the commercial introduction of such optical fiber transmission system products by companies such as Ortel (later acquired by Lucent) and Harmonic that became OEM suppliers to the then two major Cable TV equipment companies, Scientific Atlanta (SA), and General Instrument (GI). This allowed the much higher quality analog video signal transmission over fiber before they are converted back to the traditional RF signal at the optical node at the end of the optical fiber link, for further distribution over the coax tree and bus plant [12,14–16].

By combining the fiber for the feeder portion of their plant and coax for the distribution bus and drops, the Cable MSOs now have the powerful hybrid-fiber-coax (HFC) plant. The HFC architecture uses the power of optical fiber transmission technologies and shortened significantly the coax part of the video distribution plant. This eliminated many problems with the RF coax amplifiers in the feeder plant and improved greatly the bandwidth of the entire network. The quality of the video transmission and distribution is as a result improved by several orders of magnitude, transforming the entire Cable TV industry into one with telecom services capabilities. Further advances, as a results of research work at telecom research labs such as Bellcore, NTT, BT, and AT&T Labs, have also greatly helped the Cable TV industry by making it possible to use EDFAs, DWDMs and nonlinearity reduction techniques for the higher capacity and more flexible trunking and distribution of RF-video encoded optical signals over the HFC networks [14–18]. For example, the first commercial EDFA for cable TV industry video distribution application was developed by Synchronous Communications (later acquired by Motorola Broadband) following Bellcore's first research report in 1991 and NTT's report in 1992 [16–18]. During these time periods of early 1990s to the end of 1990s, Cable MSOs have invested billions of USD [19] for the network upgrade to the optical-fiber-enabled HFC based platform, with wider RF bandwidth (upgrading from 450 or 550 MHz systems to 750 and 870 MHz systems). With the HFC network upgrade and the development of advanced digital modulation (e.g., 64- and 256-QAM) and video encoding and compression technologies (e.g., MPEG 2, MPEG 4, etc.), and the coax cable and RF amplifiers having much higher bandwidth than the copper wire, the Cable TV industry's HFC network platform is ready to provide also high-speed data and voice services in addition to the analog and digital TV services [12,20,21]. The industry in this time period of 1990s also put a great deal of effort in the development and marketing of high-speed data-over-cable technologies of Cable Modem (CM) with cable modem termination systems (CMTS) in their headends (HE). Such technologies were developed by many companies as a result of the successful coordination effort by CableLabs (a joint laboratory set up by the major players in the Cable TV industry) in standardization of high-speed data-over-cable service interface specifications (DOCSIS) and interoperability testing programs [12,22].

It is important to recognize that this large-scale upgrade from the traditional all-coax plant to the fiber-rich HFC networks with a short coax-bus distribution network has dramatically transformed the entire Cable TV industry, to the extent that now they could become the full service providers, and not just the TV or video services providers, which is in their root. Some forward-looking Cable MSOs then began to call themselves telecommunication companies (e.g., Cox Communications, Charter Communications) instead of just Cable TV companies. The NCTA, the National Cable TV Association, is now the National Cable and Telecommunications Association. The SCTE, the Cable TV industry's main industry engineering association, has also changed the official name from Society of Cable TV Engineers to Society of Cable Telecom Engineers [23]. Furthermore, some of the Cable TV

HEs (Headends) now combine the functions of both HE and CO (central office) as well as Data Centers, so some of these are now called Master Communications Centers, and have transmission and switching/router equipments for massive video, telephony, and high-speed data distribution and communications.

Actually, it took more than 10 years for both xDSL and Cable Modem technologies to reach the stage of any significant scale of actual deployment after the basic technologies have been proven early in the laboratories. The strong competitions between xDSL and Cable Modem technologies for broadband access have a long history, with CM ahead of ADSL in initial deployment (though DSL probably had a longer technology history). A lot of pros and cons in comparisons between the two access technology options can be readily found in the open literatures. In general xDSL is dedicated to a particular customer with typically lower speed (1.5-6 Mb/s) data services (except until recently when VDSL comes into the picture), while Cable Modem service is shared by many users receiving data from the same optical node at the fiber end of the HFC network, typically with higher data rate (20-30 Mb/s) services shared by a group of customers (could be more than a few hundreds). So, the actual access speed in terms of bit-rate per subscriber comparison depends on many factors including how many are sharing the same CM bandwidth at a given time. One can see many advertisements from each camp trying to focus on its own unique advantage. But, even a side-by-side comparison will be non-deterministic as it depends easily on the time of access-speed testing, and the best-case and worst-case scenarios in CM sharing due to the statistical multiplexing nature of the packet data bandwidth sharing.

In the US the CM-based high-speed data access services started first before the ADSL-based services, and for many years even up to now, the CM subscribers outnumber ADSL service subscribers by about a factor of 1.6 [13], which may or may not be due to technology reasons. This popularity with CM services is in sharp contrast to the cases outside US, most likely because Cable TV companies outside US are in general not as strong as the US Cable MSOs and have yet to become strong competitors of Telcos. Note some major US Cable MSOs have formed joint ventures to establish Cable MSOs in Europe and Japan.

With major US telecom companies, such as SBC (new AT&T) and Verizon seriously planning and deploying FTTH on a large scale to achieve a competitive advantage, Cable MSOs in the US are also now evaluating their options for facing the competition. Chapter 9 by Jim Farmer of Wave 7 discusses in depth the evolution and migration from HFC to FTTH, and it presents a comprehensive view from an industry expert familiar with both platforms of broadband access. It seems that Cable MSOs are feeling the competition pressure being built up by the Telcos' FTTH visibility [23], but they believe they have an HFC network with sufficient bandwidth upgrade flexibility to meet the short-term needs, so their FTTH plan, if established, would most likely be a longer-term future plan, perhaps to be considered seriously 7–10 years from now. However, the competitive advantages of an all-fiber, all-digital, all-IP, single converged network based on FTTH could come in somewhat sooner, especially with the mandated year 2009 over-the-air all-digital broadcast TV transition, and the expected future popularity of HDTV, and HD video streaming and video communications over the broadband Internet. Therefore, it is likely that Cable MSOs' migration time table from the current HFC to a fiber deep, mini-optical-node HFC, and then to full FTTH, will need to be reevaluated as time progresses, especially if their Telco competitors are able to offer full 100–1000 Mb/s *symmetrical* high-speed access, and all-digital IP TV in the next 5–7 years or so.

Note both 100 Mb/s DSL and 100 Mb/s Cable Modem services are possible with more advanced signal processing, encoding, and modulation techniques, and in fact even Gigabit per second transmission may be possible over existing copper wires and coax cables (which have much higher intrinsic bandwidth than copper wires). But they come with limitations and cost increases. With the cost of optical fiber access dropping rapidly from large-scale deployments, and its ability to accommodate the nearly unlimited bandwidth (e.g., with 10 GbE and WDM) without access distance limitations, FTTH is the ultimate choice of broadband wireline access which ensures future upgrade ability for a long time to come.

12.3   BROADBAND FTTH DRIVERS, TRIPLE-PLAY, COMPETITION AND IPTV

As discussed in the Introduction section, the FTTH concept and trials started in the US by major telecom companies in mid-1980s. In addition to high first-installation-cost, drivers or killer applications demanding the broadband capability of FTTH-enabled services were not there to justify the deployment of FTTH during that time period. But, now in 2005–2006, as SBC team stated in the Introduction in Chapter 2, the age of FTTH has finally arrived. What have been the drivers for significant FTTH deployment over the last few years? What led many community networks to begin to establish their own broadband FTTH network independent of the major telecom operators? Why did many rural communities in the US start the FTTH deployment before the major cities? What are the real bandwidth needs in the future and how does the perceived future bandwidth requirement drive the FTTH planning?

The real drivers for FTTH include many interconnected aspects. Certainly, greatly reduced first-installation-cost (FIC) of the FSAN standard-based FTTH systems (using now low-cost high-performance optoelectronics technologies) is an important factor to consider first, where the installation cost may easily dominate over the actual system equipment cost. The increased customer demands for faster, higher-speed broadband access and lower overall triple-play services cost is another. The promise of an ultimate high-capacity ultra-broadband access with practically unlimited future upgrade capability is one key factor for deploying optical fiber-based access network, as it is the 'once-for-all' broadband access network. The long-term potential cost reduction in terms of OPEX (operation and maintenance expenses) is also a very important factor. For example, as already stated in the Introduction, NTT expects to have OPEX savings of 800 B JPY from FTTH deployment in 2005–2010. The rate of investment return critically depends on the revenue versus cost of providing FTTH-enabled broadband services. Furthermore, Verizon sees 'the FTTP deployment as an opportunity to reinvent the network and to move, from manual order taking to web-based order fulfillment; from copper-pair allocation to service/ bandwidth allocation via software; from service activation via installation and dispatch to software-based activation/deactivation; from limited reactive trouble isolation to proactive performance monitoring at the OLT and the ONT, and from manual asset inventory to auto-discovery and reporting of assets' [10].

There are also factors that are highly dependent on the national and regional telecom policy (e.g., e-Japan), regulatory conditions and additional economic incentives, as well as the specific competitors' offering or strategies. From a brief analysis of the FTTH deployment in several major countries, high-speed multimedia-based Internet access plus the voice and video services, namely the 'triple-play,' is the well-known leading service driver for FTTH. However, even NTT admitted there is no single 'killer application' that demands the high bandwidth which the current xDSL and CM access technologies cannot provide. It is the total expected high-bandwidth need (there seems to be a consensus around about 50–60 Mb/s per average residential subscriber, for now) for emerging new services including the expected HDTV, HD video streaming (on-line VOD, video on demand), two-way video communications and video conferencing, interactive on-line gaming, HD video file sharing and video 'BLOG,' etc., over the broadband Internet. Therefore, it seems that there are many emerging high-bandwidth video-centric applications but not a single 'killer application' as people used to expect. In addition, some leaders in the field of high-speed home networking have also pointed out (see a later section on Broadband Home Networks) that the potential of home networks for in-home digital cinema (home theater or home entertainment room) with large-screen displays, and personal HDTV production with low-cost HDTV video cameras will probably accelerate the bandwidth needs to the 100 Mb/s–1000 Mb/s range, leaving FTTH the only ideal access option for the future.

It can be observed that the real bandwidth needs may not be as important as the *perceived* needs of the customers. This is mainly an issue of customer's *psychology*. The new players of the broadband service providers would like very much to convince the customers that they need the advanced higher speed Internet access (and comes with it the low-cost voice service via VoIP) offered only by the new service providers. Their most obvious competitive strategy is to emphasize in their advertising their technical edge in higher and higher speed access which can be easily understood by average customers in terms of speed of Internet access, and price of service, assuming the quality and reliability of service are about the same as their competitors'. Indeed, such competition has been really one of the most important drivers in many major FTTH efforts in several countries at this stage.

Since the future high-bandwidth services will be video-centric as compared with simply voice and traditional high-speed Internet access, Telcos will need a competitive video services strategy to bring the triple-play or multiple-play services (full services) to customers. As a consequence, now IPTV or TV over IP (Video over IP as compared with VoIP, the Voice over IP) becomes *essential* for Telcos to provide, as *they do not have any choice but to provide such video services* in order to face the competitive challenges. This single most critical consideration has now been changing the inner workings of FTTH deployment plans of many Telcos worldwide, and will continue to dominate the forces behind the broadband competition for years to come [24–26]. See also Chapter 2 regarding SBC's IPTV considerations in their FTTH/FTTN deployment. This one single factor, like high-speed Internet access, will likely transform the entire telecom/cable TV industry landscape forever, and usher us into the Broadband Future.

Note that the importance of TV or video services as broadband drivers was clearly recognized by the telecom industry since the mid-1980s. Most of the early FTTH trials included video delivery strategies as well as video technology and services trials. There is also a long history in this aspect. One can remember, as a video broadband strategy, Bell Atlantic's early attempt to acquire or merge with TCI, one of the major Cable MSOs in the early 1990s. The merger did not work out. There was also the attempt in creating VOD centric consortium by several major RBOCs and hiring the top managers from the Video and Entertainment Media industry to head such a venture. It did not work out either. Then around 1998 AT&T spent about $106 B USD to acquire both TCI (Tele-Communications International, Inc.) and Media One, and called it AT&T Broadband, which became the largest Cable

TV MSO in the US [27]. This was considered centric to the new broadband strategy of AT&T at the time. AT&T Broadband was then formed with headquarters in the old TCI headquarters in Denver. This acquisition by AT&T of key video services players was in the correct strategic direction, but the purchase and selection of TCI was considered a mistake by some industry insiders, as it was too expensive, with TCI's cable plant being the least upgraded among the major Cable MSOs. AT&T has also officially stopped the Telegraph service so the official name AT&T can actually be considered to represent *American Telephone and Television* instead of the original *American Telephone and Telegraph*. In addition to the expensive purchase, the task of truly integrating TCI and Media One, now AT&T Broadband, into AT&T was a formidable one, and time was not the AT&T side. Eventually after less than 4 years of operation, in December 2001, it was announced that Comcast was to merge with (or acquire) AT&T Broadband. AT&T sold the entire AT&T Broadband, its premier broadband Cable TV division, back to the then third largest Cable MSO, Comcast Corp., for only $47 B USD, making Comcast the largest Cable MSO (AOL/Time-Warner was the second largest). This buy and sell led to a huge loss for AT&T of over $58 B in less than 4 years. This was considered a tremendous failure on the AT&T management side to properly implement or execute their video broadband strategy, which in itself could have been a correct one in the right direction. Some analysts think that this single disaster has caused AT&T such a significant financial loss that it ultimately led to the end of AT&T. In January 2005, the acquisition of the remainder of AT&T by SBC for a mere $16 B USD was announced [27]. The demise of this venerable 100+ years old pioneer telecom giant, the great Ma Bell, AT&T, was certainly unanticipated at the 1984 divestiture. This exemplified a truly significant historical event in the fundamental transformation of the telecom landscape over the last 25+ years. The gradual loss of Voice revenue to the new VoIP competition plus the failure to execute its new Video strategy led finally to AT&T's demise. AT&T's failed attempt to redefine its broadband future with its bold video broadband strategy may have been the most pivotal cause. After acquiring AT&T, SBC decided in January 2006 to rename itself the new AT&T, with a new logo. So the story continues.

More such fundamental changes are expected to come in the Broadband Future when the transition from a Narrowband World to the Broadband World goes through the complete evolution, and Television (and Video Communications) will dominate the telecom landscape just as Telephone (and Voice Communications) used to. The development of ultra-broadband FTTH optical access and high-speed in-home networking is just part of this transition process toward the Broadband Future. But, with this great broadband transition, the human society will have arrived at the era of image-and video-dominated broadband telecommunications, the era of ubiquitous Visual Communications, the most natural form of human communications.

12.4 BROADBAND COMPETITIONS WORLDWIDE: A FEW EXAMPLES

If we take a closer look at the Broadband FTTH drivers behind NTT's major FTTH initiative, and the large-scale FTTH plans of SBC (Project Lightspeed) and Verizon (Project FiOS), certainly the desire to offer advanced triple-play services of voice, video, and Internet access over a converged network is an important one, but the forces of competition are perhaps driving the rapid increase in the Internet access speed or data rate that the FTTH technology platform can easily provide. New comers certainly like to make customers believe that they offer newer, more advanced services at lower cost to consumers, thereby creating a market share for the newcomers. The easiest or most obvious advertisement focus is probably the high-speed Internet access and ability to offer some types of video services. In the following, a few examples of broadband competition will be described and the drivers behind the push to FTTH discussed. Such examples of broadband competition from Japan, US, and some parts of Europe will serve to illustrate the diversity of current and emerging broadband service providers and their offerings, and their competitive strategies. In the coming decade, these Broadband competitions worldwide will redefine the landscape of telecom and information services as well as digital media and entertainment.

12.4.1 EXAMPLES OF BROADBAND COMPETITION IN JAPAN

In Japan, NTT is now divided into NTT East and NTT West, covering different regions. The main nationwide competitors in the broadband access are: NTT East, NTT West, KDDI, Yahoo! BB (Broadband), UCOM (part of USEN, a Cable Music Broadcaster), and J-COM, formerly Jupiter Communications, the main Cable TV service provider [11]. The broadband services market competition in Japan focused on the triple-play services, with emphases on three aspects: (1) price competition for high-speed Internet access; (2) Internet access speed competition; and (3) service competition. The main broadband services include high-speed Internet access, IP video services, IP telephone services, and broadcast video services.

NTT's regional communications business is served by NTT East and NTT West, which provide telephone and broadband services to more than 60 million subscribers. Among the competitors of NTT, Yahoo!BB is a relatively new broadband service brand name but is among the most aggressive in taking away the market share from NTT in the triple-play service market in Japan. It was established by Softbank BB (Softbank is a key investor of Yahoo!). Under the Yahoo!BB brand, it offers such services as Internet connectivity, IP telephony, public wireless LAN services, video-on-demand, and online gaming. It has been aggressively competing against NTT in the higher speed Internet access offering. Yahoo!BB offers both ADSL and Hikari (Light) services, with Yahoo!BB ADSL services (8–12 Mb/s and up to 50 Mb/s in 2006) to 4.9 million users. Yahoo! Hikari services can provide symmetric 100 Mb/s high-speed Internet access. Both are packaged with IP telephone or VoIP services (brand named BB Phone by Yahoo!BB) [28].

Yahoo!BB now also offers nation-wide IP video broadcasting service called BBTV [28]. It is now working with equipment vendors such as Cisco, UTStarcom, and others to implement multi-casting technology for video services nationwide as the basis of its BBTV broadcasting service, an IPTV service which has over 5000 VOD titles, and offers programming on 36 channels, with further expansion of video content planned [29]. Such IPTV service will compete directly with J COM, the Cable TV service provider, and NTT, which is also actively establishing IPTV service platforms. OnDemandTV, a joint venture of NTT West and Itochu, has selected SkyStream's Mediaplex-20 video headend to support a nationwide rollout of IP-based broadcast television in Japan. OnDemandTV will use NTT East and NTT West's networks to deliver TV directly to consumer homes. The service will initially provide 23 broadcast channels as well as an extensive library of video on demand (VoD) movies [30]. The OnDemandTV service is currently available to NTT West subscribers in the Osaka metropolitan area. Initially, the service is being broadcast over NTT West's fiber network using MPEG-2 compression. OnDemandTV plans to expand its service on a national scale to the remainder of NTT West and all of NTT East subscribers in the near future, delivering content over its widespread xDSL network. Note that while much of the past competition has been in the ADSL offering, the newer competition arena is now in the 'Broadband FTTH Services' or 'Hikari, light-enabled services.' For FTTH services delivery, NTT, being a founding members of the FSAN, is deploying PON-based (BPON, GPON or GE-PON) system with or without RF video overlay for FTTH as well as P2P access technologies for FTTB services delivery, respectively. Yahoo! BB, on the other hand, is mostly deploying the P2P active Ethernet platform offered by Cisco without using the FSAN/ITU-T standardized passive-splitter-based PON systems, until recently. In late 2005 they started using UTStar-com's new Gigabit Ethernet PON system for supporting their 'Yahoo! BB Hikari Fiber-to-the-Home (FTTH) service,' including IPTV [31], but not without initial glitches, as is the case that many other broadband service providers are facing when introducing a new technology to speed up their competitiveness in offering advanced broadband services.

Based on the information available in the open literature and in the news releases, it is clear that most of the national broadband service providers in Japan, including NTT East, NTT West, KDDI, Yahoo! BB, UCOM, and even the Cable MSO, J-COM, are all offering or planning to offer FTTB (FTT-MDU) fiber-grade services for up to 100 Mb/s symmetric data access. Most are using or plan to use Gigabit Ethernet PON or P2P active optical Ethernet technologies to build up their broadband access networks. Some combines FTTB with new VDSL technologies within the buildings to the individual apartments. For example, UCOM focuses on the FTTx market for MDUs and offers symmetrical 100 Mbps services to customers in MDUs.

The high level of broadband competition in offering high-speed IP-based access services in Japan is responsible for the rapid revenue growth in 2005 of many core and edge router companies in the US (Cisco, Juniper, Redback, Avici, etc.). Also, most of the broadband service providers in Japan have already established plans for offering IP video broadcast or on-demand VOD services, through their fiber-deep high-speed IP-based access networks. Such high-speed broadband-access-enabled IPTV offerings compete directly with the traditional cable television services offered by the Cable TV companies such as J-COM.

J-COM is Japan's largest Cable MSO with 2 Million subscribers and was founded in 1995 as Jupiter Telecommunications as a joint venture involving Sumitomo of Japan and TCI of US. J-COM's principal investors also include Microsoft, Mitsui, and Matsushita Electric. Beginning in late July 2005, J-COM started to offer multiple-dwelling unit (MDU) customers Internet services with access speeds of up to 100 Mbps (both upstream and downstream) by extending its existing fiber-optic cable network to the outside wall of participating MDU buildings (FTTB or FTT-MDU), and using coaxial cable within the building. The 100 Mbps service is offered to MDU customers as part of a bulk package that participating building owners or managers will purchase from J-COM to serve the entire MDU building. This is no longer the regular HFC, but is FTTB as the optical node is inside the building, with coax distribution within the MDU building. This sets up the 100 Mb/s as the primary competitive offering for high-speed Internet access fro broadband service providers in Japan in the 2005–2006 time frame [32].

If over the next few years all major broadband service providers in Japan can offer triple-play services including voice, data and video; what would be the next competitive differentiator other than the traditional price and service quality? If both the inexpensive VoIP service and the high-speed Internet access at 100 Mb/s symmetric data rate become commonly available from all major service providers, the next true competitive differentiator could very well be the IPTV and IP VOD services. This is a brand new area presenting many technical and operational challenges for the broadband service providers, especially those with the telecom heritage, not only in terms of new technologies (hardware, software, and system integration) but also in terms of video or multimedia contents (movies and DVD libraries). Therefore, IPTV is indeed the next arena for keen competition over the next 10 years for all major broadband service providers in Japan, just as is the case in many other countries [33].

12.4.2 EXAMPLES OF BROADBAND COMPETITION IN EUROPE

In Europe there are now strong competitions in the broadband services market within many countries, and service providers are also reaching beyond their own countries to compete. Most of the broadband competition is still in the platform of xDSL high-speed access for the residential customers, with some Cable Modem-based high-speed broadband access where the Cable MSOs are strong, except probably in Sweden and the Netherlands where FTTH platform is starting to take effect quickly.

In UK, British Telecom (BT) is the incumbent telecom carrier with ADSL broadband service offerings, and it is facing competition from several fronts. ISP Wanado, a France Telecom's subsidiary in UK, became a strong competitor to BT with 2 Million subscribers. France Telecom owns both Wanadoo and mobile network operator Orange, and said the Wanadoo name will be rebranded Orange in 2006 in the UK and across its other European domains to create brand unity among France Telecom's on-line and mobile businesses. This interesting fact shows the nature and diversity of international competition between France Telecom and BT in UK, BT's territory, and similar competition elsewhere in Europe [34].

Furthermore, in UK, two major Cable MSOs, NTL and TeleWest Broadband (service brand name 'Blue Yonder Internet') are both offering Cable Modem-based broadband Internet access services up to 4 Mb/s, in addition to cable TV and IP telephony services, in competition with BT and others [35]. Both NTL and TeleWest have close ties to major Cable MSOs in the US due to joint venture investments in the past.

France is also one of Europe's most competitive markets for broadband Internet services with triple-play service growing rapidly, primary over the xDSL platform (ADSL and ADSL2$^+$). France Telecom has about half of the residential broadband subscribers. France Telecom is facing strong competition from Iliad, Neuf Cegetel (merger of Neuf Telecom and Cegetel, the fixed line unit of French media group Vivendi Universal), Club Internet (a German operator subsidiary), as well as Telecom Italia's subsidiary ISP Tiscali, with business in France. Many provide ADSL2$^+$ high-speed Internet access (with IP telephone service) of up to 18 Mb/s for short reach, and with Club Internet and Neuf offering digital IP TV services [36].

These two examples of competition in UK and in France simply show how competitive the broadband services market is in Europe in general, as competitors in France are companies representing French, Italian, and German interests, and competitors in UK representing British, French, and US interests. It can be expected that in the near future FTTH-enabled broadband network would be the next platform to push the speed frontier of competition in offering IP HDTV and higher-bandwidth services in Europe.

12.4.3 EXAMPLES OF BROADBAND COMPETITION IN THE US

In the US, the strong competition between telecom operators and Cable MSOs is well known. Overt the last 5 years or so, some Cable MSO have offered even the PSTN-based telephony services in addition to their broadcast analog and digital TV, and high-speed Cable Modem-based Internet access services. In the competition between Cable Modem and xDSL-based broadband Internet access services, the total number of CM subscribers has been much larger than the ADSL subscribers in the US. This may be due to many reasons including ADSL's distance limitations and later adoption. There has been the perception that Cable MSOs have been more aggressive in marketing and pushing the technology deployment, while Telcos were a little late in doing the same. Recently, it seems the growth of xDSL-based subscribers has been faster than the CM-based subscribers, as Telcos become more familiar with the technology deployment and operations, as well as the marketing strategies.

To take the initiative and prepare for higher bandwidth Internet access and triple-play competition, major telecom operators such as SBC (AT&T) and Verizon have committed to large-scale deployment of FTTH access network across their own respective regions. Verizon FiOS project is based on the BPON as specified by the ITU G.983.x. It basically consists of a 622 Mbps downstream data stream at nominally 1490 nm, while the upstream is at 155 Mbps at 1310 nm. The video is transported via an additional wavelength at 1550 nm as an RF overlay [10]. This RF overlay for video delivery is not adopted by SBC where an all IP network is envisioned even for video (IPTV) when FTTH deployment started (see Chapter 2). Verizon intends to spend $20B delivering fiber to most of his customers, the most expensive plans in the world after NTT. The broadband access service speeds are 5 and 15 Mb/s for residential customers and 30 Mb/s to business customers. Verizon passed 1 million homes in 2004 (home passed, not necessarily connected with services), and expects to pass an additional 2 million home in 2005. A good reason Verizon is moving rapidly to fiber-based FTTH access platform is that they have lagged behind in DSL. In California, mostly an SBC state, DSL yearend 2003 beat Cable Modem 2.34 to 1.93 M. In New York, mostly a Verizon State, 537000 DSL subscribers were outmatched 3-1, by 1.75 M of Cable Modem subscribers. Similarly, Cable Modem is ahead 862 K to 301 K in New Jersey, 597 K to 196 K in Virginia, and 704 K to 253 K in Massachusetts, all Verizon states. But the gap has been narrowing in the last few quarters in 2005. Verizon also faces Cablevision's intense price-cutting. With FTTH, Verizon has already rolled out video service in late 2005 and early 2006 with its RF video overlay system [37]. It also plans to deploy GPON (up to 2.4 Gb/s downstream and 1.2 Gb/s upstream) instead of BPON, starting 2006.

For SBC, its project Lightspeed has the following strategies: for the greenfield areas, they will deploy FTTH; in brownfield/overbuild areas, they will deploy a FTTN (N for node or neighborhood) platform that utilizes VDSL in the last link over copper line. SBC began deployment with BPON and expects to move to GPON as the best direction for continued full service networks supporting IP video. Figure 12.1 shows a generic PON architecture can be deployed to have FTTH and FTTN/FTTC/VDSL configurations. The networks will support switched digital video (SDV) employing IP as the end-to-end protocol. Thus, there is no RF overlay in the SBC access networks, unlike the case of Verizon. Note that even within the two major US telecom operators the FTTH strategies are quite different. SBC opted for an all-IP network with FTTH access, so they do not deal with analog TV broadcast services but offer IPTV services instead.

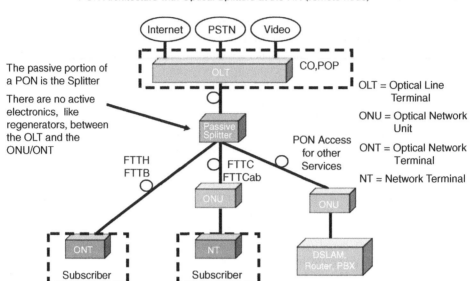

Figure 12.1 A generic PON architecture for both FTTH/FTTB (Fiber to the Building with Apartments or MDUs) and FTTN/FTTC (node/curb) where ONU is followed by xDSL over existing twisted pair copper wires to reach customer's homes.

Verizon, on the other hand, has RF video overlay for their FTTH deployment and therefore can offer video services immediately that are similar to what the Cable TV companies are offering, without the need for set-top boxes in customer's homes (special digital set top box needed for IPTV). As a consequence though, local Cable TV franchise permits are needed in local municipalities which could add to the delay and deployment cost.

The rollouts of FTTH and FTTN in SBC began in the last quarter of 2005, and SBC's goal is to expand the Project Lightspeed initiative to 18 million homes within 3 years. All of the customers will have a video service option as part of the triple-play service offering. See Chapter 2 for more details regarding SBC's FTTH strategies, technology considerations, and triple-play services including IPTV and deployment plans.

Many Cable MSOs such as Comcast and Cox are now aggressively offering higher-bandwidth services including HDTV and digital TV/VOD services in addition to the VoIP and high-speed Internet access via Cable Modems. According to market studies by Infonetics Research, the number of North American Cable VoIP subscribers jumped to about 500,000 in 2004, a 900 % increase from 2003. The penetration of VoIP service delivered directly from MSOs among cable broadband subscribers is expected to increase from 2% in 2004 to 15% in 2007 [38].

For high-speed Internet access, the number of Cable Modem subscribers has been significantly higher than the DSL subscribers in the US, as mentioned before. Since VoIP is a 'broadband phone' service over Cable Modems, it is an effective way for Cable MSOs to take some voice service market share from the incumbent Telcos. So in the triple-play, data, voice and video, Cable MSOs are ahead of Telcos and now the real arena for competition turns to the emerging video services of digital TV, VOD, and HDTV. On this account, Cable MSOs are at this stage way ahead of the Telcos as Cable MSOs not only have the heritage and broadband HFC network for TV broadcast and digital video service delivery, but may also have the upper hand in terms of *video content ownership and distribution rights*. This is because major Cable MSOs have ownership or investment in major media and video content producing companies, partly due to close historical and business ties. Furthermore, the Cable TV industry has been aggressively offering both the digital HDTV contents and digital VOD services now, perhaps preparing to preempt the next stage of competition. Telcos' recourse is then FTTH-enabled high-bandwidth high-speed Internet access, and with it, IPTV- and IP-based VOD services. They also need an effective video content acquisition and management strategy. It is therefore essential for Telcos to go with FTTH in order to compete with higher and higher speed Internet access services with an IP converged network.

SBC and Verizon's serious FTTH deployment in many regions of the country and the visibility from FTTH-enabled higher speed Internet access will be a challenge to the Cable MSOs, who believe that HFC network platform can provide sufficiently high bandwidth to compete effectively, so that FTTH for the cable network plants can be planned for a much later stage. In fact, there are currently solutions which can provide up to 100 Mb/s access services over fiber-deep HFC networks by using a 'multiple-channel bonding technique.' By combining the multiple RF channels using the familiar multi-level digital modulation techniques such as 256 QAM, downstream speed of 120–160 Mb/s (with four-channel bonding) or higher could be achieved by this channel bonding [39]. DOCSIS 3.0, a new emerging CableLabs spec. that will implement this 'channel bonding' technique, can boost downstream speeds to 100 Mbps and higher, competitive with FTTH-enabled services at this access speed, except that it is shared among the users from a fiber node at the end of the HFC network. In fact, based on the same technique, 200 Mb/s downstream services over fiber-deep HFC network is also under consideration, to fend off the all-fiber FTTH competition. It also supports IPv6, which will give MSOs the ability to better manage their IP address pools and gain more sophisticated provisioning tools. DOCSIS 3.0 is in the final review period, and a final spec could be ready by early 2006 [40], but it will take some further effort to make it a reality.

The time frame for real FTTH deployment needed by the Cable MSOs besides the HFC network is probably 7–10 years out, as the broad bandwidth of a fiber-deep HFC with coax drop access network could be utilized to its full extent with these advanced signal processing and encoding techniques. This will delay the need for full FTTH upgrade, at least for the next 5–7 years. See Chapter 9 by Jim Farmer of Wave 7 for an in-depth analysis and discussion of the evolution path from an HFC network to one with combined HFC/FTTH networks. This seems to place the Cable MSOs in a favorable position in their competition with the Telcos. Telcos have no choice but to deploy FTTH and FTTN/VDSL with IPTV, with digital video-on-demand services. Indeed, in the triple-play competition where voice is no longer centric and video is in Cable's heritage, and with Cable Modems competing effectively with DSL-based high-speed Internet access, Telcos have little choice but to expand their strategies and become successful with their new *Video-Centric* broadband services competition. In this sense, all-fiber ultimate optical broadband access with FTTH is a must for Telcos to compete, while Cable TV industry can take time to go through the gradual evolutionary upgrade toward FTTH in the future from their already upgraded fiber-deep HFC networks. The impact and the benefits of an all-IP converged FTTH network as compared with the benefits of gradual evolutionary upgrade of a hybrid high-bandwidth HFC network may take some time to be fully realized and fully appreciated.

12.5 BROADBAND COMPETITION IN HONG KONG

Hong Kong is a world-leading city in terms of high-speed mass transportation as well as advanced telecom services. It has one of the highest mobile phone and broadband access penetration rate in the world. Within this tiny special administrative region of China there are about 7 million Hong Kong people, not counting the many visitors to this city. Having world's best international airport and well known for being the Asia Pacific's center of sea-freight services and logistics facilities, Hong Kong's transportation and telecom infrastructure had to be world class to serve its commercial and logistics operations. Telecom statistics from the Hong Kong's OFTA (Office of Telecom Authority, the FCC of Hong Kong) indicated that the number of mobile phone service subscribers exceeded 8.4 million as of October 2005, representing $\sim$121% penetration of the population [41]. It is also one of the few cities in the world to have early adoption of the 3G cell phone services providing video, image and data access on mobile phones. The broadband penetration rate exceeded 60 % in 2004. In December 2001, the capacity of external telecommunications facilities recorded a tremendous increase over the year from 18 to 235 Gbps. The major increase in bandwidth was mainly attributed to the installation of new submarine optical fiber cable systems with increased capacity brought in to Hong Kong, and was made possible due to the rapid development of many high-capacity DWDM/EDFA-based undersea optical fiber systems in this time period [1].

In the broadband local access services market in Hong Kong, there are many competitors. The key players in Hong Kong's broadband market are: PCCW, HGC, HKBN, and i-Cable, to be described in more details below. An interesting aspect is that Hong Kong is well known for being a 'vertical city' with many tall high-rises (commercial and apartment buildings) in its landscape over a relatively small area of land between hills and bays. Because most of the Hong Kong residents live in one of these high-rise high-density building complex with many apartments, FTTB (Fiber-to-the-Building or MDUs) is the norm for the building access, with a relatively short distance of nonfiber (copper wire or CAT 5 cable or coax) distribution plant to the customer, making it easier for cost-effective broadband deployment. Sometimes fiber in the building riser (FTTR, R for Riser) is also deployed when necessary. Most of the buildings are residential and commercial buildings with fiber already connected to the buildings, so FTTB is almost ubiquitous in Hong Kong. Often the competitive advantage of one serviced provider against the other is whether the provider has access to a particular building or building complex. For example, HGC, being part of a conglomerate that owns many real-estates including both commercial and residential buildings in Hong Kong, has fiber (FTTB) connected to these buildings and therefore has competitive advantages in reaching the residential customers in these buildings.

The incumbent telecom operator is PCCW (which has the Hong Kong Telecom heritage). It is offering xDSL-based broadband access under the brand name of Netvigator, at 1.5, 3, and 6 Mb/s access speeds. Its broadband service is under the name of NOW Broadband [42], which is among the early pioneers in successfully offering IPTV (switched digital TV with a digital set-top box) to residential customers. FTTB/VDSL services with access speed up to 20–30 Mb/s are also being considered for IP HDTV service in 2006.

In addition to the PCCW's NOW Broadband, the main broadband service providers in Hong Kong include Hutchison Global Communications Broadband (HGC), Hong Kong Broadband Network (HKBN, a subsidiary of City Telecom, formerly a CLEC), and the Cable TV company, i-Cable. Both HKBN [43] and HGC Broadband [44] offer optical Ethernet-based high-speed access from 10 to 100 Mb/s, much faster than PCCW Netvigator's current offering of 6–8 Mb/s ADSL-based service, although VDSL-based higher speed access up to 20–30 Mb/s is being considered by Netvigator for their next-generation NOW Broadband IPTV services. The smaller Cable TV company i-Cable provides Internet access at speed of up to 10 Mb/s. It turned out that NOW Broadband has captured over 0.5 million IPTV subscribers already and compete quite well against i-Cable which so far offers traditional TV broadcasting services. Both i-Cable and NOW Broadband require set-top boxes, so customers are used to having set-top boxes for their TV set, which is good for NOW Broadband's IPTV offering. HGC Broadband and HKBN have similar optical Ethernet-based services over the access network with FTTB (fiber carrying Gigabit Ethernet to the building) plus CAT 5 cabling and Ethernet switches inside the apartment buildings. Both of them are offering video services over their Ethernet broadband IP access networks. HKBN, an aggressive broadband service provider working closely with Cisco, has been offering *bb10* and *bb100* to residential customers for 10 and 100 Mb/s access speed. HKBN also started to offer even higher speed, 1 Gb/s Internet access, under the service name of *bb 1000*. This may be the first city in the world with such high-speed access offering beside Tokyo, where Yahoo! BB also has service offering of up to 1 Gb/s access. In HKBN, this is with optical fiber all the way (FTTH) to the customer's residence or apartment [45]. HGC Broadband also offers 10 and 100 Mb/s Ethernet services with its extensive FTTB network in Hong Kong. It also offers IPTV service under the brand name of YesPlus TV [44].

Two notable facts on the broadband competition in Hong Kong are:

1. PCCW is a world leader (with Telco heritage) in the first successful IPTV rollout (NOW TV) and significant subscriber market share, using 6 Mb/s ADSL broadband access, and with an aggressive video content acquisition plan. While their incumbent telecom operator status has certainly helped, NOW Broadband did have an aggressive high-visibility marketing effort and TV content acquisition effort. To face the competition, they are also planning the VDSL offering for HDTV-based IP video services. This would be an upgrade from their ADSL service, currently at 6 Mb/s and available to over 90 % of PCCW customers. A very important experience at PCCW's NOW Broadband is that IPTV delivery appears to be more secure and piracy-proof than Cable TV or Satellite TV delivery. PCCW's NOW Broadband service, which uses three-layered content protection comprising network-based conditional access, digital copyright protection and analog copyright protection, says that this extensive security measures results in 0 % piracy, compared to the 15 % piracy other pay TV operators in Hong Kong are reporting. Furthermore, since in the digital switched video service, viewer statistics is easy to come by, the content providers have a good handle on which video programs are watched by how many viewers, which is not possible in a conventional broadcast TV mode. All of this is "music to every video content owner's ears". In PCCW's case, this has enabled NOW TV to secure premium content for their IPTV platform, some of which has never been shown before in Hong Kong and probably would never have been without IPTV [33].

2. HKBN is an aggressive all-IP Ethernet-based broadband player in Hong Kong to offer high access speed of 10–100 Mb/s (FTTB + CAT 5E cabling access) and 1000 Mb/s (FTTH) for all-IP triple-play services (telephony, broadband, and IPTV). Overall, HKBN has more than 530,000 broadband customers. HKBN installed and owns all their fibers and did not need to lease from others. HKBN also worked exclusively with Cisco's products. HKBN's 'All-IP Platform for All Services' includes the FTTH service rollout in 2005 for *bb1000*, the symmetric Gb/s services. With this, HKBN indicated that their *bb1000* is 'making Hong Kong the first market in the World whereby 30 % of the total households, approximately 800,000 households out of a total of 2.2 million households, can now enjoy World Leading Fiber-to-the-Home (FTTH) symmetric 1 Gbps Internet access service.' [45–46]. This is indeed an excellent example of the FTTH-enabled access speed in real-life broadband service offering. However, with an initial monthly subscription fee of ~$220 USD per month for *bb1000*, although it will probably have many enthusiastic early subscribers, it will take quite a few years to have significant number of common residential customers. This may have to wait until after the ultra-high-bandwidth services such as HD IPTV and HD video communications become more common or when enough computers begin to have direct Gb/s Ethernet interfaces as standard LAN interfaces instead of the current 10/100 Mb/s Ethernet interfaces.

Note that PCCW's successful IPTV deployment strategy and actual operational experience in IPTV delivery and video content management are being used as a prime example for other Telcos to learn from. Several major telecom operators worldwide have come to visit PCCW in Hong Kong to exchange information and ideas with PCCW on the various IPTV issues, and learn first hand the pitfalls and the technical and operational challenges.

The higher speed advantage of HKBN's FTTH enabled symmetric 100 and 1000 Mb/s Ethernet-based access when compared with PCCW's FTTB-enabled 6-8 Mb/s Asymmetric DSL access or even 20–30 Mb/s VDSL-based access is clearly in HKBN's favor. HGC Broadband's 10–100 Mb/s access services are also faster than PCCW's 6 Mb/s ADSL services. With FTTB already in place, the in-building wiring changeover to all fiber to the customer's apartment is in HKBN's FTTH offering. However, for PCCW, this is not a must at this stage, as long as the speed of access is sufficient for the broadband triple-pay services they offer. In a way, since FTTB in Hong Kong is almost complete and ubiquitous, the FTTH competition will come only when new high-bandwidth broadband service needs can be established. This is actually the case for HGC Broadband, which could offer FTTH 1000 Mb/s services as their platform is similar to that of HKBN, but they are not offering 1000 Mb/s to individual residential customers yet, as in their view, this can wait until there are services that really demand 1000 Mb/s access, as most of their current customers are happy with 10–100 Mb/s services. The widespread demand for 1000 Mb/s services, therefore, will take sometime, even in Hong Kong, a leading broadband city in the world. The interesting observation here is that higher and higher speed access services, while desired by many true broadband users, are not sufficient conditions to guarantee market share, as there are many other factors which influence customers' response. As in the case of PCCW NOW Broadband IP TV services, which are delivered over their comparatively slower access speed of 6 Mb/s ADSL platform, but with a good offering of original and unique video contents, a superior IP video management over this delivery platform, together with aggressive local marketing, PCCW has been able to establish a significant IPTV market share in Hong Kong.

12.6 BROADBAND OPTICAL HOME NETWORKS: THE POTENTIAL OF BROADBAND HOME NETWORKING OR 'GIGA-HOMES'

Japan is leading in R and D toward advanced home networks. Being a major supplier country of consumer electronics to the world, with many companies that are the key developers of HDTV, DVD, large-screen displays, and various digital consumer electronics for home use, it is natural that Japan sees a strong push toward advanced in-home networks interconnecting the next-generation digital electronics and optoelectronics consumer devices which are IP-network-controlled 'information appliances.' At the preeminent annual 'Consumer Electronics Show' held in January every year at Las Vegas, Japanese companies are major players and innovators of many new products for home use. The development of advanced network-capable information and entertainment appliances is a general trend. This leads to advanced home networking needs, and home networking follows naturally the widespread availability of FTTH-enabled high-speed broadband access. This section will use Japan's push in fiber-based optical home network as the prime example of what may happen in the near future.

According to Masaki Naritomi, General Manager of Lucina Division of Asahi Glass Co. of Japan, a world leader in polymer optical fibers for home networking [47], there are three-steps in the strategy to penetrate Fiber-*in*-the Home, 'FITH,' in Japan:

- 1st step: FTTH and Fiber at riser in the apartment
- 2nd step: Fiber in the wall
- 3rd step: Optical home network

These will be discussed in more details later.

In Japan there are 127 M (million) people and 48 M families. Twenty-five million subscribers use CATV but only 2.8 M use Cable Modem-based Internet service over CATV. Over 3 M subscribers use FTTH service. Price of broadband access service is among the lowest in the world: service prices of DSL (10 Mbps) and FTTH (100 Mbps) are going down to $20–35 USD/month. Yahoo! BB provides 1 Gbps FTTH service of $40/month including IP phone, broadcasting and video on demand.

Status of housing supply in Japan is as follows [47]: In the existing housing (48 M) in Japan, 52 % are detached houses and 48 % are in apartments, which means 23 M families live in apartments including condominiums. For new housing, 1.2 M new houses were completed in 2004 including 0.3 M of new condominiums.

Correspondingly, the status of FTTH in the apartment housing is as follows. Since 2001, optical fibers have been installed at riser (FTTR, R for riser) in all new tower-type condominiums. Up to 100 K subscribers use optical fiber at riser in apartments already. However, there are difficulties for rewiring the existing apartments, so that fiber to the first (ground) floor and DSL at riser using existing telephone copper wire is typical FTTH in the apartments. These numbers count as FTTH in Japan. Seventy percent of apartments are the middle and small size of less than 200 families. These apartments use UTP cables. For realizing FTTH in the apartments, other than the fascinating new high-bandwidth services and '*killer*' *applications* for over 1 Gb/s bandwidth, it is very important to have (1) ease of rewiring fibers in the apartments with (2) cost down comparable to UTP cable systems. Naritomi expects that one can achieve these two goals by installing graded-index polymer fibers, PFGI-POF at riser in Tower-type condominiums, and plans to complete one such project in Tokyo by the end of year 2007 [47].

The need for 'Giga-Homes' with Gb Ethernet access to the homes is argued strongly and convincingly by Naritomi. He made the following interesting observations on the current triple-play services and the emerging new high-bandwidth service needs [47]:

(1) Present services in condominiums include Internet access, IP phone, and HD Video on Demand; in addition, local services such as information notice, security web-camera, music and 'Karaoke,' and even TV 'channel on demand' (meaning all TV programs within 1 week are stored, so customers never miss TV programs of the week) are provided. These triple-play services plus local services will require 100 Mb/s or higher speed access, especially if one also considers also the new trends discussed below.
(2) In triple-play over a single converged network, Telephone, Internet, and TV broadcasting as well as VOD are carried over one media. Telephone (VoIP) and high-speed Internet access are already unified, but TV broadcasting over the IP network is not ready, but *Internet TV broadcasting* service has been started by CBS, ABC, and CNN in the US and will become commonplace soon. Japanese government plans to have digital TV broadcasting over FTTH access networks from 2006. This will drive up the high bandwidth needs of broadband access toward 100 Mb/s and Gb/s.
(3) Two important new trends will generate additional high-bandwidth service demands (nearly 'killer applications'?): Internet 'Blog,' and large-screen high-definition displays.

12.6.1   HD VIDEO 'BLOG'

People use 'Blog' in Japan. Blog is a personal website for "Bloggers", which is operated by ASP, ISP, etc. A personal web site is a challenge for most people to operate and maintain, but Blog is easy to join in for not only independent users but also business users. There are over 1.7 million Bloggers in Japan. Blog is changing from a personal diary and personal opinion column to an advertisement forum and a personalized *independent news agency*. Note starting mid-2005 SONY is marketing relatively inexpensive consumer-grade digital HD cameras (~$1500USD), making high-definition video quality available for individuals to make HD home videos. This could promote *individual personalized TV broadcast stations* (over the Internet) due to convincing high-quality pictures and videos for Internet audiences worldwide. This can compete with large established TV broadcast stations in a way like personal computers (PCs) competing effectively with large mainframe central computers. In this way, HD video-based Blog sites can generate (or demand) huge upstream data, so that FTTH enabled broadband access of 100–1000 Mb/s will be demanded by more and more residential customers, especially of the younger Internet-savvy generation, making Giga-Homes a necessity in the future. Optical fiber home networking can then become a natural part of meeting this service demand. Note even established company such as NISSAN now provides Blog advertisement, targeting the growing Blog audience, with significant potential commercial impact [47]. China has also seen a rapid growth of "Bloggers".

12.6.2   LARGE-SCREEN HIGH-DEFINITION DISPLAY FOR HOMES

Display is a *consumer optoelectronics* (the term consumer electronics alone does not describe it well) technology. Display market is undergoing major changes, with thin and large displays being at the center of focus for strong competition and consumer attention. The average growth rate from 2004 to 2007 of large displays is expected to be 48 % for plasma display panel (PDP), 8 % for projection TV (PTV), and 23 % for TFT-LCD. The price of thin and large displays is coming down rapidly. HDTV is picking up huge momentum in the year end of 2005, so certainly 2006 will be the year of HDTV, but there is a learning curve to follow as the HDTV contents are not yet abundant or readily available to all customers, so broadband service providers are making their best to improve the situation. In some areas in the US, the largest Cable MSO Comcast is offering over 10 INHD channels as part of their 100+ channel digital TV/HDTV services, with reasonably low subscription prices. Once users are used to HDTV it is a point of no return for them in video quality service demand, but the coexistence of both SD and HD video contents will be around for a long time before the complete transition is accomplished. Most the HDTV sets are at least 36 in or larger (e.g., 65 in) in display with thinness and lightweight being essential features. Large-screen display for built-in home theaters in newly designed upscale homes is becoming popular and could mushroom in the near future. The cost of such an additional broadband HD home entertainment center for a new-built home can be spread over years of mortgage payment so it is more acceptable to home buyers. Note the extension DVI and HDMI are needed for such HDTV sets. This means high bandwidth (3–5 Gbps) over up to 30 m distances between displays and recorders/players inside homes may be required.

These two important new trends point to the need for Giga-Homes and the external access speed of Gb/s access enabled by FTTH seems a natural demand. In fact, FTTH and FITH (Fiber-in-The-Home, with polymer optical fibers) would go together in the evolution process toward the optical broadband home networks. The following (see schematic in Figure 12.2) outlines the three steps in such an evolution from FTTH to FITH, according to Naritomi [47]:

There would be a three-step strategy to penetrate Fiber in the Home, 'FITH' in Japan:

- 1st step: FTTH outside and Fiber at riser in the apartment
- 2nd step: Fiber in the wall
- 3rd step: Optical home network

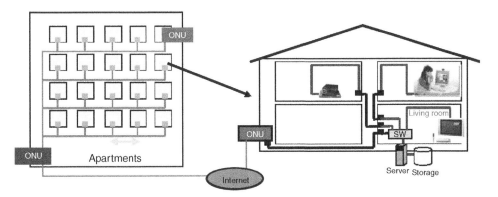

Figure 12.2 FTTH (apartments) and Fiber in the Home (FITH), with fiber to the apartments and fiber in the (apartment) homes for broadband home networks.

For FITH optical in-home wiring, large display needs DVI/HDMI extensions. Graded-index polymer optical fibers (PFGI-POF) would be a very good solution to transmit both Ethernet and DVI/HDMI data inside homes due to high-bandwidth, low cost, high-coupling efficiency, and ease of installation. Related to high-bandwidth home networking, an industry association called *Digital Living Network Association* (DLNA) has started to support triple-play services on digital consumer electronics over Ethernet, so there is no barrier between PC and digital consumer electronics/optoelectronics (information appliances) using Ethernet interfaces throughout the home. The DLNA sponsors include many major international players such as SONY, Toshiba, Sharp, Panasonic, NEC, Fujitsu, Samsung, Thomson, TI, Philips, IBM, HP, Intel, Kenwood, Microsoft, Motorola, Nokia, Huawei, Lenovo, etc. [48]. The DLNA members share a vision of a wired and wireless interoperable network of Personal Computers (PC), Consumer Electronics (CE), and mobile devices in the home enabling a seamless environment for sharing and growing new digital media and content services.

In an optimist outlook [47], it is expected that in 2007 subscribers of FTTH in Japan will increase to 30 % share of all Internet users and the price of 1 Gbps FTTH service will be the same as DSL service. In 2008 DLNA-based digital home electronics will be readily available and Blog could become a key tool for business, and FITH enabled Giga-Homes with high-speed optical home networks will likely be needed.

The examples above refer mainly to the Optical Home Network effort reported in Japan, but similar effort is being actively pursued in other regions including US and Europe. The essential point here is that, beyond the FTTH deployment, FTTH may go hand-in-hand with high-speed FITH, optical-fiber wired in-home networks, making unlimited broadband access to the homes and in the homes a complete story. Such ubiquitous, unlimited broadband access would have far-reaching impacts in human beings' daily lives in the years to come.

12.7 RESEARCH ON TECHNOLOGIES FOR NEXT GENERATION BROADBAND OPTICAL ACCESS: WDM PON ACCESS NETWORKS AND FIBER/WIRELESS INTEGRATION

Most of the FTTH systems deployed or being considered for deployment are either P2P active Ethernet systems or passive PON-based P2MP systems. In PON, FSAN/ITU-T developed standards for APON/BPON and GPON were developed for deployment at different stages. In Ethernet-based EPON, IEEE EFM group has developed GE PON-based architecture with Gigabit Ethernet capability which is taking a more popular and prominent role now. In these TDM or IP PON systems, while a few optical wavelengths are used, for example, 1310, 1490, and 1550 nm if there is RF video overlay, no real multiple-optical-channel, high-channel-count WDM systems have been deployed. Many leading players including both NTT and KT agree that some types of WDM-based access network architecture can benefit significantly from the use of CWDM and DWDM optoelectronics technologies, similar to the power of DWDM in revolutionizing the long-haul global optical fiber communications networks. So on the topic of next-generation broadband optical access networks, several research labs have been studying the various flavors of WDM PON or WDM hybrid PON/ring access architectures and associated monitoring, protection, and restoration techniques and related WDM system technologies [11,49–58], to prepare for future upgrade of the FTTH system currently being deployed.

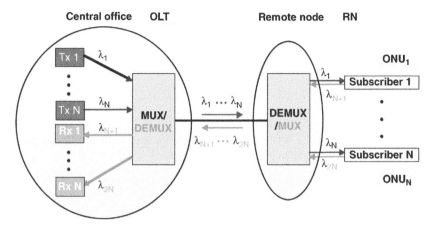

Figure 12.3 The schematic of a generic WDM PON access network architecture.

A typical WDM PON network architecture is a physical PON but a logical star, with point-to-point connection between the OLT and the ONU, as each ONU receives its own assigned wavelength. The two examples of Figures 12.3 and 12.4 basically show the similar schematic of a typical WDM PON architecture. The use of a WDM mux/dmux at the RN (remote node) constitutes the 'passive' nature of PON, as in TDM PON a passive optical splitter is used, while here the WDM mux/dmux is typically a passive WDM optical component (and if AWG is used as the WDM mux/dmux or wavelength-routing optical

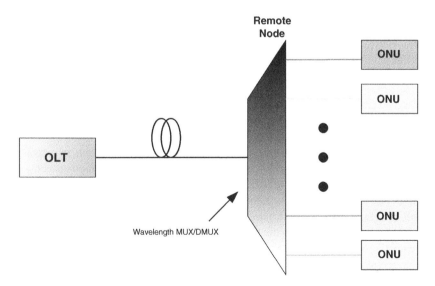

Figure 12.4 The schematic of a WDM PON access network with a star distribution architecture with a single WDM mux/dmux at the RN for distribution.

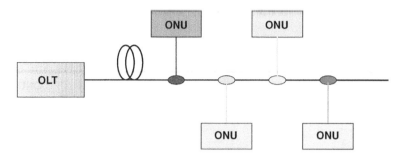

Figure 12.5 The schematic of a WDM PON access network with a bus distribution architecture where many single-wavelength add/drop multiplexers are needed.

component at the RN, the remote node, an athermal AWG can be used without any temperature control electronics at the RN). Note while Figures 12.3 and 12.4 shows the star configuration in the distribution of the wavelength-specific signal to each ONU, it could in principle also be an optical bus (tree and branch) distribution architecture (like in coax distribution bus in a typical HFC network), as shown in Figure 12.5, where an individual optical add/drop multiplexer (OADM) for each specific wavelength is needed for each ONU node, which would be much more expensive to implement.

Why is WDM PON access network considered the next-generation broadband optical access by many research labs? There are several obvious reasons [7]:

(1) Higher bandwidth: with dedicated wavelength and bandwidth for each ONU (no sharing as in TDM), each ONU can have its own maximum downstream bandwidth without sharing with other ONUs;
(2) Flexible bandwidth and format: data rate or even format on each wavelength channel can be different;
(3) Higher security due to nonsharing of signal;
(4) Eliminating the need for ranging or collision avoidance scheme for upstream path; and
(5) Higher upgrade-ability because of the dedicated link and dedicated wavelength.

However, the drawbacks of WDM-based access network are that (a) it may be of higher cost as WDM components are costly, though cost is coming down rapidly; and (b) inventory of the wavelength-specific user-site ONUs is a major issue unless an architecture with so-called 'colorless' ONUs are used. Otherwise, it will require wavelength management for wavelength-specific ONUs for specific customer sites, adding to the component cost and the operational complexity. Furthermore, if because of higher bandwidth access and higher performance of the WDM access networks, optical performance monitoring, network protection, and restoration become necessary due to the increased demand of broadband services availability and reliability, the needed technologies and management would become even more costly. Nevertheless, due to its almost unlimited access bandwidth and flexibility, as well as graceful upgrade ability, WDM PON or hybrid WDM PON/Ring architectures are considered the next-generation broadband optical access architectures, and many leading research labs have been working on these to reduce the impact of the drawbacks just mentioned. Figure 12.4 above shows a generic WDM PON architecture where the optical

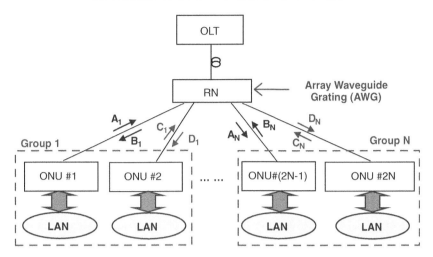

Figure 12.6 The simplified schematic of a WDM PON access network with an AWG at the RN, and with wavelength assignment such that each pair of ONUs protect each other in case of access fiber cut [for details, see 59,60].

splitter at the remote node in a TDM PON is replaced by a wavelength mux/dmux or WDM routers like AWG, so specific wavelength carries signals dedicated for a specific ONUs without the signal-sharing problems of TDM PONs.

Three main directions seem to be of great current interest in these WDM PON access network-related research efforts: (1) developing optical protection/restoration schemes in the WDM access networks [see e.g., 59,60]; (2) designing an WDM architecture which can use 'colorless' (nonwavelength-specific) ONUs [11,54–58,61]; and (3) integrating WDM PON architecture with Ring or other architectures to enhance the overall performance, upgradeability, and reliability of the network [59,60].

For (1), optical protection and restoration schemes in WDM PON networks, Figure 12.6 shows the simplified schematic of an example of a WDM PON with a self-protection technique using AWG routers and special wavelength assignment schemes to allow two ONUs as a pair to protect each other in case of fiber cut [for details, see 59,60].

In (2), the design of *colorless* ONUs usually requires the WDM channels for the ONUs to be distributed from OLT (in the CO) downstream for use by the ONUs as upstream optical channels (with remodulation at the ONUs), and these wavelength channels are being sent down together with the downstream data intended for the ONUs [50,51,54–58]. Often this is called the spectral-slicing techniques, as the broadband optical sources such as LED, ASE source, or supercontinuum source are used to generate the multiple optical channels for distribution to the ONUs, for remodulation with ONU data and sent upstream. Figure 12.7 shows just one such example of architecture and implementation, where a C-band (1540–1560 nm) supercontinuum is distributed downstream and used as the source for ONUs to send data upstream, while the downstream data from OLT to ONUs are carried over the L-band (1570–1600 nm) WDM channels [61]. Each ONU receives its own assigned wavelength channel and is remodulated with the upstream signal at the ONU, but the ONU itself

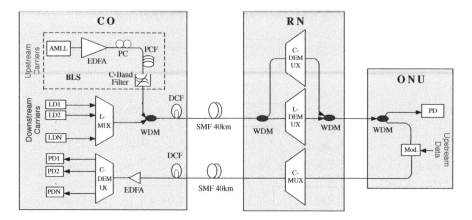

Figure 12.7 An example of WDM PON architecture with a C-band supercontinuum source distributing the optical channels downstream to all ONUs for remodulation and upstream data transmission (only one ONU shown). Downstream data transmission uses L-band optical channels. All ONUs are identical (with the same WDM coupler, a C/L band separator), and therefore 'colorless' ONUs.

does not require a wavelength-specific design. The designated optical channel for each ONU has been demultiplexed at the RN and the WDM coupler (which is a C/L band band-pass filter or band separator) is identical to each other in ONUs. This makes it a colorless ONU as each ONU is the same as the other, with no specific wavelength assigned to the ONU. There are several other variations of similar techniques [50,51,54–58]. Certainly, colorless ONU implementations such as this example and others are mostly still in the research stage and require further technology development to improve the performance and the cost-effectiveness.

Another aspect is that while WDM PON access network provides dedicated wavelength and hence bandwidth to individual ONUs, it is often desirable to also have the broadcast function in addition to the dedicated optical wavelength carrying the specific targeted data. So, a WDM access network with both dedicated optical wavelength and a broadcast wavelength would be desirable in some cases where broadcast video or data can be distributed through this wavelength-specific access network. An overlay wavelength can be added for the broadcast function provided that the WDM Mux/dmux at the RN can be bypassed.

In migrating from TDM PON to WDM PON when upgrade is needed in the future, the passive component of the optical splitter at the RN can be replaced by passive WDM mux/dmux or passive WDM routing component (such as athermal AWG), so that the passive outside plant nature of the PON is preserved and the installed fiber plant does not need to be changed. This provides a smooth upgrade at least as far as the fiber outside plant is concerned. Since fiber plant installation is the biggest part of the first installation cost, this graceful upgrade from TDM to WDM PON is an important factor for upgrade considerations. Note also for the initial WDM PON systems, CWDM technologies are much lower in cost than the DWDM technologies, so CWDM-based WDM PON deployment should be considered first. Figure 12.8 shows an example of CWDM PON architecture where specific

CWDM-PON for FTTH/FTTB Broadband Access

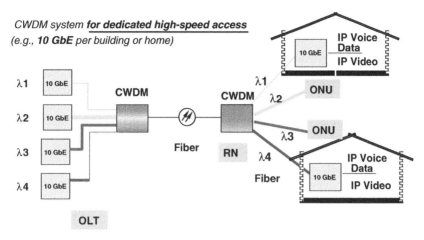

Figure 12.8 The schematic of a CWDM PON access network with CWDM mux/dmux at the RN, and with a specific wavelength targeted to each ONUs with high bandwidth. The case of 10 GbE data is shown.

wavelength can be dedicated to each ONUs (though further sharing can also be done), such that each ONU can receive its high-speed data on its own wavelength channel, making even 10 GbE data access readily achievable.

Other area of current interest is the wireline/wireless broadband network integration in a Hybrid Fiber/Wireless Access Network. The use of Wi-Fi (IEEE 802.11) based wireless access point (WAP) connected to the ONU, with wireless distribution of 11 or 54 Mb/s data over short indoor distances, proves to be a natural extension of the broadband fiber access network, for video and data distribution within a home or an apartment building. Future integration effort will also be a necessity when the broadband home networking with fiber-in-the-home (FITH) needs to integrate with the wireless broadband access technologies to achieve the flexibility of broadband reach everywhere within the home.

12.8 THE BROADBAND FUTURE, WITH IP HDTV/VOD AND HD VIDEO COMMUNICATIONS

The emergence of HDTV and digital home theaters with large-screen displays as well as HD video-based communications and video-centric media creation (HD Video Blog) will probably demand more and more in-home bandwidth for both downstream and upstream access to the outside world, mostly over IP-based converged networks. The emerging broadband society will be a video-rich society, where IPTV and IP VOD with HDTV and HD DVD will become the dominant forms of entertainment and communications. Leading telecom operators and new broadband service providers are all starting to offer IPTV services in addition to VoIP and high-speed Internet access. Verizon has already begun testing their FiOS-based IPTV, and SBC and Bellsouth have initiatives of their own [62]. In Europe, IPTV services already in place include Fastweb in Italy, HomeChoice in the UK, MaLigne and Free in France, and Telefonica in Spain [26]. In Asia, PCCW Ltd and Hong

Kong Broadband in Hong Kong, Softbank/Yahoo! BB in Japan, as well as Chunghwa Telecom in Taiwan are all offering IPTV VOD or Movie-on-Demand services [33]. Many other countries in Asia and Europe are also planning to offer similar IPTV services.

In the near future, Intelligent Homes with an in-home broadband network for Internet-connected information appliances will probably start to gain popularity when the younger generation are building their new homes, when the cost could be spread over many years embedded in the long-term mortgage payments. In addition to traditional roles of homes, homes will become workplace for many people who wish to work at home and from home, at least part of the time when commuting to work is not desirable. This will require significant upstream bandwidth from homes to the outside world. HD video-based video conferencing and video communications are not only useful for those who wish to work at home and still communicate with all coworkers worldwide, but will also be used for tele-education and tele-medicine when the user-friendly interface and infrastructure are gradually established in most of broadband homes [47,63,64].

Integrated with FTTH-enabled optical broadband access, high-speed broadband wireless technologies such as Wi-Fi and WiMAX [65], and UWB-USB for in-door wireless HDTV/ cinema distribution and video conferencing/communications will likely become common infrastructure required in the near future. The impact of such integrated broadband wireline/ wireless infrastructure, with FTTH- and FITH-enabled ultimate broadband access, on human societies and civilizations in the next 10–20 years would certainly be significant. Though difficult to fully envision at this stage, it is clear that the Broadband Era has finally come, an era of very-high-bandwidth *broadband visual communications*, and of nearly-unlimited possibilities in information and media access, which will transform all aspects of human lives in the years to come.

ACKNOWLEDGEMENTS

I thank my colleagues at CUHK's Lightwave Communications Labs., Center for Advanced Research in Photonics: Prof. Calvin C. K. Chan, Prof. Lian K. Chen, and Prof. Frank Tong, for research collaborations and stimulating discussions on the topic of next-generation WDM broadband optical access networks and technologies. Many of our graduate students in the Lightwave Communications Lab., including my Ph.D. student, Zhaoxin Wang, contributed to the WDM access network work, which have been presented at recent OFCs and ECOCs, and deserve our thanks. I also appreciate the insightful talks given at the Broadband Access Seminars in CUHK, in 2005, by Mr Ricky Wong of Hong Kong Broadband, Dr Lian Wu of PCCW, and Mr Larry Yuen of HGC. I also would like to thank Dr. Masaki Naritomi, General Manager of Lucina Division of Asahi Glass Co. of Japan, for kindly providing me his POF 2005 Plenary Paper material on Broadband Home Networking in Japan.

REFERENCES

1. Lin C (*Invited Tutorial*). Photonics for broadband optical communications – a historical perspective', 9th *OECC (Opto-Electronics and Communications Conference)*, Yokohama, Japan, July 2004.
2. Lin C (*Plenary Talk*). Photonics for broadband communications – past, present and future', 8th *OECC (Opto-Electronics and Communications Conference)*, October 14–16, Shanghai, China, 2003.

3. Payne DB, Stern JR. Transparent single-mode fiber-optic networks. *J Lightwave Tech* 1986; **LT-4**: 864–869.

4. Wagner SS et al. Experimental demonstration of a passive optical subscriber-loop architecture. *Electron Lett* 1988; **24**: 325–326.

5. Shimada S, Hashimoto K, Okada K. Fiber optic subscriber loop systems for integrated services: The strategy for introducing fibers into the subscriber network. *J Lightwave Tech* 1987; **LT-5**: 1667–1675.

6. Lin C, Kobrinski H, Frenkel A, Brackett CA. Tunable 16 optical channel transmission experiment at 2 Gb/s and 600 Mb/s for Broadband Subscriber Distribution, *ECOC'88*, Brighton, UK, September, 1988.

7. Shumate, PW, Jr. 'FTTx', *OFC* Short Courses, 2001–2003, and *ECOC* Short Course 2002.

8. For FSAN history and current activities, see http://www.fsanweb.org.

9. For ITU-T activities and some of the PON standards, see http://www.itu.int/ITU-T.

10. Abrams M, Becker PC, Fujimoto Y, O'Byrne V, Piehler D. FTTP deployments in the United States and Japan – Equipment Choices and Service Provider Imperatives. *J Lightwave Tech* 2005; **23**: 236–246.

11. Shinohara H (*Plenary Talk*). Broadband Expansion in Japan, *OFC'05*, March 8, Anaheim, CA, USA, 2005.

12. Ciciora W, Farmer J, Large D, Adams M. *Modern Cable Television Technology: Video, Voice, and Data Communications* (2nd ed), Elsevier; San Francisco, 2004

13. See information and statistics available at the NCTA webpages, www.ncta.org.

14. Way WI. Subcarrier multiplexed lightwave system design considerations for subscriber loop applications. *J Lightwave Tech* 1989; **7**: 1806–1818.

15. Mao XP, Bodeep GE, Tkach RW, Chraplyvy AR, Darcie TE, Derosier RM. Brillouin scattering in externally modulated lightwave AM-VSB CATV transmission systems. *IEEE Photonics Tech Lett* 1992; **4**: 287.

16. Way WI, Choy MM, Yi-Yan A, Andrejco M, Saifi M, Lin C. Multi-channel AM-VSB CATV signal transmission using an erbium-doped optical fiber. *IEEE Photon Tech Lett* 1989; **1**: 343.

17. Kikushima K, Yoneda E. Erbium-doped fiber amplifiers for AM-FDM video distribution systems. *IEICE Trans.* 1991; **E74**(7): 2042–2048,

18. Lin C, Way WI, (*Invited*). Optical amplifiers for multi-wavelength broadband distribution networks, *ECOC'91*, Paris, France, September 1991.

19. See NCTA webpages, e.g., http://www.ncta.com/Docs/PageContent.cfm?pageID=314

20. Lin C, Ovadia S, Eskildsen L, Andersen WT. (*Invited*). Multichannel AM/QAM video systems for hybrid fiber/coax distribution networks. *IOOC'95*, Hong Kong, July 1995; and Dai H, Ovadia S, Lin C. Hybrid AM-VSB/256-QAM multichannel video transmission over 120 km of standard single-mode fiber with cascaded EDFAs, Technical Digest, *CLEO (Conference on Lasers and Electro-Optics)'96*, paper CFB3, June 2–7, Anaheim, CA, 1996.

21. Lin C (*Invited*). Broadband optical access over hybrid fiber-coax (HFC) networks, *IEEE LEOS Summer Topical Meeting on Broadband Optical Networks*, Ft. Lauderdale, Florida, July 2000.

22. See, e.g., http://www.cablelabs.com/ for more information about CableLabs and Cable Modems based on CableModem/DOCSIS®

23. See information at SCTE website: http://www.scte.org/home.cfm.

24. SCTE Engineers in the Cable TV industry have been discussing the emerging Broadband Competition against the other players in both the IPTV and the Wireless Broadband for Cable TV delivery in SCTE's *Conference on Emerging Technologies (ET)*. See http://et.scte.org for the sessions and speakers on the Broadband Competition issues in the latest SCTE *Conference on Emerging Technologies*, Tampa, Florida, January 10–12, 2006.

25. Parandekar H (Software Development Manager, Cisco Systems, Inc.) in The Converged World of 2010 – A case for video delivery over DOCSIS®, *2006 SCTE Conference on Emerging Technologies*, Tampa, Florida, Jan. 10–12, 2006.

26. See, e.g., http://www.iptvinformation.net/IPTV + FAQ.aspx and http://www.s2data.com/iptv2006/ for IPTV industry information and conferences.

27. As reported in CNN Money, December 19, 2001, Comcast wins AT&T Cable. See http://money.cnn.com/2001/12/19/deals/att/.

28. See, for example, webpages of Softbank BB and Yahoo!BB for details (in Japanese): http://hikari.softbankbb.co.jp/ and http://bbpromo.yahoo.co.jp/promotion/usable/bbtv/

29. Softbank BB Launches BBTV Service in Japan Powered by UTStarcom's mVision IPTV System,' UTStarcom news release, July 6, 2005: http://investorrelations.utstar.com/ReleaseDetail.cfm?ReleaseID=168015

30. NTT's IPTV VOD service plan is indirectly mentioned in the press release of SkyStream Networks in April 2005: http://www.broadbandtrends.com/News%20Articles/April%202005/Skystream_NTT_04132005.htm

31. UTStarcom Signs Contract With Japan's Softbank BB to Expand GEPON Network in Support of Fiber-to-the-Home Service, UTStarcom news release, July 6, 2005: http://investorrelations.utstar.com/ReleaseDetail.cfm? ReleaseID=168013

32. See J:COM To Offer 100 Mbps Internet Access Services To Multiple-Dwelling Units, http://www.jcom.co.jp/pdf/newsrelease/en/20050622_en.pdf, June 22, 2005.

33. Soong J. Why is Asia leading the Global IPTV Revolution? in http://www.convergedigest.com/blueprints/ttp03/2005bns1.asp?ID=211&ctgy=Market.

34. See for example the webpage of Wanadoo, http://www.wanadoo.co.uk/

35. See for example, the webpage of TeleWest, http://www.telewest.co.uk/

36. See http://www.dslprime.com/News_Articles/news_articles.htm

37. See Verizon FiOS TV goes live in 7 new Texas Markets, January 5, 2006, reported in http://www.broadbandtrends.com/News%20Articles/2006/January%202006/VZ_TXTV_01052006.htm.

38. As reported by Jeff Baumgartner in CED Magazine, February 2005.

39. Brown K. Driving DOCSIS 3.0, on April 2005 CED on-line: http://www.cedmagazine.com/article/CA513765.html

40. Jeff Baumgartner, Keeping up with CableLabs, on July 2005 CED on line: http://www.cedmagazine.com/ced/2005/0705/07f.htm

41. See http://www.ofta.gov.hk/en/datastat/key_stat.html

42. See http://now.com.hk/ and http://www.nowbroadbandtv.com/eng/

43. See http://www.hkbn.net

44. See http://www.hgcbroadband.com/

45. See http://www.hkbn.net/bb1000/ and http://www.hkbn.net/bb1000/opt_conn.html for description of the 1000 Mb/s, Optical Broadband, bb1000 FTTH/FTTB access services at Hong Kong Broadband Networks.

46. Wong R. Building the Hong Kong 21st Century Network, Seminar on Broadband Access, Chinese University of Hong Kong, April 28, 2005.

47. Naritomi M (Plenary Talk). Home network in Japan from present to future, *14th International Conference on POF (Polymer Optical Fibers)*, Hong Kong, September 2005.

48. See http://www.dlna.org/about/roster for DLNA member list and activities.

49. Wagner SS et al. A Passive photonic loop architecture employing wavelength division multiplexing, *Proc. Globecom'88*, paper 48.1, 1988.

50. Zirngibl M, Doerr CR, Stulz LW. Study of spectral slicing for local access applications. *IEEE Photon. Technol. Lett.* 1996; **8**: 721–723.

51. Frigo NJ, Reichmann KC, Iannone PP, Zyskind JL, Sulhoff JW, Wolf C. A WDM-PON architecture delivering point to point and multicast services using periodic properties of a WDM router, PD-24, *Proc Conf Optic Fiber Commun OFC'97*, Dallas, TX, February 16–21, 1997.

52. Feldman RD, Harstead EE, Jiang S, Wood TH, Zirngibl M. An evaluation of architectures incorporating wavelength-division-multiplexing for broad-band fiber access. *J Lightwave Tech* 1998; **16**(9): 1546–1559.

53. Park SJ, Kim S, Song K-H, Lee J-R. DWDM-based FTTC access network. *J Lightwave Tech* 2001; **9**: 1851–1855.

54. Jung DK, Shin SK, Lee C –H, Chung YC. Wavelength-division-multiplexed passive optical network based on spectrum-slicing techniques. *IEEE Photon Technol Lett* 1996; **10**(9): 1334–1336.

55. Wroodward SL, Iannone PP, Reichmann KC, Frigo NJ. A spectrally sliced PON employing Fabry-Perot Lasers. *IEEE Photon Technol Lett* 1998; **10**(9): 1337–1339.

56. Healy P, Townsend P, Ford C, Johnston L, Townley P, Lealman I, Rivers L, Perrin S, Moore R. Spectral slicing WDM-PON using wavelength-seeded reflective SOAs. *IEE Electron Lett* 2001; **37**(19): 1181–1182.

57. Kim HD, Kang SG, Lee CH. A low cost WDM source with an ASE injected Fabry-Perot semiconductor laser. *IEEE Photon Technol Lett* 2000; **12**: 1067–1069.

58. Akimoto K, Kani J, Teshima M, Iwatsuki K. Gigabit WDM-PON system using spectrum-slicing technologies, in *ECOC'03*, Remini, Italy, September 2003, Paper Th2.4.6.

59. Chan TJ, Chan CK, Chen LK, Tong F. A self-protected architecture for wavelength-division-multiplexed passive optical networks. *IEEE Photon Technol Lett* 2003; **15**: 1660–1662.

60. Wang ZX, Sun XF, Lin C, Chan CK, Chen LK. A novel centrally controlled protection scheme for traffic restoration in WDM passive optical networks. *IEEE Photon Technol Lett* 2005; **17**(3): 717–719.

61. Zhang Bo, Lin C, Huo L, Wang Z, Chan C-K. A simple high-speed WDM PON utilizing a centralized supercontinuum broadband light source for colorless ONUs, paper presented at *OFC'06*, Anaheim, CA., March 2006.

62. SBC Confirms Project Lightspeed IPTV Field Trial, http://www.tvover.net/SBC + Confirms + Project + Lightspeed + IPTV + Field + Trial.aspx, 3 November 2005.

63. Lin C, Chen LK, Chan CK.(*Invited*). Broadband Access and Fiber-to-the-Home–Current Status, *10th OECC (Opto-Electronics and Communications Conference)*, September 4–6, 2005, Seoul, Korea.

64. IEEE-USA's Committee on Communications and Information Policy, Accelerating Advanced Broadband Deployment in the US, http://www.ieeeusa.org/forum/positions/broadband.html, Feb. 2003.

65. See IEEE 802.11 Wireless LAN, http://grouper.ieee.org/groups/802/11/ and IEEE 802.16 Wireless MAN, http://grouper.ieee.org/groups/802/16/

Index

Printed and bound by CPI Group (UK) Ltd, Croydon, CR0 4YY

16/04/2025

14658555-0001